ECONOMIC IMPACT AND CONTROL OF SOCIAL INSECTS

CONTRIBUTORS

R. D. Akre
Department of Entomology
Washington State University
Pullman, WA 99164

J. M. Cherrett
Department of Applied Zoology
University College of North Wales
Bangor, Gwynedd, LL57 2VW, Wales

J. P. Edwards
Ministry of Agriculture
Fisheries and Food Pest Infestation
  Control Laboratory
Slough, Berks, England

H. G. Fowler
3103 McCarty Hall
Gainesville, FL 32611

J. R. J. French
Division of Building Research
CSIRO
Box 56
Highett, Victoria, Australia

J. E. Gillaspy
Department of Biology
Texas A&I University
Kingsville, TX 78363

W. H. Gotwald, Jr.
Department of Biology
Utica College of Syracuse University
Utica, NY 13502

L. Greenberg
Department of Entomology
Texas A&M University
College Station, TX 77843

R. A. Johnson
Center for Overseas Pest Research

College House, Wrights Lane
London W8 5SJ, England

C. S. Lofgren
U.S. Department of Agriculture
Box 14565
Gainesville, FL 32604

J. F. MacDonald
Department of Entomology
Purdue University
West Lafayette, IN 47907

J. D. Majer
School of Biology
Western Australian Institute of
  Technology
Kent Street
Bently, Western Australia 6102

J. K. Mauldin
U.S. Department of Agriculture
Forest Service
Southern Forest Experiment Station
Box 2008
GMF, Gulfport, MS 39503

E. A. McMahan
Department of Biology
University of North Carolina
Chapel Hill, NC 27514

P. K. Sen-Sarma
Forest Research Institute and College
P.O. New Forest
Dehra Dun, India

S. B. Vinson
Department of Entomology
Texas A&M University
College Station, TX 77843

T. G. Wood
Center for Overseas Pest Research
College House, Wrights Lane
London, W8 5SJ, England

ECONOMIC IMPACT AND CONTROL OF SOCIAL INSECTS

edited by
S. Bradleigh Vinson

PRAEGER SPECIAL STUDIES • PRAEGER SCIENTIFIC

New York • Philadelphia • Eastbourne, UK
Toronto • Hong Kong • Tokyo • Sydney

Library of Congress Cataloging in Publication Data
Main entry under title:

Economic impact and control of social insects.

Bibliography: p.
Includes index.
1. Insect pests. 2. Insect societies. 3. Insect
pests—Control. 4. Beneficial insects. I. Vinson,
S. Bradleigh, 1938–
SB931.E26 1985   628.9′657   85-16715
ISBN 0-03-063208-0 (alk. paper)

Published in 1986 by Praeger Publishers
CBS Educational and Professional Publishing, a Division of CBS Inc.
521 Fifth Avenue, New York, NY 10175 USA

© 1986 by Praeger Publishers

Printed in the United States of America on acid-free paper

INTERNATIONAL OFFICES

Orders from outside the United States should be sent to the appropriate address listed below. Orders from areas not listed below should be placed through CBS International Publishing, 383 Madison Ave., New York, NY 10175 USA

Australia, New Zealand
Holt Saunders, Pty, Ltd., 9 Waltham St., Artarmon, N.S.W. 2064, Sydney, Australia
Canada
Holt, Rinehart & Winston of Canada, 55 Horner Ave., Toronto, Ontario, Canada M8Z 4X6
Europe, the Middle East, & Africa
Holt Saunders, Ltd., 1 St. Anne's Road, Eastbourne, East Sussex, England BN21 3UN
Japan
Holt Saunders, Ltd., Ichibancho Central Building, 22-1 Ichibancho, 3rd Floor, Chiyodaku, Tokyo, Japan
Hong Kong, Southeast Asia
Holt Saunders Asia, Ltd., 10 Fl, Intercontinental Plaza, 94 Granville Road, Tsim Sha Tsui East, Kowloon, Hong Kong
Manuscript submissions should be sent to the Editorial Director, Praeger Publishers, 521 Fifth Avenue, New York, NY 10175 USA

CONTENTS

Contributors   ii

Introduction   vii

1   The Biology, Physiology, and Ecology of Termites   1
T. G. Wood and R. A. Johnson

2   Economically Important Termites and Their
Management in the Oriental Region   69
P. K. Sen-Sarma

3   Termites and Their Economic Importance in Australia   103
J. R. J. French

4   Economic Importance and Control of Termites
in the United States   130
Joe K. Mauldin

5   Beneficial Aspects of Termites   144
Elizabeth A. McMahan

6   The Economic Importance and Control of
Leaf-Cutting Ants   165
J. M. Cherrett

7   The Biology, Physiology, and Ecology of
Imported Fire Ants   193
S. Bradleigh Vinson and Les Greenberg

8   The Economic Importance and Control of
Imported Fire Ants in the United States   227
Clifford S. Lofgren

9   The Biology, Economic Importance, and Control of the
Pharaoh's Ant, *Monomorium pharaonis* (L.)   257
J. P. Edwards

10   Biology, Economics, and Control of Carpenter Ants   272
Harold G. Fowler

11   The Beneficial Economic Role of Ants   290
        William H. Gotwald, Jr.

12   Utilizing Economically Beneficial Ants   314
        Jonathan D. Majer

13   *Polistes* Wasps: Biology and Impact on Man   332
        James E. Gillaspy

14   Biology, Economic Importance, and Control of
        Yellow Jackets   353
        Roger D. Akre and John F. MacDonald

        Index   413

        About the Editor   422

INTRODUCTION

The purpose of this book is to bring together information concerning the problems that social insects cause to man, his structures, food, and treasured artifacts. In addition to information concerning the problems caused by social insects, current control technologies are discussed. In order to provide some background, several chapters on the biology, physiology, and ecology of several select social insects are included.

Drs. T. G. Wood and R. A. Johnson contribute a chapter that provides a general view of the biology of termites. They bring together a diverse group of biologies and provide a good foundation both for understanding the problems caused by termites and for understanding control techniques.

The economic importance and control of termites varies with the geographic region. Dr. P. K. Sen-Sarma discusses the termite problem in the Oriental Region, where termites are not only a problem to buildings, but a major crop pest. The termite problem in the Australian region, where tropical, temperate, and desert species exist, is discussed by Dr. J. R. J. French. The challenges to people living with termites in North America are discussed by Dr. J. K. Mauldin. As he points out, in the United States, termites are primarily a problem to structures people build. These problems are largely preventable but too many structures are built without proper protection.

However, termites are not all bad. As Dr. E. A. McMahan points out, they are essential components of the ecosystem, and our world would be uninhabitable without them. Her chapter points out one of the major problems with social insects: they are essential to the ecological health of the environment, but they may cause direct problems to people.

Ants are another important group of social insects. Ants cause problems to our food, destroy structures, and cause direct problems to people. Unlike termites, some ants sting and then become nuisance pests. However, for some people insect stings are life-threatening.

Dr. J. M. Cherrett discusses one of the more important economically important ant pests, the leaf-cutter ants. As he points out, leaf-cutters are a major pest in tropical areas of Central and South America. He discusses the history of their control and the problems that may be facing tropical agriculture in Central and South America due to changes in available control methods.

The biology and spread of the fire ant into the United States is discussed by Drs. S. B. Vinson and L. Greenberg. The spread of the imported fire ant represents a classic example of what can happen when a species of relatively little concern in its original range is accidently introduced into another area. The economic importance of this ant is presented by C. S. Lofgren. The difficulty in placing a monetary figure on the economic importance of some social insects is demonstrated in his chapter. Not only does the fire ant pack a painful sting, but, as Lofgren points out, the imported fire ant also causes some problems with crops, wildlife, and some of our engineering efforts.

The biology and economic importance of carpenter ants are discussed by Dr. H. G. Fowler. Like termites, carpenter ants primarily are associated with destruction of structures erected by people, but, as Fowler notes, the economic importance of carpenter ants is difficult to determine. The Pharoah ant represents a different aspect. This species, discussed by Dr. J. P. Edwards, does not destroy structures or crops, nor does it interfere with people by stinging. However, it is an increasing hazard because of its ability to contaminate food and sterile materials used in medical care.

Like termites, the ants represent an important component of our ecosystem. The beneficial role that ants play is discussed by Dr. W. H. Gotwald. Again, like the termites, we find that the ants are essential to the health and well-being of our environment. The biology, economics, and control of arboreal ants are discussed by Dr. J. Majer. His chapter also reviews the beneficial role that ants play, but Majer emphasizes the beneficial role of ants more from the point of view of manipulation for the benefit of people. His examples using the arboreal ants could well be extended to other species.

The remaining chapters concern the wasps. Dr. J. E. Gillaspy discusses the biology and economic importance of *Polistes*. He notes the aspects of *Polistes* that have the potential to benefit man, rather than recommending the species be destroyed because of its sting. The concluding chapter, by Drs. R. D. Akre and J. F. MacDonald discusses the biology, behavior, ecology, and economic and medical importance of vespids. Their chapter points out the conflict social insects can represent in their interaction with people. The difficulties in determining their economic impact, and the problems encountered in the control of these species that came into conflict with people are also discussed.

Hopefully, the book will provide insight into the benefits that social insects provide to the human race and their essential role in the ecosystem. Certain social insects, such as bees, represent a significant importance to man. Only the limits of our technology and understanding of the biology of social insects hinder our ability to use these species to protect our food supply.

While it cannot be denied that social insects are essential to the balance of an ecosystem, certain species come into conflict with the interests of man or represent a more direct medical threat to certain people. Often it is difficult to assess the costs from destruction due to social insects in terms of dollars, while a more accurate assessment can be determined for insect pests in terms of food or fiber. Hopefully, this book will provide a clearer understanding of the economic cost of social insects.

In addition, the mutual challenges various researchers encounter in the control of social insects and the approaches taken to overcome these difficulties will perhaps be useful to others involved in the control of social insects where they cause a problem.

I would like to give special thanks to Dr. J. K. Mauldin who helped me with the chapters dealing with termites. I also wish to thank my wife, Patricia Vinson, for her editorial help and patience.

S. Bradleigh Vinson

Chapter 1

THE BIOLOGY, PHYSIOLOGY, AND ECOLOGY OF TERMITES

T. G. Wood and R. A. Johnson

INTRODUCTION

This chapter attempts to provide a basic account of termites that will facilitate the understanding of ensuing chapters on their economic importance. In the space available it is impossible to cover all aspects of their biology comprehensively, or equally, and we have been selective in that aspects directly related to their economic significance have been emphasized.

The fascination of termites for biologists (Hegh 1922) and our concern with their damage to our property (Kofoid 1934) has developed rapidly. Compilations of papers dealing solely with termites include UNESCO and Zoological Survey of India (1962), which emphasized economic aspects and Bouillon (1964), which included several quantitative studies of the composition and size of colonies. Reviews of termite morphology, biology, and zoogeography were included in Krishna and Weesner (1969, 1970) and behavioral aspects of biology were emphasized by Howse (1970). A popular account of their biology was presented by Skaife (1955). Termite damage to crops, trees, and buildings was covered on a worldwide basis by Harris (1971) and for tropical southern Asia by Roonwall (1979). The ecological importance of termites was emphasized by Lee and Wood (1971b) and reviews of research in this field were included in Brian (1978). Recent compilations of papers on social insects include several contributions on termite biology, ecology, physiology, and economic importance (Howse and Clement 1981, Hermann 1979, 1982a,b,c).

At present, termite control depends on the use of a very limited number of insecticides, all of which present health and environmental hazards. Recent research on termites has confirmed that their dependence on microbial symbionts for digestion and a complex chemical-based communication system for colony integration opens possibilities for the development of alternative and selective methods of control. For these possibilities to be realized, these largely laboratory-based research programs need to be integrated with ecological studies of the major pest species, many of which are of greatest significance in tropical regions remote from specialized laboratories.

CLASSIFICATION AND WORLD DISTRIBUTION

Termites belong to the order Isoptera, consisting of 2,231 living and over 60 fossil species. Several versions of their higher classification exist, with the number of families ranging from five to nine. The following arrangement is based on that originally proposed by Hagen (1858), who divided the 60 then known species into four genera, *Calotermes* (= *Kalotermes*), *Termopsis, Hodotermes* and *Termes*. The first three genera form the basis of separate families (Kalotermitidae, Termopsidae, Hodotermitidae) and to these have been added two families derived from Hagen's *Termes* (Rhinotermitidae and Termitidae) and two represented by single species (Mastotermitidae, Serritermitidae) (Grassé 1949, Snyder 1949, Emerson 1955, 1965, 1968). These seven families are subdivided into 15 subfamilies as follows:

Family: Mastotermitidae
             Kalotermitidae
             Termopsidae

  Subfamily   Termopsinae
                 Porotermitinae
                 Stolotermitinae

             Hodotermitidae
                 Cretatermitinae (fossil)
                 Hodotermitinae

             Rhinotermitidae
                 Psammotermitinae
                 Heterotermitinae
                 Stylotermitinae
                 Coptotermitinae
                 Termitogetoninae
                 Rhinotermitinae

             Serritermitidae

Termitidae
  Apicotermitinae
  Termitinae
  Macrotermitinae
  Nasutitermitinae

The first 6 families are referred to collectively as the lower, and the seventh family (Termitidae) as the higher termites, a division based on the dependence of the lower termites on symbiotic protozoa for digestion (discussed under "Food Resources"). Termite phylogeny has been summarized by Krishna (1970), Wilson (1971), and Ampion and Quennedey (1981). The fossil subfamily Cretatermitinae contains the earliest described termite (and social insect), *Valditermes brenanae* (Jarzembowski 1981) from southern England, estimated to be around 120 million years old. The fossil species are closely related to the living forms, indicating little change over many millions of years. In the family Termitidae, genera originally included in the subfamily Amitermitinae (Kemner 1934, Harris 1971) are divided between the Apicotermitinae and the Termitinae (Grassé and Noirot 1954, Sands 1972a).

The number of genera and species allocated to the different families is shown in Table 1.1. The number of species in each genus ranges from one to over 200 (e.g., *Nasutitermes*). The order is dominated by the higher termites, with over 80% of the genera and 74% of the species. In the lower termites, the Kalotermitidae is the largest family, with 21 genera and 350 species. Of the 236 known genera, less than 10% are of economic importance (Harris 1971).

The distribution of termites (including selected pests) in relation to the major zoogeographical regions is shown in Fig. 1.1 and tabulated in detail by Wood (1979). They are entirely terrestrial and within their latitudinal limits are restricted by arid areas (Johnson and Wood 1980a) and are rarely found at altitudes above 3,000 m. Even so they occur over more than half of the world's land surface, an area occupied by well over a hundred countries with a combined population exceeding three billion people. Many of the countries affected are extremely poor and almost half have a gross national product of less than US $500 per capita (1980). Some pest species have a very restricted distribution (e.g., *Cryptotermes*, *Reticulitermes*); others have well-defined geographical limits (e.g., *Microtermes*), while others (e.g., *Coptotermes*) are pantropical.

The number of different species is greatest in the tropical rain forests close to the equator and decreases as the latitude increases. Only a few primitive species extend into the cool temperate regions, the limit of their distribution corresponding approximately to the 45°N and S latitudes. Primitive forms living on as relics in what are now cold climates include

TABLE 1.1
Numbers of Living Genera and Described Species in the
Seven Termite Families

Families	Number of genera	Number of species	
Lower termites			
Mastotermitidae	1	1	
Kalotermitidae	21	350	(2 subspecies)
Termopsidae	5	17	
Hodotermitidae	3	17	
Rhinotermitidae	14	206	(17 subspecies)
Serritermitidae	1	1	
Higher termites			
Termitidae			
Apicotermitinae	40	169	(3 subspecies)
Termitinae	71	639	(21 subspecies)
Macrotermitinae	13	288	(22 subspecies)
Nasutitermitinae	67	543	(10 subspecies)
TOTALS	236	2231	(75 subspecies)

several species belonging to the Termopsidae. *Archotermopsis wroughtoni* (Desneux) is found up to 3,000 m in the sub-Himalayas, *Zootermopsis angusticollis* (Hagen) and *Z. nevadensis* (Hagen) in British Columbia, and the genus *Porotermes* with three species in the tips of the southern hemisphere continents; in Chile, in Cape Province, and in Tasmania and south-eastern Australia. In the northern temperate region, the genus *Reticulitermes* (Rhinotermitidae) is well adapted to the cold winters, although its survival has been aided by the favorable environment created by man. The heated soil beneath buildings not only allows termites to remain active throughout the year, thereby increasing the damage they cause, but has facilitated their spread into colder areas.

The high proportion of endemic genera occurring in the Ethiopian, Neotropical, and Oriental regions, particularly those belonging to the four subfamilies of the Termitidae which evolved during the Cretaceous and Tertiary periods (Emerson 1955), indicates that these three regions are the origin of many of the existing genera.

MORPHOLOGY

Unlike most other insects, the termite parents, the so-called king and queen, live together with their offspring (Fig. 1.2A) in a communal nest. The functional offspring (workers, soldiers) are effectively sterile indi-

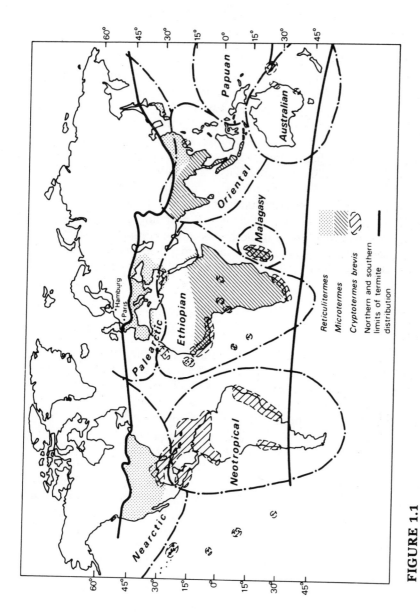

FIGURE 1.1

Limits of termite distribution in relation to zoogeographical regions and distribution of some economically important species.

viduals that assist in the care of the brood and perform most other jobs essential for the success of the community. This reproductive division of labor helps to increase the feeding and reproductive efficiency of the colony and is reflected in the development of morphologically distinct forms or castes (polymorphism). The reproductive castes are either primary or secondary forms. The primary reproductives are the adult winged termites (imagos or alates), which leave the nest at certain times of the year, mate, and establish new colonies. The secondary or supplementary reproductives develop functional reproductive organs without leaving the nest. They can act as replacement reproductives should one or both of the original parents die, or supplement the egg-laying capacity of the queen should this fall below the level necessary to maintain the colony. Termites are most readily recognized as small, pale, wingless insects exposed to view when a nest is disturbed, or a plant or piece of timber is attacked. These are the nonreproductive castes which have been arrested in their development to act as food gatherers, nest builders, nurses (workers), or guards (soldiers), to suit the needs of the community (Fig. 1.2). The behavioral division of labor exhibited by members of this community is known as polyethism.

Reproductives

External Morphology

These are, morphologically, the least modified of the different castes. They vary in size according to the species, from the large African *Macrotermes* (Macrotermitinae) with a body length of 20 mm and a wing span of 90 mm, down to a *Afrosubulitermes* (Nasutitermitinae) from Africa, which is some 4–5 mm long, with a wing span of 9–10 mm. The body is divided into three regions: head, thorax, and abdomen.

Strong triangular mandibles project in front of the head and are of major taxonomic importance, since the teeth on their inner margins have undergone changes as the termites have evolved (Ahmad 1950, Fig. 1.2B and 1.3). The molar plates (Fig. 1.2B) possess a number of transverse ridges in the wood and grass-feeding species, but are modified into a crushing cusp in soil-feeding species (Sands 1965c, Deligne 1966). A small pore of varying size and shape, the fontanelle, is found in the frontal area in the Rhinotermitidae and Termitidae. It is the external opening of the frontal gland (Fig. 1.4A). The function of this exocrine gland in the imago is not known, but in the soldiers of some species it has become modified for defense. A second cephalic exocrine gland, located at the base of the mandibles, has been observed in most termite families (Brossut 1973). The mandibular gland of cockroaches is concerned with aggregation, but its

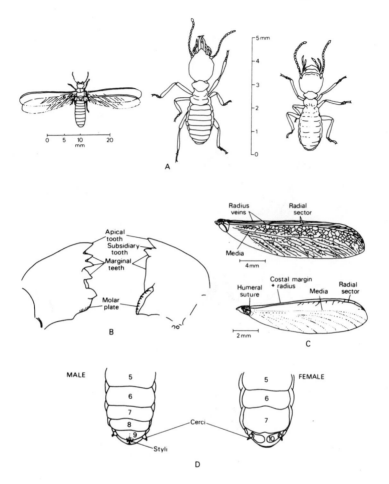

FIGURE 1.2
A: *Ancistrotermes latinotus* (Holmgren), left to right alate imago, major soldier, major worker
B: *Reticulitermes lucifugus*, imago mandibles
C: fore-wing of *Zootermopis angusticolus* (above), *Coptotermes pacificus* Light (below)
D: terminal tergites of male and female *Coptotermes formosanus*.
(*Sources:* A, after Harris 1971; C, D, modified from Weesner 1969)

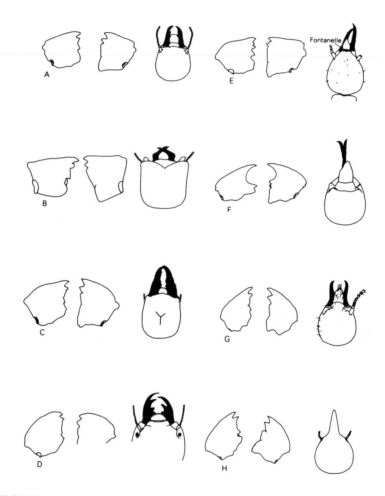

FIGURE 1.3

Imago-worker mandibles and soldier heads.

A: *Mastotermes darwiniensis*, Mastotermitidae;
B: *Cryptotermes havilandi* (Sjöstedt), Kalotermitidae;
C: *Zootermopsis angusticollis*, Termopsidae;
D: *Hodotermes mossambicus* (Hagen), Hodotermitidae;
E: *Coptotermes formosanus* Shiraki, Rhinotermitidae;
F: *Serritermes serrifer*, Serritermitidae;
G: *Microtermes najdensis* Harris, Termitidae-Macrotermitinae;
H: *Nasutitermes arborum* (Smeathman), Termitidae-Nasutitermitinae.

function in termites is not known (Noirot 1969a). Large compound eyes are prominent on each side of the head. In all families except Termopsidae and Hodotermitidae, there are, in addition, two simple eyes or ocelli situated near the upper border of the compound eyes. The number of antennal segments varies from 15 in some of the higher termites to 32 in the most primitive (*Mastotermes*).

The thorax consists of the prothorax, bearing the first pair of legs, and the mesothorax and metathorax, each with a pair of wings and the second and third pair of legs, respectively. In most termites a distinctive

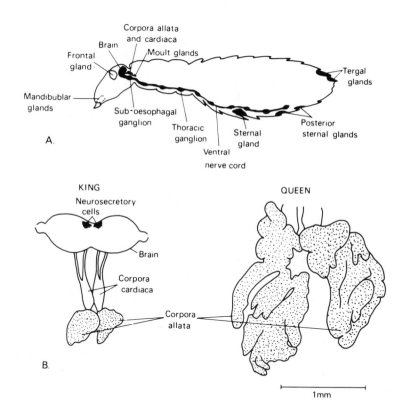

FIGURE 1.4

A: Location of important endocrine and exocrine glands in a generalized termite

B: Cerebral endocrine system of the king and the hypertrophied corpora allata of the mature queen *Macrotermes subhyalinus* (Rambur).

(*Source:* Modified from Lüscher 1976)

number of sclerotized spurs are present at the distal end of the tibia, the number on the fore, mid and hind tibia, respectively, being expressed as a tibial spur formula (e.g., 2 : 2 : 2, 3 : 3 : 3). A tibial-tarsal exocrine gland has been observed on the legs of some Rhinotermitidae, but its function remains unknown (Bacchus 1979). The tarsal segment is five jointed in the Mastotermitidae, Termopsidae, and Hodotermitidae and four jointed in the vast majority of the other termites. The wings do not differ greatly except in the Mastotermitidae, where the hind wings have an enlarged anal lobe, providing a link with a cockroach-like common ancestor of termites (McKittrick 1965, Krishna 1970). The venation is simple, with only the subcostal and radial sector veins being strongly sclerotized (Fig. 1.2C). Cross veins are a character of the primitive families. In the Kalotermitidae the medius varies in form and position relative to the other veins according to the genus (Krishna 1961). A feature, unique to the Isoptera, is the presence of a line of weakness (humeral suture) near the base of the wing, which allows them to be shed after the mating flight, leaving behind a triangular stump or scale attached to the thorax.

The abdomen consists of ten segments, the dorsal plates (tergites) being similar in both sexes, but ventrally dissimilar in the terminal segments (Fig. 1.2D). There are no external genitalia. There are two abdominal exocrine glands (Fig. 1.4A), which play an important role in social behavior. The sternal glands are found in all castes of the termite families so far investigated (Ampion and Quennedey 1981). In *Mastotermes* there are three sternal glands, one each in the middle of the third, fourth, and fifth sternites. In the other familiies, there is a single gland located in the anterior of either the fourth (Termopsidae, Hodotermitidae) or fifth sternite (Kalotermitidae, Rhinotermitidae, Termitidae). Posterior sternal glands (Ampion 1980), whose function is unknown, have been found in the terminal sternites of the imagos of several of the lower termites. The tergal glands, which appear to be restricted to the imagos, are lacking in several subfamilies. In the primitive *Mastotermes*, there are eight glands (tergites 3 to 10), but this number is reduced in the other families. In the Kalotermitidae, there are generally two glands on tergites 9 and 10; they are lacking in the Termopsinae, the Hodotermitidae, and certain Rhinotermitidae but occur in representatives (females only) of all subfamilies of the Termitidae (Ampion and Quennedey 1981). In the Macrotermitinae, for example, they are only found in *Macrotermes* (tergites 3 to 10 of *M. subhyalinus*; Ampion 1980) and *Microtermes* (tergites 9 and 10).

Unlike the primary reproductives, the secondary forms do not leave the parental nest and their offspring therefore add to the size of the colony. Their appearance can take different forms, depending on the species. They may develop from winged forms (Termitidae), larvae or nymphs (see "Caste Differentiation").

Reproductive System

The male and female reproductive systems have been described by Weesner (1969). The number of panoistic ovarioles per ovary ranges from seven to eight in some primitive (e.g., *Kalotermes*) to several thousand in the higher termites (e.g., *Macrotermes*). These differences are reflected in the egg-laying capacity of the mature females (Nielsen and Josens 1978). Figures range from eight eggs per day in *Cryptotermes havilandi* (Sjöstedt) (Wilkinson 1962), 1,000 per day in *Coptotermes formosanus* (King and Spink 1974) up to 86,400 per day in *Odontotermes obesus* (Rambur) (Roonwal 1960). In many species, egg production is continuous throughout the year, although there may be weekly or seasonal fluctuations. In termites living in temperate regions (e.g., *Reticulitermes*), egg production is often suspended during the colder months. Intraspecific variations in egg production depends on several interrelated factors, including colony size (Sieber 1982), food (LaFage and Nutting 1978), and temperature (Garcia and Becker 1975, Steward 1983). Copulation occurs many times throughout the life of the reproductives and sperm is stored in the spermatheca, where it can remain viable for several months.

The male remains relatively unaltered throughout its life, whereas in the female, as egg-laying capacity increases to meet the growing needs of the community, there is an enlargement of the abdomen by the continuous growth of the intersegmental membrane. This abdominal swelling, known as physogastry, is found in other insects, but to a less spectacular degree. In *Mastotermes*, the Kalotermitidae, and Termopsidae, the females develop only slight physogastry. In the other families, the growth of the female abdomen is more noticeable, reaching its maximum in the African *Macrotermes*, with a six-fold increase in length, from 20 mm to over 120 mm. Physogastry is generally less marked in the secondary reproductives.

Workers

Form and Function

This is the most numerous caste and they are responsible for nest construction, foraging, feeding, and, in the soldierless termites, for defense. In the higher termites the trend is toward a permanently sterile worker caste, incapable of further development. However, in lower termites the worker caste is less well defined, develops gradually over several molts, and, although possessing specific worker characteristics, retains the potential for further development into reproductives or soldiers (Miller 1969). To take into account this developmental flexibility, terms such as pseudergate (Grassé and Noirot 1947) and pseudoworker (Williams 1973) have been used to describe worker equivalents in this group. Since the pseudergate of *Kalotermes flavicollis* (Fabricius) (Fig. 1.5) appears to be a highly specialized form, it has been proposed that the term worker is used for all termites to describe an individual's behavior, irrespective of its developmental potential (Watson and Sewell 1981).

Workers are wingless immature individuals, which otherwise resemble the adults. When fully functional, they have strongly sclerotized mandibles, lightly sclerotized heads, and except in a few species, very lightly sclerotized bodies. In most cases, compound eyes and ocelli are lacking. The number of antennal segments is usually smaller than the other castes. In the lower termites, the size of the workers can vary considerably, depending on their stage of development. In some of the Termitidae there are two distinct sizes based on sex and referred to as majors and minors. In the Macrotermitinae, the major workers are males (Fig. 1.6) and forage outside the nest, whereas the minors are females and work in the nest (Noirot 1974). Task bias may also be age related (temporal polyethism) and be reflected in minor morphological differences and physiological changes, such as gland activity. In general, the younger individuals tend to remain in the nest caring for the brood, while the older individuals initiate nest repair and foraging (McMahan 1979).

Most of the African soldierless termites have a line of weakness behind the metanotum, and when molested by predators the abdominal muscles contract, the abdominal wall ruptures, and the intestine bursts, scattering its contents (Sands 1972a, 1982). This phenomenon (autothysis) has also been observed in some South American Apicotermitinae (Mathews 1977). Use of rectal contents for defense has been observed in *Skatitermes* (Coaton 1971) and *Speculitermes* (Sands 1982).

The Alimentary System

The alimentary system occupies a large part of the abdomen as is usual with animals consuming materials that are difficult to digest. Noirot and Noirot-Timothée (1969) gave an excellent review of the anatomy of the alimentary system. There are variations between taxonomic and feeding groups but in all termites there are three major regions: foregut (stomodeum), midgut (mesenteron), and hindgut (proctodem) (Fig. 1.5).

Food entering the buccal cavity is subjected to secretions from the paired salivary glands before passing into the foregut. This consists of a short esophagus leading to its posterior dilation, the crop, and a muscular gizzard where food is comminuted by cuticular armature of the internal epithelial folds. It terminates in an esophageal valve that penetrates into the midgut. The latter has a peritrophic membrane and in lower termites its simple junction with the hindgut is marked by a proctodeal valve and the Malpighian tubules, which are excretory organs. The midgut-

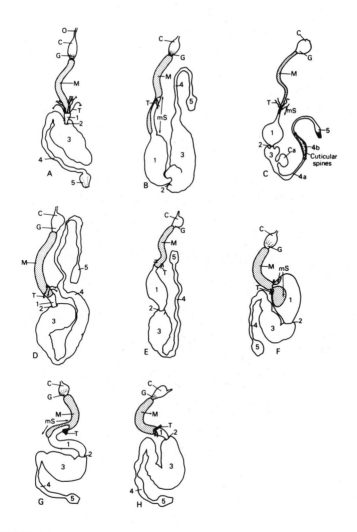

FIGURE 1.5

Major regions of the alimentary system.

    A: Lower termites, *Kalotermes flavicollis*;

    B&C: Termitinae, B: *Drepanotermes rubriceps* (Froggatt), (herbivore and detritivore); C: *Cubitermes severus* Silvestri, (soil-feeder);

    D,E: Apicotermitinae, D: *Allognathotermes hypogeus* Silvestri (soil-feeder); E: unidentified 'soldierless' species (soil-feeder);

    F,G,H: Nasutitermitinae, *Syntermes dirus* (Burmeister), (grass-feeder); G: *Nasutitermes arborum* (wood-feeder); H: *Postsubulitermes parvicontrictus* Emerson (soil-feeder).

Notation: O, esophagus; C, crop; G, gizzard; M, midgut; T, Malpighian tubules; mS, mixed segment; 1, 1st proctodeal segment; 2, enteric valve; 3, paunch (Ca, caecum); 4, colon; 5, rectum.

(*Sources:* A, B, D, E, F, G, H, after Noirot and Noirot-Timothée 1969; C, after Bignell, Oskarsson, and Anderson 1980b)

13

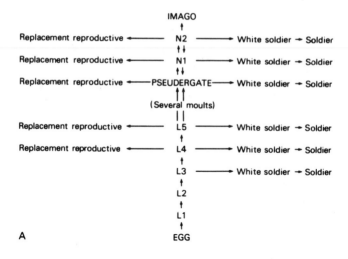

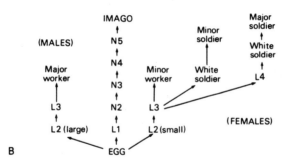

FIGURE 1.6
Generalized scheme of caste development:
A: *Kalotermes flavicollis* (modified from Lüscher 1961a);
B: *Macrotermes michaelseni* (modified from Okot-Kotber 1981).
Notation: L, larva; N, nymph.

hindgut junction varies in the Termitidae from a simple junction in the Macrotermitinae to a complex mixed segment consisting of mesenteric epithelium bearing Malpighian tubules on one side and proctodeal epithelium on the other (Fig. 1.5B, C, F, G). The mixed segment typically harbors numerous bacteria between the gut wall and the peritrophic membrane. The hindgut consists of five parts (P). P_1 may be very short, long, and tubular or dilated. P_2 is a short, muscular funnel containing the enteric valve which is of considerable taxonomic importance in certain termites (Sands 1972a). P_3, the paunch, is considerably dilated, occasionally possesses a lateral diverticulum (the caecum), and typically contains the majority of the intestinal symbionts. P_4, the colon, is a narrow, often contorted, tube that in certain soil-feeding termites is divided into anterior and posterior regions, with the latter containing numerous elongated, internal cuticular filaments that serve as sites for the attachment of symbiotic bacteria, such as actinomycetes (Fig. 1.5C). P_5, the rectum, is short and muscular.

Soldiers

The soldier caste is characterized by specialization of the head for defense of the colony. In general, two types of soldiers can be distinguished: mandibulate types, which rely on large, prominent mandibles; and nasutoid types, in which the front of the head is modified for chemical warfare and the mandibles are often atrophied. The two categories are not mutually exclusive (Deligne et al. 1981).

Mandibulate soldiers occur in all families, and defense is achieved by the biting, pincer-like, or snapping action of the mandibles. These are generally robust structures, with or without lateral teeth, whose diverse and sometimes grotesque appearance is of taxonomic importance (Fig. 1.3). The labrum is also modified from the adult form. It may have a cartilaginous, characteristically shaped tip and be considerably elongated. The other mouthparts vary little from the basic adult pattern. The head capsule is often elongated to accommodate the muscles necessary for the large mandibles. The enlarged head is used by some termites (e.g., *Cryptotermes*) to effectively seal off threatened galleries and entrance holes, a habit known as phragmosis.

Soldiers employing chemical defense are confined to *Mastotermes*, the Rhinotermitidae, and Termitidae. In *Mastotermes* and a few other species (e.g., *Globitermes sulphureus* (Haviland)) and many Macrotermitinae, saliva is used in defense and the salivary glands may occupy a large part of the abdomen. In *Mastotermes*, the modified salivary secretion acts as an irritant and has a dark, rubber-like appearance when rubbed onto an opponent. In *Macrotermes carbonarius* John, saliva is discharged into the open wound caused by the mechanical action of the mandibles (Maschwitz et al. 1972). The Rhinotermitidae and Termitidae possess a frontal gland opening at a pore, the fontanelle, which may appear as a conspicuous hole (e.g., *Coptotermes*, Fig. 1.3E). Some soldiers have a frontal gutter for controlling the flow of the frontal gland secretion. The modification of the head for chemical warfare reaches its peak of specialization in the Nasutitermitinae, where the frontal pore opens at the tip of a snoutlike projection, the nasus (Fig. 1.3H). When disturbed, the modified mandibular muscles compress the frontal gland and its secretion can be expelled for several centimeters, thereby avoiding any physical contact with the enemy.

The main enemies of termites are predatory ants. Defense secretions can act in several ways, either as an irritant, repellent, toxicant, mechanical immobilizer, or a combination of these. Their chemical composition has been examined in a number of species and more than 70 compounds have so far been identified, many of which incorporate chemical structures hitherto unknown. The salivary defense secretions of *Mastotermes* and several Macrotermitinae contain quinones, that have general toxicity and irritant properties. Terpenes are common components in the Nasutitermitinae. Several monoterpenes and diterpenes with insecticidal properties have been identified (Prestwich 1979, Deligne et al. 1981).

Compound eyes are present in the Mastotermitidae and most Hodotermitidae, progressively reduced in the Termopsidae, and absent in the rest. Traces of the ocelli are present only in the most primitive species. The number of antennal segments varies from 11 to 29 in different species, but is liable to greater interspecific variation than in the imago. The rest of the body is changed little from the adult form. Dimorphism is common and some species have trimorphic soldiers. The forms differ in size and behavior; minor soldiers accompany foraging parties while major soldiers are largely employed in defense of the nest.

The Endocrine System

Many aspects of larval and adult development (see "Caste Differentiation") are under endocrine control (Noirot 1977). The anatomy and histology of the endocrine system (Fig. 1.4) has been described by Noirot (1969b) and is essentially similar to that observed in the nonsocial insects.

THE NERVOUS AND SENSORY SYSTEMS

The termite nervous and sense organs have been described by Richard (1969). The general plan of the central nervous system is similar to that of other insects, with a cerebral (the brain) and subesophageal ganglion, and three thoracic and six abdominal ganglia (Fig. 1.4A).

The nervous system is the target of the insecticides most commonly used for termite control, the organochlorine compounds (e.g., aldrin, chlordane). They may be ingested or, since most are lipophylic, absorbed through the cuticle. They act by disrupting the functioning of the nervous system, resulting in hyperexcitability and muscular spasms, followed by paralysis and death.

ESTABLISHMENT, GROWTH, AND INTEGRATION OF THE COLONY

Colony Foundation

Foundation by Alate, Reproductive Pairs

The dispersal of winged reproductives is common to all termites and is the most important means of establishing new communities. Imagos are only produced by relatively mature colonies (see "Colony Growth," Noirot 1969b) and the number produced ranges from 1% to 43% of the total individuals (Wood and Sands 1978) representing over 50% of the energy of the society.

Detailed observations on swarming have only been made for a few species (Nutting 1969, Nutting and Haverty 1976, Nel and Hewitt 1978, Nutting 1979a). In general, the primitive pattern may be summed up as "little and often." In *Cryptotermes* (Kalotermitidae) and *Zootermopsis* (Termopsidae), for example, the alates tend to be released in small numbers over a period of several months, but often show seasonal peaks. In the higher termites, swarming becomes increasingly restricted to certain times of the year and geared to specific times of the day, with the result that much larger numbers emerge together than in the more primitive families. Swarming tends to synchronize with the regional weather pattern and occurs during the warmer months in temperate regions and largely during the wet season in the tropics. In localities where there are several species belonging to the same genus, swarming occurs either at different times of the year, or at different times of the day (Wood 1981). In addition to acting as a species-isolating mechanism, it seems likely that different flight times during the day are a behavioral adaptation for avoiding at least some of the many predators (Nutting 1969, 1979a). Other defense mechanisms include "jumping" in *Zootermopsis*, caused by a rapid contraction of the flight muscles (Stuart 1969), and feigning death, as in *Trinervitermes occidentalis* (Sjöstedt) (Sands 1965d). In spite of these behavioral defense strategies, however, usually over 99% of the swarming termites are lost to predators (Wood and Sands 1978, Deligne et al. 1981).

In view of the high predation pressures, survival is aided by a short flight period and the rapid location of a mate. In some of the lower and most of the higher termites, rapid pairing is achieved by means of a chemical sex attractant released from either the tergal or sternal glands (Fig. 1.4A). In most cases, the female attracts the male, but in *Hodotermes mossambicus* the roles are reversed (Leuthold 1977).

The Kalotermitidae seek out cracks or holes in timber above ground to establish their colonies. In other termites, including those that eventually build mounds or nest in trees, the founding pair burrow into the soil and establish their colonies below ground.

Foundation by Budding

Examples of colony foundation by the isolation of part of an existing one (budding) have been reported for all termite families, except the Serritermitidae (Nutting 1969). However, it is more common in the lower termites since the essential prerequisites include a large, diffuse nest system and the potential for isolated groups to develop supplementary reproductives. It may constitute the principal method of colony foundation in some species. For example, colonies headed by primary reproductives are extremely rare in *Mastotermes darwiniensis* (Gay 1970) and have never been observed in *Reticulitermes lucifugus* (Rossi) from Italy (Grassé 1949). Budding ensures that new colonies are quickly established and avoids the high energy losses resulting from predation. In addition, recent studies have shown that, at least in some species, the primary and secondary reproductive forms have different optimum climatic requirements for colony foundation. For example, in *Cryptotermes*, the imagos require higher humidities for reproduction (Steward 1983). Colony foundation by budding could, therefore, increase the climatic range of a species and may be an important adaptation in the successful establishment of introduced species (Lenz and Barrett 1982). Budding, particularly in the Rhinotermitidae (e.g., *Coptotermes*, *Reticulitermes*), can result in difficult postconstruction building protection, since the usual soil treatments may only succeed in isolating any subsidiary nests within the building from the main subterranean nest. If left untreated, these isolated nests become new colonies and the damage to the property continues.

Foundation by Social Fragmentation

The third method of colony foundation, referred to as social fragmentation or sociotomie (Grassé and Noirot 1951) has only been observed in a few species of the Termitidae (Nutting 1969). This unusual, risky and exceedingly rare behavior involves the migration of the reproductives, accompanied by workers and nymphs, to a new nesting site, leaving the orphaned community to develop secondary reproductives. At least in some cases, this migration has resulted from attack by subterranean doryline ants (Harris and Sands 1965), and extraneous pressures may be the primary cause of this behavior.

Colony Growth

Colony Establishment

The details of colony establishment and subsequent development have largely been obtained by direct observation of laboratory cultures (Becker

1969). Copulation occurs within several hours to several days from establishment of the copularium (Nutting 1969). Oviposition begins from within five days to two to three weeks. After the first batch of eggs is laid oviposition ceases until these have hatched or until the first foraging workers have developed. Populations at the end of one year vary: 10–60 in various Kalotermitidae, where the food (sound wood) is not very nutritious; several hundred to 1000 in certain Hodotermitidae, Rhinotermitidae, and Termitidae, where the larval food is more nutritious (Nutting 1969, King and Spink 1974).

In the economically important subfamily Macrotermitinae the survival of young colonies is dependent on the early successful inoculation of the fungus comb with *Termitomyces* (Sands 1969) and it is clear that two quite different methods are used (Johnson et al. 1981b). In *Microtermes*, the first workers appear in the nest approximately seven weeks after pairing and foraging commences some two to three weeks later (Johnson 1981). A primordial fungus comb appears in the nest shortly afterward and within a few days, fungal mycelia are clearly visible. Food passed through the guts of the foraging termites is deposited onto the comb and its digestion by the fungus takes from five to six weeks. It therefore takes 16 to 17 weeks before the fungus comb begins to contribute to the nourishment of the colony. During this critical period, the energy requirements for colony establishment are provided by the two reproductives which, since they are unable to feed, suffer a 75% loss in dry weight. The critical inoculation of the new comb is achieved by the female reproductive which, before leaving the parental nest, ingests asexual fungal spores from the fungus comb. These survive in her digestive tube for about ten weeks, before being deposited on the new, potential comb. The only other member of the subfamily known to successfully inoculate the fungus comb using this method is *Macrotermes bellicosus* Smeathman, where the male carries the inoculum. The second method appears to depend on the first foraging workers picking up basidiospores at or close to the soil surface, produced by the *Termitomyces* fruiting bodies (basidiocarps) (Sieber 1983).

Colony Growth

All colonies go through the three growth phases (juvenile, mature, and senile) recognized by Noirot (1970) but it is the rates that appear to differ in different species rather than the life span. Bodenheimer (1937) fitted logistic curves to Kalshoven's (1930) data on the growth and senescence of colonies of *Neotermes tectonae* (Dammerman). Most colonies reached their maximum size (844 individuals) after 5½ years and died after ten years while larger colonies reached their maximum size (2,844 individuals) after eight years and died after 15–16 years. Other species have more rapid growth rates (e.g., *Coptotermes formosanus* with populations exceeding 5×10^4 after four years, King and Spink 1974). In the mound-building *Macrotermes bellicosus* Smeathman in West Africa the juvenile period of rapid growth in numbers and mound size lasts for four to six years and populations of $4-8 \times 10^5$ are attained. During the mature period of four to six to 10–12 years, populations stabilize and alates are produced; it is not clear what induces the senescent phase but maximum longevity appears to be 20–25 years (Collins 1981a). A very similar time span was indicated for the more populous colonies (maximum exceeding 5×10^6 individuals) of *M. michaelsoni* (Sjöstedt) in East Africa (Darlington 1982a).

Although the ability to produce supplementary reproductives endows some colonies with potential immortality there are no reliable estimates of colonies living beyond the 50 years estimated for *Nasutitermes exitiosus* (Hill) (Ratcliffe et al. 1952) and *Amitermes hastatus* (Haviland) (Skaife 1955). The maximum recorded size of a single colony appears to be 7×10^6 for *Mastotermes darwiniensis* in northern Australia (Spragg and Paton 1980).

Caste Differentiation

Developmental Pathways

Termite development is a complex phenomenon, particularly in the lower termites. This is due, in part, to the production of three morphologically distinct functional forms, but also to the ability of some species to alter a particular line of development, according to the changing needs of the community. It is not always possible therefore to describe development in precise terms, such as the number and duration of larval and nymphal stages.

Although relatively few species have been studied, the development of many of the economically important species is known, at least in outline. These include:

Mastotermitidae
   Mastotermes darwiniensis   (Watson et al. 1977)
Kalotermitidae
   Kalotermes flavicollis   (Lüscher 1952)
   Kalotermes spp.   (Sewell and Watson 1981)
   Neotermes jouteli (Banks)   (Nagin 1972)
   Neotermes tectonae   (Kalshoven 1930)
Termopsidae
   Zootermopsis spp.   (Miller 1969, Lüscher 1974)

Hodotermitidae
    Anacanthotermes ahngerianus (Jacobson)   (Zhuzhikov et al. 1972)
    Hodotermes mossambicus   (Watson 1973)
Rhinotermitidae
    Reticulitermes santonensis de Feyteaud   (Buchli 1958)
    Reticulitermes flavipes (Kollar)   (Howard et al. 1981)
    Reticulitermes virginicus (Banks)   "
    Coptotermes formosanus   (King and Spink 1974)
Termitidae
    Amitermes evuncifer Silvestri   (Noirot 1955)
    Macrotermes michaelseni   (Okot-Kotber 1981)
    Microtermes sp.   (Johnson 1981)
    Trinervitermes spp.   (Sands 1965d)

(a) Imago. Termite development is hemimetabolous, or an incomplete metamorphosis. Postembryonic development of the winged adult includes several larval and nymphal (possessing wing pads) instars. In the European drywood termite *Kalotermes flavicollis*, development follows a linear pathway (Watson and Sewell 1981). There are between five and eight larval instars leading to the fully grown larva or pseudergate (Fig. 1.6). In all other termites studied, development of the imago is evident at the first or second molt. In *Mastotermes darwiniensis*, for example, although nymphs do not appear until the fifth instar, the reproductive line can be distinguished from the worker line after the first molt (Watson et al. 1977). There are eight nymphal instars leading to the winged adult. In the higher termites, development is usually much more predictable (e.g., *Macrotermes michaelseni*, Fig. 1.6). Nymphs are formed at the first molt and there are usually five nymphal instars leading to the imago.

(b) Secondary Reproductives. These forms can be divided into three basic types; adultoids develop from imagos, nymphoids develop from nymphs of various ages and ergatoids develop from workers. Adultoid reproductives are only found in certain Termitidae (Noirot 1956, 1969b). In some species, their development depends on the time of year when the colony is orphaned. In *Macrotermes michaelsoni*, for example, they are only produced if nymphs or alates are present in the nest when the primary reproductives die or are removed (Sieber and Darlington 1982). The other secondary reproductives are neotenics and develop functional reproductive organs, without becoming alates. The nymphoid or brachypterous neotenics have been recorded in numerous species (e.g., *Drepanotermes perniger* (Froggatt), Watson 1974) and often appear in large numbers. Ergatoid or apterous neotenics are apparently less common. Field colonies of *Mastotermes* are normally headed by numerous apterous neoten-

ics (Gay 1970), developed from workers following a single molt (Watson et al. 1977). Ergatoids have been observed in a few species of the higher termites, including *Neotermes corniger* (Thorne and Noirot 1982).

(c) Workers.   In the higher termites, there is a trend toward a permanently sterile worker caste, stabilized with one particular adult feature, namely the shape and hardness of the mandibles. They are formed after the second or third molt, depending on the subfamily (Noirot 1969b). In the Macrotermitinae, for example, workers appear after three larval instars and are dimorphic (Fig. 1.6).

A fixed worker caste rarely exists in the lower termites, since, should the need arise, individuals assuming the role of workers retain the ability to develop into soldiers or neotenics. In addition, the distinction between larva and worker may be poorly defined, since the former may assist in performing certain tasks, such as the care of the brood. In *Kalotermes flavicollis* (Fig. 1.6), the developing larvae are able to feed themselves and participate in the care of the younger larvae and parents after the third instar. The worker stage or pseudergate, formed by progressive larval molts or nymphal reversionary molts, retains extreme developmental flexibility.

(d) Soldier.   The molt to the soldier caste is preceded by a characteristic, obligatory larval instar (white or presoldier) which lacks pigment and is unsclerotized, but otherwise resembles the mature soldier. In the lower termites, they can develop from workers (*M. darwiniensis*), pseudergates (*K. flavicollis*; Fig. 1.6) or rarely from nymphs (*Zootermopsis*) and appear as early as the third or fourth instar in newly founded colonies and instars five to ten in mature colonies. In the higher termites, as with the workers, soldier development is generally more predictable and the stage at which they appear is often very precise for each species. The dimorphic soldiers of the Macrotermitinae, for example, develop after four or five instars, depending on their size. In most of the Macrotermitinae and Termitinae, soldiers are females. In most Nasutitermitinae they are males and in a few species of the Termitinae (e.g., *Amitermes evuncifer*) there are equal proportions of male and female soldiers.

Caste Determination and Regulation

In mature termite colonies, with few exceptions, the production of imagos is seasonal, whereas the neuter castes are produced more or less continuously throughout the year. For each species, the relative proportions of workers and soldiers are maintained at fairly constant levels (Haverty 1977, 1979) and these ratios are reestablished after disturbance (Lenz

1976). Caste determination is, therefore, regulated and geared to the needs of the community, allowing it to maintain optimum feeding and reproductive efficiencies in a highly competitive environment. The mechanisms involved in caste regulation are complex and many details have yet to be worked out. It is clear, however, that both extrinsic and intrinsic factors are involved.

Caste regulation is best understood in the lower termites. In this group, when the founding reproductives are removed from a colony, neotenics quickly develop to replace them. Since neotenics are not produced in intact colonies, it was postulated that the primary reproductives are in some way inhibiting their development. Subsequent observations and experimentation on *Kalotermes flavicollis* and *Zootermopsis* have developed and refined this inhibition hypothesis and recognized the essential role played by hormones (Miller 1969, Stuart 1969, Lüscher 1972, 1973, 1977, Greenberg and Stuart 1982). These developments have recently been summarized by Brian (1979), Stuart (1979), and de Wilde and Beetsma (1982). The hormonal mechanisms thought to be involved in caste regulation can be divided into two aspects: an extrinsic factor involving pheromones and an intrinsic factor-juvenile hormone.

Pheromones are compounds that, when released by one individual, affect the behavior or physiology of other individuals of the same community (Karlson and Lüscher 1959). Several pheromones produced by the male and female reproductives have been implicated in the control of neotenic formation in *K. flavicollis* (Lüscher 1961a). Perhaps the one most studied is the proposed inhibitory pheromone produced by the female, which acts to prevent the production of more female secondary reproductives, but does not affect the development of nymphs to alates. It appears to be circulated through the larval population by anal trophallaxis. The males produce both an inhibitory pheromone preventing the development of male reproductives and a stimulatory pheromone that promotes the formation of female reproductives. However, none of these pheromones have been isolated and the evidence for their existence is based on indirect experimentation. Even so, the pheromone-mediated hypothesis proposed for *K. flavicollis* has been widely accepted as a basis for explaining caste regulation in the lower termites, although the results of more recent studies have questioned certain aspects of its general application (Stuart 1979, Greenberg and Stuart 1982).

Juvenile hormone (JH) is produced by the corpora allata (Fig. 1.4) and is well known in insects for its role in the control of metamorphosis and reproductive development. It is a terpenoid; four forms are known, the type isolated from the higher termites (*Macrotermes*, Meyer et al. 1976) is JH-III (C-16 juvenile hormone). Its involvement in caste differentiation is now well established. In many species, the hormone or its analogs, when administered to larvae topically, by feeding, via injection, or in the vapor form, increases soldier production (Lüscher 1969, Wanyoni 1974, Howard and Haverty 1979, Hrdy 1981, Okot-Kotber 1980a).

In the lower termites, developmental pathways can be altered according to the needs of the community (Fig. 1.6). Yin and Gillot (1975), studying the endocrine control of caste differentiation in *Zootermopsis angusticollis*, have proposed that a particular developmental pathway is determined by the levels (titers) of JH in the hemolymph at different times during the instar. Since developmental competence is highest early in the instar, the JH titer during this period determines the identity of the following instar. High titers result in soldier development, with ever-decreasing levels giving rise to apterous larvae, brachypterous larvae, and neotenics, respectively, with very low or zero titers preceding the imaginal molt.

In the lower termites, a constant soldier ratio is maintained by the existing soldiers inhibiting the formation of additional forms (Lenz 1976). However, whether soldiers produce inhibitory pheromone, which lowers the activity of the larval corpora allata, or employ some other method for maintaining a low JH titer, is not known.

In the higher termites, caste development is less flexible and in some species is sex related. For example, in the Macrotermitinae the major workers are males and the minor workers and soldiers are females (Fig. 1.6). However, as in the lower termites, the overriding influence imposed by JH is indicated by the fact that soldier characters have been induced in the male larvae of *Macrotermes* following treatment with hormone analogs (Okot-Kotber 1980b). In addition, in the third instar female larvae, low corpora allata activity determines worker development, medium activity determines minor soldier development, and high activity determines major soldier development (Okot-Kotber 1980b).

The early separation of the reproductive and neuter lines in these termites has also raised the possibility of blastogenic caste formation. The hypertrophied corpora allata of physogastric *Macrotermes* queens, which may be thirty times as large as that of the male partner (Fig. 1.4) produce large quantities of JH (Lüscher, 1976). This is reflected not only in high titers of JH-III (Lanzrein et al. 1978), but high levels have also been recorded in the eggs and the anal fluid. Varying levels of JH have been found in the eggs at different times of the year and appear to be correlated with nymph formation; low levels giving rise to nymphs and high levels to neuters. Alternatively, some of the eggs may be eaten (oophagy) and used to feed the developing larvae. Such a trophic pathway could be expected for the rich source of extrinsic JH contained in the anal fluid, since it is produced almost continuously and eagerly sought after by the nurse workers.

Other caste regulating factors include caste elimination and colony nutrition. Caste elimination has been observed in several of the lower termites (Lenz 1976). In *K. flavicollis*, for example, several neotenics are formed following the removal of the primary reproductives, but only one pair are retained (Ruppli 1969). Fights occur between neotenics of the same sex and wounded or supernumerary individuals may be eaten by the larvae or pseudergates. The important influence of nutrition on caste development is poorly defined. It may affect an individual's physiological state, which in turn influences developmental competence (Miller 1969, Esenther 1977, LaFage and Nutting 1978, Stuart 1979).

Knowledge of the mechanisms involved in caste regulation has important practical implications in termite control. The treatment of termite colonies or groups of termites in the laboratory has in some cases resulted in the overproduction of soldiers (Haverty and Howard 1979), defaunation (Howard and Haverty 1978), the inhibition of feeding (French and Robinson 1978), and increased egg mortality (Howard 1980). However, JH analogs are unstable when exposed and this approach to control requires further study and has yet to be tested in the field (Howard and Haverty 1979, Hrdy 1981).

COLONY INTEGRATION

Termites owe much of their success and consequent economic importance to their highly developed social organization (Stuart 1969, Howse 1970, Wilson 1971, Hermann 1979). Since they spend most of their lives in total darkness, they rely almost entirely on tactile and chemical stimuli for communication and are well equipped with the appropriate receptor organs.

The intimate body contact maintained within the nest makes tactile communication almost inevitable. Sudden jerking movements of the body are used by the larvae to attract the attention of the nurse workers and by workers and soldiers to spread alarm or excitement. Termites also respond to vibrations transmitted through the nest material, which are detected through the subgenual organs in their legs. The characteristic tapping noise produced when a nest or foraging party is disturbed is an alarm signal created by the soldiers striking their heads against the floor of the nest or the food source.

Chemical communication is of primary importance in the integration of the termite society and appears to be involved in almost every aspect of colony life. However, it is a complex phenomenon and, as yet, few examples have been conclusively demonstrated. It is thought that termites recognize members of their own community by their so-called colony odor. This is a mixture of cuticular secretions and the odor absorbed onto the cuticle from the nest material. Subtle differences in the composition of the cuticular hydrocarbons may also be used to identify the various castes within the community (Howard and Blomquist 1982).

Termites produce a number of pheromones and some have already been mentioned in relation to postflight pairing and caste differentiation. Two types are distinguished, releaser pheromones, which evoke an immediate behavioral response (e.g., sex attractant) and primer pheromones, which stimulate or inhibit some physiological function (e.g., female inhibitory pheromone). They may be airborne or circulated by direct contact during mutual grooming and feeding (see "Trophallaxis").

In addition to the sex attractant produced by the tergal and sternal glands, releaser pheromones are also used for regulating different aspects of behavior associated with foraging, nest building, and defense. A trail pheromone produced by the sternal gland is particularly important in those species that forage above ground and outside the protection of runways and sheeting. It can function both as a stimulus to recruit foragers to the food source and to guide the termites between the nest and their food (Oloo and Leuthold 1979). The same pheromone may also influence building and alarm behavior. The trail pheromone has been isolated and identified in a few species (Moore 1974). Pheromones involved in nest construction in *Macrotermes* have been reported to be present in the saliva of the worker termites and produced by the fat body of the queen (oleopalmitic acid, Bruinsma and Leuthold 1977). In some of the higher termites alarm and aggressive behavior is initiated by a pheromone produced by the soldier frontal glands (Deligne et al. 1981).

THE NEST SYSTEM

Spatial Organization

The nest, comprising the royal cell, brood chambers, food stores, and associated foraging galleries and runways, is a more or less closed system. Peripherally, its boundaries are constantly changing as foraging parties exploit different areas of their environment. Many species have a distinct, centralized nest system with eggs, juveniles, and food stores concentrated in a hive or habitacle ("endoecie," Grassé 1949) around the royal cell so that food and nest-building materials have to be transported some distance. There is every gradation from centralized (monocalic) nest systems to extremely diffuse (polycalic) nests. In the latter the royal cell is isolated from widely dispersed, interconnected brood chambers, galleries, and specialized structures containing stores of food and even water (Leprun 1976).

The variety of nest systems have been discussed by Emerson (1938), Grassé (1949), Noirot (1970), Howse (1970), and Lee and Wood (1971b). Six categories were recognized by Baroni-Urbani et al. (1978): (a) Associated with soil, (i) subterranean with no external indication of their presence (e.g. many soil-feeders), (ii) subterranean with surface holes for the passage of foragers (e.g. those feeding on above ground vegetation), (iii) partly subterranean and partly epigeal (mound builders); (b) Associated with living or dead vegetation above ground, (iv) within the substrate with no external indication of their presence (e.g., dry-wood termites, Kalotermitidae), (v) within the substrate with surface holes for the passage of foragers (e.g., many lower termites that nest in logs or trees), (vi) attached externally to the plant (e.g., several Nasutitermitinae).

The cryptic nature of types (i), (ii), (iv), and (v) make nest location difficult if not impossible, and presents greater difficulties in control compared with (iii) and (vi), where the nest centers can be located and destroyed.

Composition and Construction

The materials used for construction were reviewed by Lee and Wood (1971b), Wood and Sands (1978), and Roy-Noel (1979) and consist of soil, feces and saliva in varying proportions. Soil carried in the mandibles or crop of the workers is deposited at the site of construction and cemented into place with saliva or with a thin layer of liquid feces. New constructions have a "pelleted" appearance and are often filled in later with soil to make a hard, compact structure. Kalotermitidae have no contact with soil and their nests are simple excavations within their food source. Their feces are dry, pellet-like and accumulate as a loose frass within the nest. Soil-feeding termites, which also largely excavate nests within their food source, make their nest structures from a mixture of transported and fecal soil. The Macrotermitinae deposit their feces on special structures, fungus combs, within the nest and the rest of the nest system (mounds, chambers, runways) is constructed entirely from soil cemented with saliva.

In many species there is no precise selection of soil particles and the nests reflect the particle-size composition of the soil over a wide range of values, with preferential selection of clay on sandy soils and of sand on clayey soils. Some species are more consistently selective. Many Macrotermitinae, particularly those species of *Macrotermes* and *Odontotermes* that construct large mounds, select clay-rich subsoil, and the exterior of the mounds of several other species (e.g., *Coptotermes* in Australia) is also rich in clay. Parts of the subterranean nests of soil-feeding *Apicotermes* consistently consist of 81–83% fine sand (Stumper 1923, Noirot 1970). The composition of different regions of the nest system is often very variable. The royal cell of *Macrotermes* is often much richer in clay than other regions of the nest. The central hive of certain wood-feeding Nasutitermitinae is sometimes constructed entirely from liquid feces; associated structures are often variable mixtures of soil and feces and the external wall of the nest may be constructed largely from soil (e.g., *Nasutitermes exitiosus* in Australia, Gay and Calaby 1970). Structures dominated by feces are collectively known as carton, and thin fecal layers are often used to line the internal walls of galleries and runways.

The chemical composition of termite nests, particularly mounds, was reported by Lee and Wood (1971a, 1971b), Wood and Sands (1978), and Sheikh and Kayani (1982). Boyer (1976) made particularly detailed studies on African *Macrotermes* mounds. The most significant changes are in the composition of the organic (fecal carton) structures, which generally have high carbon/nitrogen ratios (occasionally exceeding 100) and high lignin/cellulose ratios (up to 15.7, compared with values of less than 1.0 in wood). The addition of clay and organic matter to the constructional soil results in an increase in the total exchangeable cations. The accumulation of calcium at the base of *Macrotermes* mounds in certain localities has been reported by several authors (e.g., Pendleton 1941, Hesse 1955, Watson 1962, 1974). It is generally thought to be due to the large evaporating surface of the mounds, supplemented by an internal ventilation system which, in association with calcium-rich ground waters, results in the accumulation of salts near the base of the mound. These calcium carbonate concretions may amount to 2,000 kg per mound (Milne 1947).

Dimensions

Nests vary from simple, small excavations 1–2 m in length (e.g., some Kalotermitidae and termites that live in nests of other termites) to huge subterranean and epigeal structures up to 9 m above ground. Where mounds are abundant they commonly occupy 1% of the surface area and contain from 5×10^3 to 45×10^3 kg ha^{-1} of soil (Lee and Wood 1971b). Exceptionally their impact is more spectacular, as in the case of the old *Macrotermes* mounds in the Congo, which occupied 30% of the surface area and contained 24×10^5 kg ha^{-1} of soil (Meyer 1960). The amount of soil in above ground runways in African savannas was estimated to be 250–950 kg ha^2 (Wood and Sands 1978). The peripheral network of galleries has been traced for up to 50 m from mounds of *Coptotermes* and *Macrotermes* (see "Foraging") and there are records of galleries penetrating to depths of 70 m, although such depths are exceptional.

There are few estimates of the size of subterranean nests. Haverty et al. (1975) estimated nests of *Heterotermes aureus* (Snyder) occupied surface areas of 12.5 m^2, *Microtermes* in West Africa occupied 240 m^2 with

560 fungus combs per colony (Josens 1971) while Coaton's (1958) estimate of 6.5 ha for *Hodotermes mossambicus* in South Africa is almost certainly an overestimate.

Architecture and Internal Microclimate

Small nests are subject to the temperature and moisture regimes of the environment. Larger colonies in centrally concentrated nests produce large amounts of metabolic heat and CO_2 that must be dissipated. However, the production of heat, combined with elaborate internal and external structural modifications and site selection, contributes to varying degrees of microclimate regulation within the nest. Selected examples are shown in Fig. 1.7. The nest of *Amitermes evuncifer* is small, thin walled, exposed to the sun with no temperature regulation. *Thoracotermes macrothorax* (Sjöstedt) is thin walled but, being built in the shade, is subject to smaller temperature fluctuations. *Cephalotermes rectangularis* (Sjöstedt) is thick-walled, in the shade and has more constant above-ambient temperatures. In *Macrotermes bellicosus* the temperature was constant around 30°C, and the CO_2 produced is dissipated through the thin, external mound walls. Singh and Singh (1981) found that temperatures within the large mounds of two *Odontotermes* species in India varied between 23.5°C and 28°C, whereas atmospheric temperatures ranged from 13.7°C to 39°C. A remarkable modification of the basal region of *M. bellicosus* nests is seen in parts of West Africa where the base of the hive consists of a thick, circular plate of soil balanced on a solid soil pillar (Collins 1979). Thin vanes of soil hang down in a spiral from the underside of the plate which forms the roof of a large chamber. In old mounds the vanes are encrusted with salts indicating that one of their functions may be that of a water-cooled air conditioner. *Amitermes meridionalis* (Froggatt) in northern Australia builds thin-walled, wedge-shaped mounds which are oriented with their long axis north-south. The mounds are always in seasonally waterlogged areas (Gay and Calaby 1970), which means that the termites cannot escape into the cooler depths of the soil during the hot, wet season. *Amitermes laurensis* Mjoberg and *Amitermes vitiosus* (Hill) build similar shaped mounds in seasonally waterlogged areas. This orientation presents the minimal surface area to the sun during the hottest part of the day and when a mound was cut off at the base and reorientated east-west (Grigg 1973) temperatures rose to 6°C higher than in undisturbed mounds.

The dissipation of excess heat and CO_2 from large nests of certain African *Macrotermes* is occasionally aided by surface vents, which connect with and stimulate airflow through tunnels in the base of the mound (Weir 1973). In the Indian species *Odontotermes wallonensis* (Wasmann)

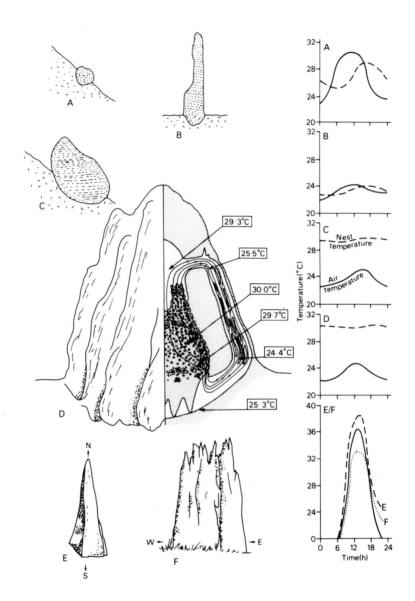

FIGURE 1.7

Nest architecture and thermoregulation.

A: *Amitermes evuncifer;*

B: *Thoracotermes macrothorax;*

C: *Cephalotermes rectangularis;*

D: *Macrotermes bellicosus;*

E: *Amitermes laurensis,* undisturbed N-S orientation;

F: *A. laurensis,* cut off at base and reorientated E-W.

(*Sources:* A–D, modified from Lüscher 1961b; E, F, modified from Grigg 1973)

these vents lead directly to the hive and are open and closed by the colony depending on the time of year (Rajagopal and Veeresh 1981).

Maintenance of a suitable moisture level in the nest is important as termites are generally very susceptible to desiccation (Collins 1969, Moore 1969). The retention of moisture within the nests is aided by the use of absorbent materials (clay, carton, fecal fungus combs) for constructing parts of the nest (Lee and Wood 1971b). Sources of water are the moisture in soil used for nest construction and in food supplies, water transported from deep soil layers, subterranean water storage chambers (Leprun 1976), metabolic water (Fyfe and Gay 1938), and natural rainfall. In high rainfall areas excess water may be a hazard and several species construct arboreal nests or mounds with water-shedding devices (Emerson 1938, 1956).

Several species have adaptable nest-building behavior. In West Africa *Amitermes evuncifer* builds subterranean nests in dry savanna, epigeal mounds in wet savanna and semideciduous forest and occasionally arboreal nests in rain forest (Johnson et al. 1980). In northern Australia *Nasutitermes triodiae* (Froggatt) builds a wide variety of mounds, each characteristic of certain localities (Ratcliffe et al. 1952). Such variations can be explained as adaptations to local environmental conditions, but it is more difficult to explain the two contrasting types of mound built by *Macrotermes bellicosus* in the same locality in Nigeria (Collins 1979).

The most intricate nests of all are built by the subterranean, soil-feeding *Apicotermes* in tropical Africa. Noirot (1970) and Grassé (1982) have reviewed the considerable literature on their structure. There is a progression from simple to complex structures involving isolation from the surrounding soil, division into a series of superimposed, intercommunicating floors, and complexity of the external wall with precisely constructed slits, pores, and channels. The adaptive significance of these incredible structures is unknown, but prevention of water seepage, defense and aeration have been suggested (Howse 1970).

FOOD FORAGING AND UTILIZATION

Food Resources

This subject was reviewed by Adamson (1943), Noirot and Noirot-Timothée (1969), Lee and Wood (1971b), and Wood (1978). The basic food is plant material: living, fresh dead, dead but in various stages of decomposition (including dung), and soil rich in organic matter (so-called humus). Specialized or incidental foods include fungi, algae, lichens, organic-rich portions of termite nests, members (including eggs) of their own colony and skins or other parts of vertebrate corpses. Occasionally, other substances such as leather or plastics are attacked.

The majority of the lower termites are wood feeders and this habit, although regarded as primitive, is a general habit of many Macrotermitinae and a specialized habit of many Nasutitermitinae and Termitinae. Most wood feeders attack dead wood, but some of those that attack living wood are potential or notorious pests of forest or plantation trees (e.g., some *Coptotermes* in Australia and Malaysia). Grass, herbs, and plant litter are resources for many termites. Some are specialists, such as the grass-harvesting *Hodotermes* (southern Africa), *Anacanthotermes* (northern Africa, Arabia, and arid regions of Asia), *Drepanotermes* (Australia), *Trinervitermes* (Africa and India), *Nasutitermes, Tumulitermes* (Australia), and *Syntermes* (South America). Others are almost polyphagous herbivores and detritivores, such as *Mastotermes* (northern Australia), *Psammotermes* (northern Africa and Arabia), and many Macrotermitinae. The latter include several species that attack living plants and are agricultural and forestry pests (Harris 1969, 1971). Very few termites have adapted to feeding on the fallen leaves of deciduous trees, but for *Macrotermes bellicosus* in West African savanna they are a specialized diet for part of the year (Collins 1981b) and throughout the year for *Macrotermes carbonarius* and *Longipeditermes longipes* (Haviland) in Malaysian rain forest (Matsumoto and Abe 1979). Soil feeding is a specialized habit found in many Termitinae, several Nasutitermitinae, and most Apicotermitinae, which collectively comprise approximately 60% of all the higher termite genera. They appear to utilize noncellular organic material intimately mixed with mineral material (Wood et al. 1983, Anderson and Wood, 1984). The lack of knowledge of their feeding behavior is a major gap in termite ecology.

The most widespread example of specialized feeding is on fungi, where Macrotermitinae ingest asexual spores of their symbiotic *Termitomyces*, which plays an essential role in digestion. Fungi are also an important dietary item for many species feeding on rotting wood (Sands 1969) and an incidental component for other species. Extreme specialization is rare but there are a few examples. *Hospitalitermes* in southeast Asia (Kalshoven 1958) mounts special expeditions in search of lichens to supplement the basic diet of bark and mosses (Collins 1980b). *Ahamitermes* and *Incolitermes* feed exclusively on the carton of *Coptotermes* nests in Australia (Gay and Calaby 1970) and the dry-wood termite, *Incisitermes banksi* (Snyder) has only been found nesting, and therefore feeding, in the common mesquite in Arizona (Nutting 1979b).

Food Selection

Food selection is a function of the workers, although drywood termites would appear to have little choice (except for a choice over heartwood or sapwood) as the nest develops within the site selected by the reproductive pair. Although the olfactory sense appears to operate only over a few centimeters (Abushama 1967) it may be of particular importance for some epigeal foragers. For instance, *Coptotermes* construct long, covered runways over nonwoody material and it would obviously be advantageous to follow traces of washed-wood extractives to their source (Williams 1977). Interestingly, Han and Yan (1980) found that extracts of fungus-infested wood induced attractive and trail-following behavior in *Coptotermes formosanus*. Subterranean foragers appear to locate potential food sources on the surface by sensing the thermal shadow (i.e., slightly lower temperatures) beneath them (Ettershank et al. 1980).

Selection for a particular species of plant is not common but its corollary, the avoidance of certain plants, is widespread and is of immense practical importance in the use of tropical timbers (Fougerousse 1969, Sen-Sarma and Gupta 1978, Nour 1979, 1980, Usher and Ocloo 1979, Carter and Dell 1981). Resistance of timber occasionally is due to specific chemicals that repel certain termites. For example, *Pinus* in Australia is attacked by many termites, but not by *Nasutitermes exitiosus* because essential oils in the timber contain α- and β- pinenes which are a dominant constituent of the termites' alarm pheromone. More often, resistance is due to a complex of chemical substances (Rudman and Gay 1967, Saeki et al. 1971, Carter and Huffman 1982). Interestingly, Carter, Mauldin and Rich (1981) showed that certain nonpreferred species, consumed in "no-choice" experiments, had the effect of eliminating all gut protozoa of *Coptotermes formosanus*. It is not known whether previous feeding experience, which has been shown to affect food selection in the laboratory (McMahan 1966), influences behavior in the field. Physical factors, particularly hardness, also influence food selection.

The grass-feeding *Hodotermes mossambicus* is influenced by thickness and width of the leaves, their physiological state (green or dry), and the presence of toxic or repellent chemicals (Hewitt and Nel 1969, Nel et al. 1970). Sands (1961b) showed that the grass-feeding *Trinervitermes geminatus* preferred smaller, fine-leaved species to larger, coarse-leaved species. In Africa and Indo-Malaya certain Macrotermitinae attack a wide range of crop plants (Harris 1969) but attack on living plants is very largely limited to nonindigenous varieties. For example, in Nigeria, ground-nuts, wheat, maize, and sugarcane are susceptible, whereas the indigenous cowpea, sorghum, and millet are largely resistant (Johnson and Wood 1979, 1980b). There may also be seasonal variations in food selection as in the grass and litter-feeding *Gnathamitermes* in Texas (Allen et al. 1980).

Selection for the state of decomposition is an important factor for many wood-feeding termites (Abe 1980). Some wood-rotting fungi produce substances that either attract termites, enhance feeding rates, or degrade repellent to toxic constituents of the wood (Lund 1969, Sands

1969, Ruyooka and Edwards 1980, French 1981). This phenomenon has been put to practical use by inoculating baits for termite control with wood-rotting fungi (Esenther and Beal 1979, Beal and Esenther 1980). Dung-feeding termites appear to prefer old, dry dung to fresh, moist dung (Ferrar and Watson 1970, Coe, 1977). One would also expect the extent or type of decomposition to be an important factor affecting food selection in soil-feeding termites, but this has yet to be demonstrated. However, in Nigerian savanna we observed that where *Adaiphrotermes cuniculator* (Sands) and *Anenteotermes polyscolus* (Sands) occurred in the same soil, the gut of the former was full of organic-rich topsoil and the latter of predominantly mineral subsoil.

The size of the food source and its location also influence foraging and are, therefore, relevant to "graveyard" trials of timber resistance (Usher and Ocloo 1974). In South Australia *Nasutitermes exitiosus* is the dominant consumer of *Eucalyptus* logs and stumps but does not attack small *Eucalyptus* twigs that are exploited by *Heterotermes ferox* (Froggatt) and *Microcerotermes* (Wood 1978). Ettershank, Ettershank, and Whitford (1980) showed that *Gnathamitermes tubiformans* (Buckley) and *Amitermes wheeleri* (Desneux) located food on the surface but not buried food. In contrast, our observations in Nigeria indicate that *Microcerotermes parvulus* (Sjöstedt) foraged on buried food more readily than on surface food.

Foraging

Certain lower termites, notably the wood-feeding Kalotermitidae and Termopsidae, excavate their nest and galleries entirely or largely within their food supply, and among the Termitidae many soil feeders exhibit similar behavior. These species usually have small colonies (less than 10,000 individuals) and a compact nest and gallery system. In contrast most termites construct special foraging galleries or runways, either as part of a diffuse nest system, as in the subterranean *Microtermes* and *Ancistrotermes* (Josens 1977), or radiating out from a centrally concentrated hive. Excavation of large, centrally concentrated nest systems of *Coptotermes lacteus* (Froggatt), *Nasutitermes exitiosus* (Ratcliffe and Greaves 1940), *Coptotermes brunneus* (Gay), *C. acinaciformis* (Froggatt) (Greaves 1962) and *Macrotermes michaelseni* (Darlington 1982b) showed that foraging galleries could extend for up to 50 m from the nest. The nest system of *M. michaelseni* was estimated to cover 8,000 m^2, in which there were 6 km of subterranean galleries and 72,000 food-storage pits.

Subterranean foraging parties enter the food source through underground galleries either below ground or at its point of contact with the surface. They may emerge on the surface under covered runways or as columns of workers foraging in the open. They may excavate the food, so that all foraging is cryptic, or the food may be covered with a protective layer of soil. Internal foraging within living plants or timber is destructive and its extent may not be apparent until the object wilts, dies, or falls. In contrast, external foraging is obvious even to casual observers, as objects are often encased in a layer of soil to heights of several meters. This type of foraging is typical of detritivorous termites that feed on sparsely distributed food such as the bark of trees, grasses, herbs, and their litter on the soil surface. Some species that feed on sparse food resources do so in open foraging columns. For the most part these are nocturnal or crepuscular, but some (e.g., *Hodotermes mossambicus*, Nel 1969) forage during daylight and are darkly pigmented for protection against ultraviolet light.

Difference in daily and seasonal foraging periods is one of the ways that several species living in the same habitat exploit a variety of food resources. Foraging may occur at characteristic times of the day and occasionally there may be two daily foraging periods as for the grass-feeding *Trinervitermes geminatus* (Wasmann) (Ohiagu and Wood 1976) which emerged from 6:00 to 9:30 AM and from 4:30 to 6:30 PM. In the same locality *Macrotermes subhyalinus*, predominantly a grass-feeder, never emerged before 10:00 PM. *T. geminatus* foraged only in the dry season. Morning activity was curtailed by temperatures lower than 20°C and evening activity by temperatures greater than 35°C. Haverty et al. (1974) and Collins et al. (1973) observed that *Heterotermes aureus* and *Gnathamitermes perplexus* (Banks) in the Arizona desert had foraging limits from 7.6°C to 47°C and 9°C to 49°C, respectively. Between soil temperatures of 20°C and 33°C slight rainfall increased foraging activity, but outside these extremes foraging decreased and rainfall had no effect. Demands of the colony also influence seasonal foraging as Lepage (1981a) showed that there was an annual three-peak foraging cycle in *Macrotermes michaelseni* in semi-arid Kenyan rangeland with the larger peak being from May to July when alate nymphs were in the nest. Other observations in African savannas (Bodot 1967b, Lepage 1974a, Buxton 1981, Collins, 1981b, Ferrar 1982a) illustrate the marked seasonality in foraging patterns particularly of those termites that can store food.

Food Storage

The seasonal availability of food and environmental restrictions on foraging activity have made food storage a characteristic trait of many abundant and widespread species of termites. The most general is storage of feces, either as fungus combs in Macrotermitinae or as carton in other herbivorous and detritivorous species. In Nigerian savanna, *Microtermes* spp. forage in the wet season on roots and plant debris near or at the surface and build subterranean fungus combs that reach total dry weights of 3.3 g/m^2 in woodland and 8.5 g/m^2 in cultivated land (Wood and Johnson 1978). During the dry season the termites retreat to lower soil depths where foraging is restricted to roots. Roots are more or less absent from the cultivated land and by the end of the dry season the weight of fungus comb is reduced almost to zero and in woodland, where some roots are available, to 1.6 g/m^2. In other localities we recorded maximum fungus comb weights for *Microtermes* of 27 g/m^2. Collins (1977) recorded an average for *Macrotermes bellicosus* of 2.6 g/m^2 with an individual colony maximum of 62.63 kg. Some Macrotermitinae store finely macerated food mixed with saliva in loose aggregations (food stores) near the fungus combs. Their function is unknown, but in *M. bellicosus* the turnover time is rapid (one to two weeks) and the amounts are small (0.07–4.23 kg per nest, Collins 1981c) compared with the amount of fungus comb.

Storage of unprocessed food is commonly practiced by grass-feeding termites. In Nigerian savanna the grass-harvesting *Trinervitermes geminatus* is polycalic, with five to six mounds per colony (Sands 1961a, Ohiagu 1979a). Grass is stored in the mounds in amounts varying from 15 g to 143 g per mound, equivalent to 0.3 to 2.9 g/m^2 (Wood 1978). Darlington's (1982b) data for *Macrotermes michaelseni* indicate that 1.98 g/m^2 of grass was stored overnight in subterranean pits for removal and processing the following day. Grass is often stored in particular regions of the nest (e.g., around the nursery chamber of *Nasutitermes magnus*) and may therefore fulfill a secondary role in providing insulation (Lee and Wood 1971b).

Quantity of Food Eaten

Laboratory and field estimates for various species were reviewed by Wood (1978). Recent publications by Buxton (1981), Collins (1981b), and Lepage (1981b) have confirmed the view that Macrotermitinae consume considerably more food per unit of body weight than other termites. This is because, first, their production per biomass ratios are greater than those of other termites by two to four times (Wood and Sands 1978) and second, their fungus combs metabolize up to 80% of the food consumed by the termites (Wood and Sands 1978, Collins 1977). The considerable range in recorded feeding rates of 27.7–565.0 mg/g/day for Macrotermitinae and 2.0–90.8 mg/g/day for other herbivorous and detritivorous termites is probably more a reflection of methodology than reality. Probably the most accurate figure for Macrotermitinae is that recorded for *Macrotermes bellicosus*, where field studies of removal in Nigerian savanna indicated consumption rates of 132 mg/g/day, calculations from the turnover rates of food stores indicated rates of 66–131 mg/g/day and consumption by nests isolated in specially constructed tanks indicated 94–148 mg/g/day (Collins 1981c). In the same locality Ohiagu (1979b) calculated a rate of 7.2 mg/g/day for the grass-harvesting *Trinervitermes geminatus*. Probably the most complete data for a non-fungus-growing member of the Termitidae is Bodine and Ueckert's (1975) study of *Gnathamitermes tubiformans* in west Texas rangeland. Laboratory feeding trials indicated consumption rates of 23.8 mg/g/day and measurements of termite biomass and standing crop of grasses, forbs, roots and litter on termite-free and control plots indicated consumption rates of 47.6 mg/g/day.

The significance of consumption in relation to net primary production was reviewed by Wood and Sands (1978). They highlighted three West African savannas where termites consumed from 10% to 35.9% of the available plant debris, equivalent to 12.5–192 g/m², respectively. Studies on individual species in semiarid grasslands and rangelands in Africa indicate a wide range in annual consumption rates: 1.5 g/m² (equivalent to 1.0% of the NPP) for *Hodotermes mossambicus* in South Africa (Nel 1970) and 27.4 g for the same species (Basson 1972); 8.1 g/m² (equivalent to 2.7% of the NPP) for *Trinervitermes geminatus* in Nigeria (Ohiagu 1979b) and 0.6–4.4 g/m² for the same species in the Ivory Coast (Josens 1972). Much larger quantities, 80–150 g/m², were consumed by *Macrotermes michaelseni* in East Africa (Lepage 1981b).

Digestion and Utilization of Food

Microbial Symbionts and Their Environment

The digestive processes of termites are based largely on symbiotic relationships with various microorganisms (O'Brien and Slaytor 1982, Breznak 1983). Four groups can be recognized:

(i) *Lower Termites.* These are largely wood feeders, where the symbiotic microbiota is dominated by cellulolytic oxymonad, trichomonad, and hypermastigote protozoa with populations of $3–4 \times 10^4$ per gut and comprising up to one third of the total body weight. The protozoa occupy the lumen of the paunch, but there are also considerable numbers of facultative or obligate anaerobic bacteria ($10^6–10^7$ per gut), mainly associated with the epithelium. Spirochaetes are present, either free in the gut fluid or attached to the surface of protozoa where they form complex associations in which bacteria may also be involved (e.g., as in *Mastotermes darwiniensis*, Cleveland and Grimstone 1964). Elimination of protozoa in *Reticulitermes flavipes* by offering food treated with chlorotetracycline results in death of the termites (Maudlin and Rich 1980) and therefore this and similarly effective compounds may have potential use in bait-block methods of control. In the lower termites, more is known about protozoan-termite relationships (Honigberg 1970) than is known with other termite groups.

(ii) All Termitidae (Except Macrotermitinae and Soil-Feeding Termites). These include a variety of wood, grass, and herb feeders where bacteria are the dominant microorganisms in the gut. Protozoa may be present in small numbers but are noncellulolytic. In the past it has been assumed that the bacteria play a role in cellulose digestion comparable to that of protozoa in the lower termites. However, attempts to isolate cellulolytic bacteria from guts of *Nasutitermes* (O'Brien et al. 1979; Breznak 1983) have failed and the importance of intestinal bacteria in this group of termites remains to be demonstrated.

(iii) Macrotermitinae. These are generally polyphagous herbivores and detritivores where for each termite species there is a single, major symbiont, a basidiomycete fungus *Termitomyces*, which is cultivated on fecal-based fungus combs in the nest system. The fungus does not occur outside the nest (Thomas 1981) and the association is an obligate symbiosis (Sands 1969). A notable exception is *Sphaerotermes sphaerothorax* (Sjöstedt) which has no association with *Termitomyces*. However, the presence of cellulolytic microbes in the gut of *Sphaerotermes* (Pochon et al. 1959) does not necessarily indicate that they form part of an essential symbiosis; there is a similar lack of understanding of the significance of bacteria (some of which are cellulolytic) and protozoa in the guts of other Macrotermitinae (Rohrmann and Rossman 1980).

(iv) Soil-Feeding Termites. Their food is noncellular organic matter and until recently nothing was known about their intestinal microorganisms, although Honigberg (1970) noted the presence of protozoa in *Subulitermes* (Nasutitermitinae) and *Cubitermes* (Termitinae). The investigations of Bignell et al., (1980a, 1980b, 1982) showed that large numbers of bacteria occurred throughout the gut of *Procubitermes aburiensis* (Sjöstedt). Filamentous forms, supposedly actinomycetes, were more abundant, relative to nonfilamentous forms, than in freshly ingested soil. They were intimately associated with the epithelium of the midgut and mixed segment in this species and also in *Cubitermes severus* (Bignell et al. 1980c). In the posterior colon of *P. aburiensis* they were attached to prominent cuticular spines elaborated from the gut wall. Shorter filamentous forms included those with several characteristics of spirochetes (D. E. Bignell, personal communication) and there were many nonfilamentous bacteria of various morphologies (Bignell et al. 1980b).

Digestive Processes

Termites, with the possible exception of soil feeders, derive most of their energy from the breakdown of plant structural polysaccharides, cellulose, hemicelluloses, and lignin, which comprise 70–90% of the dry weight of wood (LaFage and Nutting 1978). Lee and Wood (1971b) and Wood (1978) reviewed the available data on the extent to which these constituents were degraded and quoted figures ranging from 74% to 99% for cellulose, 65% to 87% for hemicellulose, and 0.3% to 83% for lignin. Other sources of energy are soluble carbohydrates that are hydrolyzed by enzymes (e.g., amylases) found in the salivary glands and foregut. Other enzymes have been isolated from the midgut and hindgut but little is known of the relative importance of intestinal bacteria as opposed to epithelial secretions in their production (LaFage and Nutting 1978).

Our knowledge of digestive processes is most complete for cellulose digestion in lower termites, based on Hungate's (1938, 1943) work on the role of protozoa and recent work on bacteria, which has been summarized by Breznak (1983). In the paunch, protozoa endocytose wood particles that are degraded by anaerobic fermentation ($C_6H_{12}O_6 + 2H_2O \rightarrow 2C_2H_4O_2 + 2CO_2 + 2H_2$). Acetate is absorbed by the gut and provides the main oxidizable source of energy. Mono-, di-, and soluble oligosaccharides released from the protozoan cells are fermented by bacteria to formate and lactate; formate could be used by methanogenic bacteria and lactate could be fermented to acetate, which is absorbed. Methanogenic bacteria utilize CO_2 and H_2 to produce methane. The quantities evolved are less than would be expected on a molar basis and some H_2 could be utilized by Acetobacterium-type bacteria that reduce CO_2 to acetate ($2CO_2 + 4H_2 \rightarrow C_2H_4O_2 + 2H_2O$), which is absorbed. Many of the bacteria in the hindgut are common heterotrophs, and it is possible that they utilize soluble intermediates of cellulose hydrolysis released by protozoa (Yamin 1980) and are in turn consumed by the protozoa.

Virtually nothing is known about the processes involved in the digestion of hemicelluloses and lignin. The evidence for lignin degradation, summarized by Lee and Wood (1971b), was recently reviewed by Breznak (1982) and O'Brien and Slaytor (1982). Kovoor (1964) indicated that degradation was accomplished by demethoxylation, which would require aerobic conditions, yet the gut is generally regarded as being anaerobic. Butler and Buckerfield (1979) used ^{14}C-labeled lignin to demonstrate that degradation occurred within the termite body (i.e., not in the feces) and suggested that degradation may proceed by depolymerization in the gut, absorption of lower-molecular-weight derivatives, and the aerobic oxidation of these derivatives by termite tissues.

Very little is known about digestive processes in the higher termites belonging to group (ii) (i.e., all except Macrotermitinae and soil feeders).

It has generally been assumed that intestinal bacteria are primarily responsible for cellulose digestion. However, this has yet to be demonstrated and it is apparent that several unrelated genera (Microcerotermes, Trinervitermes and Nasutitermes) are capable of synthesizing their own cellulases (Kovoor 1970, McEwen et al. 1980, Potts and Hewitt 1973). However, recent reviewers of this subject (O'Brien and Slaytor 1982, Breznak 1983) believe that bacteria are also involved in digestive processes.

Recent work on the Macrotermitinae (group (iii)) has highlighted the key role of the Termitomyces fungus in the degradation of cellulose and lignin. The feces, which contain a large proportion of structurally unaltered plant material, are used to construct fungus combs that support the growth of the Termitomyces. The termites reingest old parts of the comb and continually add fresh feces, resulting in a "turnover" time of five to eight weeks (Josens 1971). Cellulases and enzymes capable of degrading lignin are produced by Termitomyces in the comb (Rohrmann and Rossman 1980, Thomas 1981). The presence of cellulases in the termite gut partly results from the consumption of "mycotêtes" which are aggregates of asexual spores on the comb surface known to contain C_1 cellulases (Martin and Martin 1978, 1979). The latter authors showed that C_x cellulase and β-glucosidase in the gut were derived partly from ingested fungal material and partly from the midgut epithelium and salivary glands. Cellulolytic bacteria are present in the gut but their significance is unknown (Rohrmann and Rossman 1980). Soluble carbohydrates in primary food sources or reingested comb and the chitin of fungal cell walls can be utilized by enzymes in the gut. However, it is not certain whether these enzymes are derived from gut bacteria or synthesized by the termites.

Probably even less is known about digestive processes in soil-feeding termites (group (iv)) than in the Termitidae in group (ii). The high pH and oxygen deficit in the P1 and P3 part of the hindgut (see Fig. 1.5) (Bignell and Anderson 1980) combined with the high microbial populations (Bignell et al. 1980a, c) may promote solubilization or hydrolysis of recalcitrant fractions of soil organic matter. However, nothing is known of their enzyme systems and the importance of their varied and numerous gut microbes in digestion.

Nitrogen Economy

Termite tissues contain approximately 11% nitrogen, whereas the diet of species feeding on fresh, decay-free wood (e.g., dry-wood termites and many Macrotermitinae) contains less than 0.1% nitrogen and has C/N ratios of 350–1000 per liter (La Fage and Nutting 1978). The metabolism of wood-rotting fungi and other microorganisms modifies sound wood in several ways but one end result, a lowering of the C/N ratio, produces substrates which, along with grass, leaves and similar diets, contain up to 1% nitrogen. The acquisition and conservation of nitrogen by termites (reviewed by Collins 1983) has ranked as an important question in termite biology since Cleveland (1925) demonstrated that *Zootermopsis* could survive and multiply on a diet of cellulose. Mauldin and Smythe (1973) recognized four routes: (i) extraction from the diet (including termite corpses and individuals superfluous to the needs of the colony (Dhanarajan 1978), (ii) fixation of atmospheric nitrogen, (iii) digestion of symbionts, (iv) utilization of microbial and their own waste products.

The first is accomplished by proteolytic enzymes produced by the termites and their intestinal symbionts resulting in the elaboration of termite and microbial tissue and their waste products, uric acid from termites and ammonia from protozoa. Processes (i) and (iii) appear to operate particularly efficiently in fungus-growing termites where plant tissues appear to be transformed into termite and fungal tissue (Wood 1976). In *Macrotermes bellicosus* comminuted food (0.28% N) is left for approximately one week in the food store. The modified food (0.58% N) is then ingested and deposited as fecal pellets on the fungus comb, which after five to six weeks (0.82% N) are reingested along with groups of asexual spores (mycotêtes, 6.68%N). The latter are particularly favored by young workers with greater demands for nitrogen (Thomas 1981). The ability of other termites to elaborate nitrogen by digesting their intestinal microbes is unknown (Breznak 1983).

The fixation of atmospheric nitrogen by bacteria in the hindgut was demonstrated by Benemann (1973) and Breznak et al. (1973) and recent discoveries were summarized by Breznak (1983). The time taken to double body nitrogen content varies from 2 to 2,000 years. The lower rates are obviously ecologically significant as annual production/biomass ratios of non-Macrotermitinae range from 1.5 per liter to 3.9 per liter (Wood and Sands 1978). Nitrogen fixation has been demonstrated to occur in representatives of all the major taxonomic or ecological groups of termites except Macrotermitinae.

Leach and Granovsky (1938) were the first to suggest that uric acid, stored in the fat body and transported and excreted by the Malpighian tubules into the hindgut, could be elaborated into a form of nitrogen that could be used by the insects. Breznak (1983) has summarized recent work which demonstrates that uricolysis occurred in *Reticulitermes flavipes* with the aid of anaerobic bacteria in the hindgut resulting in the production of NH_3, CO_2, and acetate. The form in which the uric acid nitrogen is absorbed by the termites is not known, both NH_3 or organic nitrogen being possibilities. Potrikus and Breznak (1981) calculated that 30% of the nitrogen requirements of *R. flavipes* could be provided in this way.

Trophallaxis

Trophallaxis involves the exchange of nutrients, either reciprocally or unilaterally, among members and guests of social insect colonies (La Fage and Nutting 1978) and is the key feature of colony nutrition. Typically, the independent castes, that is, workers, feed the dependent castes, larvae, soldiers, nymphs, alates, and reproductives. However, some larvae consume raw food (e.g., larvae of Macrotermitinae ingest mycotêtes of *Termitomyces*) and some workers are themselves dependent (e.g., old workers of *Hodotermes mossambicus*, Nel et al. 1969). The food they pass to the dependent castes is either saliva or regurgitated (stomodeal) food or liquid contents (proctodeal food) of the hindgut. Proctodeal feeding occurs only in lower termites and is the means by which young larvae and recently molted individuals (which lose their intestinal protozoa at ecdysis) acquire their gut symbionts. Saliva appears to be a high-energy food and is given to larvae and functional reproductives and in some species to soldiers. In Termitidae the nutrition of the first brood is largely dependent on salivary secretions from the royal pair elaborated from nutritional reserves in their fat body and flight muscles. Stomodeal food is fed to soldiers of many Termitidae (Noirot and Noirot-Timothée 1969) but appears to be rarely exchanged in lower termites.

Trophallaxis is accompanied by ritualistic behavior between donor and recipient, which extends to interactions with well-integrated termitophiles. It is also important in maintaining colony integration by rapid transmission of pheromones.

POPULATION ECOLOGY

Abundance in Different Ecosystems

Reviews of Lee and Wood (1971b) and Wood and Sands (1978) have been supplemented by several recent studies of the abundance of termite populations in semiarid rangeland in Texas (Ueckert et al. 1976), savanna in South Africa (Ferrar 1982b), riparian forest in Nigeria (Wood et al. 1982) and various tropical forests in Malaysia (Abe and Matsumoto 1979, Collins 1980a). Methods of sampling populations have been discussed by Lee and Wood (1971b), Sands (1972b), and Baroni-Urbani, Josens, and Peakin (1978). Arboreal and epigeal colonies with concentrated nests can be destructively sampled and the individuals counted in a series of colonies selected to be representative of the population as a whole; with large colonies fumigation may be useful in preventing escape of individuals while mounds are being excavated (Darlington 1982a). Subterranean populations can be sampled by extracting termites from cores of soil, usually at least 60 mm diameter and 40–50 cores on each occasion (Wood et al. 1977, Ferrar 1982b). The considerable vertical penetration of soil by many termites leads to variable proportions of the population remaining below the sampling depth. For example, Ueckert et al. (1976) found that members in the upper 30 cm fluctuated between zero and over 9,000 m^2 and Wood et al. (1977) found that in the dry season 50% of the *Microtermes* and *Ancistrotermes* populations in the upper 2 m were at 1–2 m.

There is as wide a range in abundance within similar ecosystems as between different ecosystems. Semiarid tropical savanna in Senegal (Lepage 1974b) supported 229 m^2 (fresh weight biomass of 1.0 g/m^2) whereas in temperate semiarid sites in North America abundance ranged from 431 m^2 (Haverty et al. 1975) to 2139 m^2 (Ueckert et al. 1976). The latter site was only sampled to a soil depth of 30 cm and the seasonal maximum of 9127 m^2 (22.2 g/m^2) is possibly the highest figure ever recorded. Abundance in tropical savannas varies from 861 m^2 (1.7 g/m^2) in the Ivory Coast (Josens 1972) to 4402 m^2 (11.1 g/m^2) in Nigeria (Wood and Sands 1978). Tropical forest populations vary from zero in montane (> 1970 m) and 38 m^2 in lower montane rain forest in Sarawak (Collins 1980a) to 4450 m^2 in rain forest in Trinidad (Strickland 1944).

Man's activities have a drastic effect on termite populations, through disturbance of nest systems, destruction or modification of food resources, and, in urban areas, through the construction of new environments. There are, however, few well documented examples. In general, the number of species and overall abundance is reduced by human disturbance, for example, various localities in Senegal (Roy-Noel 1978), dipterocarp forests in Sarawak (Collins 1980b) and riparian forest in Nigeria (Wood et al. 1982). However, there are instances where one or more species has been able to withstand the disturbance, adapt to the man-modified environment and, in the absence of potential competitors, increase their abundance and become significant pests. In northern Nigeria *Microtermes lepidus* (Sjöstedt) is a serious pest of groundnuts (Johnson et al. 1981). *Microtermes* spp. are moderately abundant (819 m^{-2}) in undisturbed savanna woodland (Wood et al. 1977), where they are associated with 23 other species at a total abundance of 3472 m^2. Clearing and use of the land for grazing reduced the number of species to 20 and overall abundance to 2010 m^2. Cultivation had a more drastic effect with the number of species reduced to eight and overall abundance to 1553 m^2 in the first year. However, *Microtermes*, by virtue of their deep nest system and ability to utilize crop residues, increased with successive years of cultivation to over 6825 m^2 in 24-year-old fields.

Dry-wood termites are not normally abundant in natural vegetation but wooden structures in buildings provide them with a potentially new environment, which several exploit successfully (Williams 1976, 1977), particularly "tramp" species such as *Cryptotermes brevis* (Fig. 1.1). The ability of people to modify their domestic environment by heating has enabled *Reticulitermes* to colonize localities (e.g., Paris, Hamburg) well to the north of its natural distribution (Fig. 1.1). Similarly, air conditioning in tropical coastal regions could lead to *C. brevis* extending its distribution.

Niche Exploitation

The majority of termites live in tropical or subtropical environments and as with many other animals the number of species increases with decreasing latitude (Wood 1976) and decreasing altitude (Kemp 1955, Kayani et al. 1979, Collins 1980a). In five localities along a latitudinal gradient from 16.5°N (semiarid savanna) to 4.5°N (rain forest) in West Africa the number of species per locality increased from 19 to 43. The number of species feeding on fresh wood and litter was more or less constant (8–11) whereas those feeding on decomposing wood increased from zero to eight and soil feeders increased from three to 31 (Wood 1976).

From these and other comparisons it appears that semiarid and savanna ecosystems, whether tropical or temperate, are inhabited mainly by species that exploit freshly dead or, occasionally, living vegetation. In tropical Africa and India, Macrotermitinae are the dominant herbivores and detritivores and with the aid of their symbiotic *Termitomyces* are able to process large amounts of plant material. Their role in equivalent ecosystems in tropical America and Australasia appears to be carried out, albeit less spectacularly, by various Termitinae and Nasutitermitinae (Araujo 1970, Gay and Calaby 1970). It is difficult to generalize on termite populations in tropical forests (Wood et al. 1982). Certainly some African forests appear to be dominated by soil-feeding species but equally there are forests where wood feeders are dominant. Soil feeders appear to be absent from Australia (Lee and Wood 1971b) and possibly less numerous in the Oriental region than in the Ethiopian and Neotropical regions.

The possibilities for exploiting subterranean, epigeal, and hypogeal nest sites and a variety of food sources give a range of opportunities for adaptation and evolution within different regions of the world. This is illustrated by a comparison of the termite fauna in three forests in Brazil, Nigeria, and West Malaysia (Fig. 1.8). In all three forests arboreal nesting is dominated by the wood-feeding *Microcerotermes* and *Nasutitermes*. In the Matto Grosso forest (Brazil) the dominant mound builder was the herbivorous and detritivorous *Cornitermes*; in Rabba (Nigeria) and Pasoh (Malaysia) forests the dominant mound builders were various Macrotermitinae or mainly soil-feeding Termitinae. Soil feeders, mainly Apicotermitinae and Termitinae, dominated the subterranean groups found at Rabba and also, mainly Nasutitermitinae, in the Matto Grosso; in contrast, Macrotermitinae were the dominant subterranean group in Pasoh.

At the community level there is little understanding of niche partitioning among groups of species with similar feeding habits. In African savannas the complex of species attacking woody and herbaceous litter are partioned partly according to food selection (wood or grass) and partly according to diurnal or seasonal activity (Bodot 1967b, Collins 1981b, Ferrar 1982a). However, the five species of subterranean *Microtermes* that have overlapping nest systems and often forage simultaneously in the same food item (Wood 1981) obviously have more complex interrelationships. Sands (1961b) studied the co-occurrence of five species of grass-feeding *Trinervitermes* in Nigerian savanna. Three species stored grass in their nests and ceased foraging for part of the year, while the other two did not store grass and foraged almost throughout the year. There were also differences in survival on different types of food and nesting habits that enabled them to exploit the mosaic of conditions available. Similarly subtle niche partitioning is likely among the 20 species of soil feeders coexisting in the riparian forest at Rabba (Wood et al. 1982).

Factors Affecting Distribution and Abundance

Soils, Vegetation and Climate

The interaction of soils, vegetation and climate in determining the distribution and abundance of termites was discussed by Bouillon (1970) and Lee and Wood (1971b). It is particularly well illustrated by the distribution of *Cubitermes* spp. in East Africa (Williams 1966). The interactions are poorly understood but occasionally a single factor may be of such overriding importance that the operation of some of the factors can be appreciated. For instance, Ratcliffe et al. (1952) noticed the absence of termite mounds on cracking clays in Australia. These soils have a high clay content, are self-mulching, subject to seasonal wetting and crack deeply in the dry season. It is likely that the seasonal physical disturbance prevents the construction of stable nest systems, although Lee and Wood (1971b) and Holt and Coventry (1982) noted the rare occurrence of mounds of *Nasutitermes longipennis* on some cracking clays.

The structure of vegetation would appear to be of greater importance than the species composition, although there are some rare examples of affinities between termites and particular plant species (e.g., *Incisitermes banksi* and mesquite in Arizona). The relative amounts of shaded and open areas were shown to be important in determining the relative abundance of different species of mound-building *Trinervitermes* and

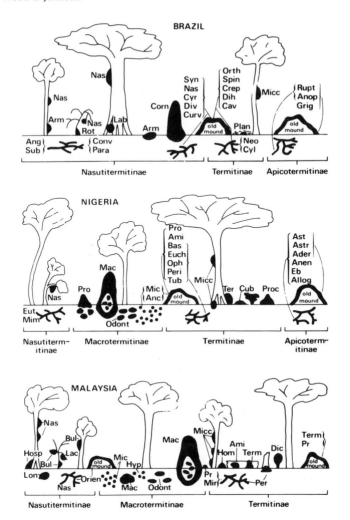

FIGURE 1.8

Spatial distribution of termite (Termitidae) nests in forests in Neotropical, Ethiopian and Oriental regions (after Wood 1979).

(* indicates soil feeders)

Brazil: riparian forest, Mato Grosso (Mathews 1977).

    Apicotermitinae: *Ruptitermes**, *Anoplotermes**, *Grigiotermes**. Termitinae: *Orthognathotermes**, *Spinitermes**, *Crepititermes**, *Dihoplotermes**, *Cavitermes**, *Planicapritermes**, *Neocapritermes**, *Cylindrotermes**, *Microcerotermes*. Nasutitermitinae: *Nasutitermes*, *Armitermes*, *Angularitermes**, *Subulitermes**, *Convexitermes**, *Paracornitermes**, *Rotunditermes*, *Labiotermes**, *Cornitermes*, *Syntermes*, *Cyranotermes**, *Diversitermes*, *Curvitermes**.

subterranean Macrotermitinae in West Africa (Sands 1965b). Excess shade, provided by exotic conifer plantations, appears to cause the decline of *Nasutitermes exitiosus* in southern Australia (Lee and Wood 1971b). The distribution of African Nasutitermitinae is closely related to vegetation type (Sands 1965d), with some genera being characteristic of savanna (e.g., *Trinervitermes*) and others of forest (e.g., *Nasutitermes*). Climatic factors obviously interact with vegetation in determining the distribution of the African Nasutitermitinae and many other termites also. However, climate and in particular temperature and air humidity, appear to be the main factors in limiting the distribution of certain *Cryptotermes* (Williams 1976). This author used the upper limits of saturation deficit and wood equilibrium moisture content to predict the occurrence of three building pests, *Cryptotermes brevis*, *C. dudleyi*, and *C. havilandi*, in Africa.

Other Animals

The effect of other animals on termite populations was discussed by Lee and Wood (1971b) and Wood and Sands (1978). The two most important groups are predators and termites themselves. Termites are subject to a vast array of predators, from opportunistic predation on alate reproductive colony-founding swarms to opportunistic and specialized predation on foraging sterile castes and specialized predation on colony brood centers. Predation appears to be particularly significant in the tropics. From Ward's (1965) and Lepage's (1974b) data, Wood and Sands (1978) calculated that the alate flights of *Odontotermes smeathmani* in Senegal would be capable of sustaining a population of eight *Quelea* birds per hectare over a period of 31 days. In Africa the Ponerine ant, *Megaponera foetens*

Nigeria: riparian forest, Rabba (Wood et al. 1982).
    Apicotermitinae: *Astalotermes**, *Astratotermes**, *Aderitotermes**, *Anenteotermes**, *Eburnitermes**, *Allognathotermes**. Termitinae: *Promirotermes**, *Amitermes*, *Basidentitermes**, *Euchilotermes**, *Ophiotermes**, *Pericapritermes**, *Tuberculitermes**, *Microcerotermes*, *Termes*, *Cubitermes**, *Procubitermes**. Macrotermitinae: *Protermes*, *Macrotermes*, *Odontotermes*, *Microtermes*, *Ancistrotermes*. Nasutitermitinae: *Nasutitermes*, *Eutermellus**, *Mimeutermes**.
Malaysia: rain forest, Pasoh (Abe and Matsumoto 1979).
    Termitinae: *Microcerotermes*, *Procapritermes**, *Homallotermes**, *Mirocapritermes**, *Amitermes*, *Termes*, *Pericapritermes**, *Dicuspiditermes**. Macrotermitinae: *Microtermes*, *Macrotermes*, *Odontotermes*, *Hypotermes*. Nasutitermitinae: *Nasutitermes*, *Hospitalitermes*, *Longipeditermes*, *Bulbitermes*, *Lacessititermes*, *Oriensubulitermes**.

(Fabr.), is a specialized predator on foraging parties of Macrotermitinae, mainly Macrotermes spp. In Nigerian savanna the annual predation of 141 Macrotermes bellicosus per m² represented 2.7 times the standing crop of workers/soldiers (Longhurst et al. 1978). Lepage (1981c) found lower annual predation rates on Macrotermes sp. near subhyalinus in Kenya of 51 termites m², representing 0.1–0.15 times the standing crop of workers or soldiers. In both areas Odontotermes and, more rarely, other species were captured. A much smaller ant, Decamorium uelense (Santchi), is a specialized predator on foraging Microtermes and annual predation rates of 632 termites m² represented 0.74 times the standing populations (Longhurst et al. 1979). Doryline ants are specialized predators on colony brood centers and Bodot (1967a) and Harris and Sands (1965) noted the drastic effects of their raids on Macrotermes and Trinervitermes colonies.

The importance of termites as a source of food for predators can be seen from Wood and Sands' (1978) comparison of the return to the ecosystem of energy and nutrients via termite feces or termite tissue. In southern Guinea savanna in Nigeria, termites consumed 35.9% of the annual litter production, representing 3500 kJ m² a. Returns via feces were 233 kJ m² a and returns via termite tissue were 277 kJ m² a; however, the nitrogen content of feces is 0.5%, whereas that of termite tissue is approximately 11% and in these areas herbivorous and detritivorous termites stimulate a predator network rather than a decomposer network.

An equally important category of animals that affect termite populations are competitors. Due to their specialized nesting and feeding habits their most important competitors are termites of the same or different species (Lee and Wood 1971b). Competitive processes involve "interference," "exploitation," and "avoidance." Indirect evidence for competition was provided by Wood and Lee's (1971) demonstration that colonies of many Australian species were overdispersed. Nel (1968) demonstrated that aggressive behavior among colonies of Hodotermes mossambicus lead to well-defined territories, and Darlington (1982a) maintained that nest density in Macrotermes michaelseni was self-regulated through the establishment of defended foraging territories. Fewer than 1% of new colonies of M. michaelseni reached maturity, whereas in an area which was regularly cleared of nests, recolonization and the survival of nests was much higher. Some species appear to avoid each other (e.g., Hodotermes mossambicus and Trinervitermes trinervoides, Nel 1968) and this is probably the normal reaction of sympatric species that may be found foraging in the same food resource (e.g., logs), but that fight if forcibly mixed (Noirot 1959). Direct interference appears to be rare but was observed by Bodot (1967a) where Amitermes evuncifer entered mounds of

Cubitermes and *Trinervitermes* eventually displacing the original inhabitants. The upsurge in populations of *Microtermes* following the decline of other herbivorous and detritivorous species resulting from clearing and cultivation is further evidence of competitive interactions.

Termites feeding on fresh woody litter would appear to have few competitors other than themselves. In contrast the possible competitive interactions among grass-feeding termites and grazing mammals, leading to denudation of pastures, have been noted by many authors in South Africa (e.g., Coaton 1954, Hartwig 1955) and Australia (e.g. Watson and Gay 1970, Lee and Wood 1971b). Studies by Nel and Hewitt (1969) on *Hodotermes* in South Africa, Ohiagu (1979b) on *Trinervitermes geminatus* in Nigeria, Lepage (1981b) on *Macrotermes michaelseni* in Kenya and Watson et al. (1973) on various species in Australia, indicate the preference of termites for dead, dry forage compared with the mammalian preference for live forage. Reduction of resources by drought or overgrazing could lead to competition by the two groups for dwindling resources. Nothing is known of the interactions between soil-feeding termites and their most likely competitors, soil-feeding earthworms.

Phenology of Populations

The homeostatic nature of termite societies regulates both the numbers in the colony and the abundance of colonies to the extent that overall abundance is subject to much smaller fluctuations than in nonsocial insects. This is particularly true for the mound-building *Macrotermes* in Africa. Darlington (1982a) showed that populations of *Macrotermes michaelseni* in Kenya appeared to be stable and self-regulating. Normal seasonal weather patterns (two wet and two dry seasons) had no effect on nest populations and there was no change in the number of larvae or weight of fungus comb when a growing brood of reproductives was present. Although rainfall affected the daily and seasonal foraging activity of this species (Lepage 1981a), the consistency in annual foraging activity over a period of three years was a further indication of colony stability. In Nigeria, Collins (1981a) estimated the abundance of *Macrotermes bellicosus* in three successive years to range from 111 to 164 m².

Studies on subterranean populations of *Microtermes* in Nigerian savanna (Wood and Johnson 1978) and *Gnathamitermes tubiformans* in semiarid rangeland in North America (Bodine and Ueckert 1975, Ueckert et al. 1976) appeared to indicate marked seasonal fluctuations in numbers. However, in both cases this was due to the downward movement of termites to below the sampling depth during unfavorable conditions, seasonal aridity in Nigeria and a combination of aridity and low temperatures in North America. As a consequence foraging activity was curtailed during these unfavorable periods (see "Foraging" and "Food Storage").

Species with less well-regulated societies than *Macrotermes* display distinct seasonal changes in caste composition. These changes can be related to the foraging, nest-building and reproductive cycles (Sands 1965a, Bouillon 1970, Lafitte de Mosera et al. 1979, Howard and Haverty 1981, Ferrar 1982a). Periods of foraging generally result in a reduction in the number of workers and soldiers due to predation. Egg production is geared to replace both these individuals and the primary reproductives that leave mature colonies *en masse* each year. The annual swarming of alates is the most spectacular seasonal event and a considerable proportion of the energy resources of the colony is channeled into alate production. Wood and Sands (1978) reviewed data that showed that the energy loss via swarming alates represented 44%–233% of the energy content of the sterile castes in seven species of Termitidae and 4% (admittedly underestimated) in one of the lower termites, *Psammotermes hybostoma* (Desneux).

The amount of energy diverted toward the production of sterile castes generally exceeds that of alates. It is particularly high in the Macrotermitinae, possibly in response to heavy predation. The data reviewed by Wood and Sands (1978) indicated that the production per biomass ratios of Macrotermitinae (5.4–10.2) (based on energy equivalents) were significantly higher than those of other termites (1.5–3.9). Confirmation is given by recent data obtained for *Macrotermes bellicosus* (7.5, Collins 1977), *Macrotermes michaelseni* (4.7, Darlington 1982) and *Trinervitermes geminatus* (1.0, Ohiagu 1979b, 2.6, Josens 1973).

An important seasonal event for some species of Macrotermitinae is the production of basidiocarps ("mushrooms") by their *Termitomyces*. These appear at approximately the same time of year as the first foraging workers are produced by new colonies (Johnson et al. 1981) and it is thought that these workers bring the basidiospores back to the colony, thereby inoculating the fungus comb.

Environmental Impact of Termite Populations

The movement of soil to build runways and nests and the remarkably efficient conversion of large quantities of vegetable matter into fecal residues and termite tissue means that, where they are abundant, termites have a marked effect on soils, nutrient cycling, growth of vegetation, and wildlife. Reviews by Lee and Wood (1971b), Wood and Sands (1978), and Roy-Noel (1979) have been supplemented by recent publications (Trapnell et al. 1976, Tinley 1977, Mielke 1978, Schaefer and Whitford 1981, Arshad 1982). These influences are relevant to this and subsequent chapters in that conventional chemical control is just as effective in killing non-harmful species as the damaging species. The protection and control of termites in structural timbers is obviously essential and if carried out properly there should be no harmful side effects. In agriculture and forestry the problems are rarely so clearly defined (Johnson and Wood 1979). There is also a much greater risk to the environment and public health when toxic chemicals are used, particularly by untrained and occasionally illiterate people in rural communities. In these situations it is important that before control measures are advocated there should be clear indications as to whether or not the apparent damage has a significant effect on yield (Wood et al. 1980a, Johnson et al. 1981a) and that control will be economically rewarding (Verma et al. 1976, Wood et al. 1980a, b). Very often this information is lacking and chemical control is recommended on an *ad hoc* basis. Ideally extension workers should be educated to assess the economic viability of control methods and the risks inherent in their use.

REFERENCES

Abe, T. 1980. Studies on the distribution and ecological role of termites in a lowland rain forest of W. Malaysia: (4) The role of termites in the process of wood decomposition in Pasoh. *Rev. Ecol. Biol. Sol.* 17: 23–24.

Abe, T. and Matsumoto, T. 1979. Studies on the distribution and ecological role of termites in a lowland rain forest of west Malaysia: (3) Distribution and abundance of termites in Pasoh Forest Reserve. *Japan. J. Ecol.* 29: 337–351.

Abushama, F. T. 1967. The role of chemical stimuli in the feeding behaviour of termites. *Proc. Roy. Entomol. Soc. London (A)* 42: 77–82.

Adamson, A. M. 1943. Termites and the fertility of soils. *Trop. Agr.* 20: 107–112.

Ahmad, M. 1950. The phenology of termite genera based on imago-worker mandibles. *Bull. Amer. Mus. Nat. Hist.* 95: 43–86.

Allen, C. T., Foster, D. E., and Veckert, D. N. 1980. Seasonal food habits of a desert termite *Gnathotermes-tubiformans* in West Texas USA. *Environ. Entomol.* 9: 461–466.

Ampion, A. and Quennedey, A. 1981. The abdominal epidermal glands of termites and their phylogenetic significance. In, *Biosystematics of social insects* (P. E. Howse and J-L. Clement, eds.), pp. 249–261. London and New York: Academic Press.

Ampion, M. 1980. *Les glandes tergales des imagos de termites: Etude comparative et signification évolutive.* Thèse, Université de Dijon.

Anderson, J. M. and Wood, T. G. (1984). Mound composition and soil modification by two soil-feeding termites (Termitinae, Termitidae) in a riparian Nigerian forest. *Pedobiologia* 26: 77–82.

Araujo, R. L. 1970. Termites of the neotropical region. In, *Biology of termites* (K.

Krishna and F. M. Weesner, eds.), Vol. 2, pp. 527–576. New York and London: Academic Press.

Arshad, M. A. 1982. Influence of the termite *Macrotermes michaelseni* (Sjost.) on soil fertility and vegetation in a semi-arid savannah ecosystem. *Agro-Ecosystems* 8: 47–58.

Bacchus, S. 1979. New exocrine gland on the legs of some Rhinotermitidae (Isoptera). *Int. J. Insect. Morphol. Embryol.* 8: 135–142.

Baroni-Urbani, C., Josens, G., and Peakin, G. J. 1978. Empirical data and demographic parameters. In, *Production ecology of ants and termites* (M. V. Brian, ed.), pp. 5–44. Cambridge: Cambridge Univ. Press.

Beal, R. H. and Esenther, G. R. 1980. A new approach to subterranean termite control—The bait block method. *Sociobiology* 5: 171–174.

Becker, G. 1969. Rearing of termites and testing methods used in the laboratory. In, *Biology of termites* (K. Krishna and F. M. Weesner, eds.), Vol. 1, pp. 351–385. New York and London: Academic Press.

Benemann, J. R. 1973. Nitrogen fixation in termites. *Science* 181: 164–165.

Bignell, D. E. and Anderson, J. M. 1980. Determination of pH and oxygen status in the guts of lower and higher termites. *J. Insect Physiol.* 26: 183–188.

Bignell, D. E., Oskarsson, H., and Anderson, J. M. 1980a. Specialization of the hindgut wall for the attachment of symbiotic micro-organisms in a termite *Procubitermes aburiensis* (Isoptera, Termitidae, Termitinae). *Zoomorphology* 96: 103–112.

Bignell, D. E., Oskarsson, H., and Anderson, J. M. 1980b. Distribution and abundance of bacteria in the gut of a soil-feeding termite *Procubitermes abunensis* (Termitidae, Termitinae). *J. Gen. Microbiol.* 117: 393–403.

Bignell, D. E., Oskarsson, H., and Anderson, J. M. 1980c. Colonisation of the epithelial face of the peritropic membrane and the ectoperitrophic space by Actinomycetes in a soil-feeding termite. *J. Invert. Pathol.* 36: 426–428.

Bignell, D. E., Oskarsson, H., and Anderson, J. M. 1982. Formation of membrane bounded secretory granules in the mid gut epithelium of a termite *Cubitermes severus* and a possible inter-cellular route of discharge. *Cell Tissue Res.* 222: 187–200.

Bodenheimer, F. S. 1937. Population problems of social insects. *Biol. Rev.* 12: 393–430.

Bodine, M. C. and Ueckert, D. N. 1975. Effect of desert termites on herbage and litter in a shortgrass ecosystem in west Texas. *J. Range Man.* 28: 353–358.

Bodot, P. 1967a. Etudes écologiques des termites des savanes de Basse Côte d'Ivoire. *Insectes Soc.* 14: 229–258.

Bodot, P. 1967b. Cycles saisonniers d'activité collective des termites des savanes de Basse Côte d'Ivoire. *Insectes Soc.* 14: 359–388.

Bouillon, A. (ed.) 1964. *Etudes sur les termites africains.* Paris: Masson et Cie.

Bouillon, A. 1970. Termites of the Ethiopian region. In, *Biology of termites* (K. Krishna and F. M. Weesner, eds.), Vol. 2, pp. 153–280. New York and London: Academic Press.

Boyer, P. 1976. Les differents aspects de l'action de certain *Bellicositermes* sur l'évolution des sols des savanes Oubanguiennes (République Centrafricaine) Extrait, *Ann. Sci. Nat. Zool.* Paris: Masson.

Breed, M. D., Michener, C., and Evans, H. E. (eds.) 1982. *The biology of social insects.* Boulder, Colo.: Westview.

Breznak, J. A. 1982. Intestinal microbiota of termites and other xylophagous insects. *Ann. Rev. Microbiol.* 36: 323–343.

Breznak, J. A. 1983. Biochemical aspects of symbiosis between termites and their intestinal microbiota. In, *Invertebrate-Microbial interactions* (J. M. Anderson, A. D. Rayner, and D. H. Walton, eds.), pp. 173–203. Cambridge: Cambridge Univ. Press.

Breznak, J. A., Brill, W. J., Mertins, J. W., and Coppel, H. C. 1973. Nitrogen fixation in termites. *Nature* (London) 244: 577–580.

Brian, M. V. (ed.) 1978. *Production ecology of ants and termites.* Cambridge: Cambridge Univ. Press.

Brian, M. V. 1979. Caste differentiation and division of labour. In, *Social insects* (H. R. Herman, ed.), Vol. 1, pp. 121–222. New York and London: Academic Press.

Brossut, R. 1973. Evolution du système glandulaire exocrine céphalique des Blattaria et des Isoptera. *Int. J. Insect Morphol. Embryol.* 2: 35–54.

Bruinsma, O., and Leuthold, R. H. 1977. Pheromones involved in the building behaviour of *Macrotermes subhyalinus* (Rambur). In, *Proceedings of the eighth international congress of the International Union for the Study of Social Insects, Wageningen, The Netherlands, September 5–10, 1977,* pp. 257–258. Wageningen: Centre for Agricultural Publishing and Documentation.

Buchli, H. 1958. L'origine des castes et les potentialités ontogéniques des termites européens du genre *Reticulitermes* Holmgren. *Ann. Sci. Nat. (Zool.)* Sér. 11, 20: 263–429.

Butler, J. H. A. and Buckerfield, J. C. 1979. Digestion of lignin by termites. *Soil Biol. Biochem.* 11: 507–513.

Buxton, R. D. 1981. Changes in the composition and activities of termite communities in relation to changing rainfall. *Oecologia* 51: 371–378.

Carter, F. L. and Dell, T. R. 1981. Screening selected American hardwoods for natural resistance to a native subterranean termite, *Reticulitermes flavipes* (Kollar). *United States Department of Agriculture Forest Service Research Paper,* no. 50–176: 1–10.

Carter, F. L. and Huffman, J. B. 1982. Termite responses to wood and extracts of melaleuca. *Wood Science* 14: 127–133.

Carter, F. L., Mauldin, J. K., and Rich, N. M. 1981. Protozoan populations of *Coptotermes formosanus* exposed to heart wood samples of 21 American species. *Mater Org.* (Berl.) 16: 29–38.

Cleveland, L. R. 1925. The ability of termites to live perhaps indefinitely on a diet of pure cellulose. *Biol. Bull. Mar. Biol. Lab. (Woods Hole)* 48: 289–293.

Cleveland, L. R. and Grimstone, A. V. 1964. The fine structure of the flagellate *Mixotricha paradoxa* and its associated micro-organisms. *Proc. Roy. Soc. Lon.* (Ser. B) 159: 668–686.

Coaton, W. G. H. 1954. Veld reclamation and harvester termite control. *Farming S. Afr.* 29: 243–248.

Coaton, W. G. H. 1958. *The Hodotermitid harvester termites of South Africa.* Union of South Africa Department of Agriculture Bulletin, 375.

Coaton, W. G. H. 1971. Five new termite genera from South West Africa. (Isoptera: Termitidae) *Cimbebasia* (A) 2: 1–34.

Coe, M. 1977. The role of termites in the removal of elephant dung in the Tsavo (East) National Park, Kenya. *E. Afr. Wildl. J.* 15: 49–55.

Collins, M. S. 1969. Water relations in termites. In, *Biology of termites* (K. Krishna and F. M. Weesner, eds.), Vol. 1, pp. 433–458. New York and London: Academic Press.

Collins, M. S., Haverty, M. L., Lafage, J. P., and Nutting, W. L. 1973. High-temperature tolerance in two species of subterranean termites from the Sonoran desert in Arizona. *Envir. Entomol.* 2: 1122–1123.

Collins, N. M. 1977. The population ecology and energetics of *Macrotermes bellicosus* (Smeathman), Isoptera. Thesis, London University.

Collins, N. M. 1979. The nest of *Macrotermes bellicosus* (Smeathman) from Mokwa, Nigeria. *Insectes Soc.* 26: 240–246.

Collins, N. M. 1980a. The distribution of soil macrofauna on the West Ridge of Gunung (Mt.) Mulu, Sarawak. *Oecologia* 44: 263–275.

Collins, N. M. 1980b. The effect of logging on termite (Isoptera) diversity and decomposition processes in lowland dipterocarp forests. In, *Tropical ecology and development* (J. I. Furtado, ed.), pp. 113–121. Kuala Lumpur: International Society of Tropical Ecology.

Collins, N. M. 1981a. Populations, age structure and survivorship of colonies of *Macrotermes bellicosus* (Smeathman) (Isoptera: Macrotermitinae). *J. Anim. Ecol.* 50: 293–311.

Collins, N. M. 1981b. The role of termites in the decomposition of wood and leaf litter in the southern Guinea savanna of Nigeria. *Oecologia* 51: 389–399.

Collins, N. M. 1981c. Consumption of wood by artificially isolated colonies of the fungus-growing termite *Macrotermes bellicosus*. *Entomol. Experimentalis et Applicata* 29: 313–320.

Collins, N. M. 1983. The utilisation of nitrogen resources by termites (Isoptera). In, *Nitrogen as an ecological factor* (J. A. Lee, S. McNeill and I. H. Rorison, eds.), pp. 381–412. Oxford: Blackwell Scientific.

Darlington, J. P. E. C. 1982a. Population dynamics in an African fungus growing termite. In, *Biology of social insects* (M. D. Breed, C. D. Michener, and H. E. Evans, eds.), pp. 401–402. Boulder, Colo.: Westview Press.

Darlington, J. P. E. C. 1982b. The underground passages and storage pits used in foraging by a nest of the termite *Macrotermes michaelseni* in Kajiado, Kenya. *J. Zool. (London)* 198: 237–247.

Deligne, J. 1966. Caractères adaptif au régime alimentaire dans la mandibule des termites (Insectes, Isopteres). *C. R. Hebd. Séanc. Acad. Sci. (Paris)* 263: 1323–1325.

Deligne, J., Quennedey, A., and Blum, M. S. 1981. The enemies and defense mechanisms of termites. In, *Social insects* (H. R. Hermann, ed.), Vol 2, pp. 1–76. New York and London: Academic Press.

de Wilde, J. and Beetsma, J. 1982. The physiology of caste development in social insects. *Adv. Insect Physiol.* 16: 167–246.

Dhanarajan, G. 1978. Cannibalism and necrophagy in a subterranean termite. *Malay Nat. J.* 31: 237–251.

Emerson, A. E. 1938. Termite nests. A study of the phylogeny of behaviour. *Ecol. Monog.* 8: 247–284.

Emerson, A. E. 1955. Geographical origins and dispersions of termite genera. *Fieldiana: Zool.* 37: 465–521.

Emerson, A. E. 1956. Regenerative behaviour and social homeostasis of termites. *Ecology* 37: 248–258.

Emerson, A. E. 1965. A review of the Mastotermitidae, including a new fossil genus from Brazil. *Amer. Mus. Novit.* 2236: 46 pp.

Emerson, A. E. 1968. A revision of the fossil genus *Ulmeriella*. *Amer. Mus. Novit.* 2332: 22 pp.

Esenther, G. R. 1977. Nutritive supplement method to evaluate resistance of natural or preservative-treated wood to subterranean termites. *J. Econ. Entomol.* 70: 341–346.

Esenther, G. R. and Beal, R. H. 1979. Termite control: Decayed wood bait. *Sociobiology* 4: 215–222.

Ettershank, G., Ettershank, J. A., and Whitford, W. G. 1980. Location of food sources by subterranean termites. *Envir. Entomol.* 9: 645–648.

Ferrar, P. 1982a. Termites of a South African savanna. III. Comparative attack on toilet roll baits in subhabitats. *Oecologia* 52: 139–146.

Ferrar, P. 1982b. Termites of a South African savanna. IV. Subterranean populations, mass determinations and biomass estimations. *Oecologia* 52: 147–151.

Ferrar, P. and Watson, J. A. L. 1970. Termites (Isoptera) associated with dung in Australia. *J. Aust. Entomol. Soc.* 9: 100–102.

Fougerousse, M. 1969. Methods of field tests in West Africa to assess the natural resistance of woods or the effectiveness of preservative products against attack by termites. In, *Termite symposium: Proceedings of the meetings held at Cambridge, June 23–24, 1969*, p. 35–55. London: British Wood Preserving Association.

French, J. R. J. and Robinson, P. J. 1978. Feeding inhibition of *Mastotermes darwiniensis* (Froggatt) [Isoptera] on J.H.A. surface-treated plywood blocks. *Z. Angew. Entomol.* 85: 360–364.

French, J. R. J., Robinson, P. J., Thornton, J. D., and Saunders, F. W. 1981. Termite fungi interactions: Response of *Coptotermes acinaciformis* to fungus decayed softwood blocks. *Mater Org.* (Berl.) 16: 1–14.

Fyfe, R. V. and Gay, F. J. 1938. The humidity of the atmosphere and the moisture conditions within mounds of *Eutermes exitiosus* Hill. *Council for Scientific and Industrial Research*, Melbourne, Pamphlet No. 82, 1–82.

Garcia, M. L. and Becker, G. 1975. Influence of temperature on the development of incipient colonies of *Nasutitermes nigriceps* (Haldemann). *Z. Angew. Ent.* 79: 291–300.

Gay, F. J. 1970. Isoptera. In, *Insects of Australia*. CSIRO publn., Melbourne Univ. Press.

Gay, F. J. and Calaby, J. H. 1970. Termites from the Australian region. In, *Biology of termites* (K. Krishna and F. M. Weesner, eds.), Vol. 2, pp. 393–448. New York and London: Academic Press.

Grassé, P. P. 1949. Ordre des Isoptères ou termites. In, *Traité de zoologie*, (P. P. Grassé, ed.), Vol. 9, pp. 408–544. Paris: Masson et Cie.

Grassé, P. P. 1982. *Termitologia, Anatomie, Physiologie, Biologie, systematique des termites. Tome 1. Anatomie, Physiologie, reproduction.* New York and Paris: Masson.

Grassé, P. P. and Noirot, C. 1947. Le polymorphisme social du termite a cou jaune (*Cal. flavicollis* F.). Les faux-ouvriers ou pseudergates et les mues regressives. *C. R. Hebd. Séanc. Acad. Sci. Paris* 224: 219–220.

Grassé, P. P. and Noirot, C. 1951. La sociotomie: Migration et fragmentation de la termitière chez les *Anoplotermes* et les *Trinervitermes*. *Behaviour* 3: 146–166.

Grassé, P. P. and Noirot, C. 1954. *Apicotermes arquieri*: Ses constructions, sabiologie. Considerations sur la sous-famille des *Apicotermitinae*. *Ann. Sci. Nat. Zool.* 16: 345–388.

Grassé, P. P. and Noirot, C. 1958. La meule des termites champignonnistes et sa signification symbiotique. *Ann. Sci. Nat. (Zool.). (Sér. 11)* 20(2): 113–128.

Greaves, T. 1962. Studies of the foraging galleries and the invasion of living trees by *Coptotermes acinaciformis* and *C. brunneus* (Isoptera). *Aust. J. Zool.* 10: 630–651.

Greenberg, S. L. W., and Stuart, A. M. 1982. Precocious reproductive development neoteny by larvae of a primitive termite *Zootermopsis angusticollis. Insectes Soc.* 29: 535–547.

Grigg, G. C. 1973. Some consequences of the shape and orientation of "magnetic" termite mounds. *Aust. J. Zool.* 21: 231–237.

Hagen, H. 1858. *Catalogue of Neuropterous Insects in the collection of the British Museum. Part 1. Termitina.* London: Taylor & Francis.

Han, M. Z. and Yan, F. 1980. A preliminary report on the comparative tests of termite trail-following pheromone analogues from fungus-infested wood. *Acta Entomol. Sinica* 23: 260–264.

Harris, W. V. 1969. *Termites as pests of crops and trees.* London: Commonwealth Institute of Entomology.

Harris, W. V. 1971. *Termites. Their recognition and control,* 2nd ed. London: Longman.

Harris, W. V. and Sands, W. A. 1965. The social organization of termite colonies. *Symposia of the Zoological Society of London,* 14: 113–131.

Hartwig, E. K. 1955. Control of snouted harvester termites. *Fmg. S. Afr.* 30: 361–366.

Haverty, M. I. 1977. The proportion of soldiers in termite colonies: A list and a bibliography. *Sociobiology* 2: 199–216.

Haverty, M. I. 1979. Soldier production and maintenance of soldier proportions in laboratory experimental groups of *Coptotermes formosanus* (Shiraki). *Insectes Soc.* 26: 69–84.

Haverty, M. I. and Howard, R. W. 1979. Effects of insect growth regulators on subterranean termites: induction of differentation, defaunation, and starvation. *Ann. Entomol. Soc. Amer.* 72: 503–508.

Haverty, M. L., La Fage, J. P., and Nutting, W. L. 1974. Seasonal activity and environmental control of foraging of the subterranean termite, *Heterotermes aureus* (Snyder), in a desert grassland. *Life Sci.* 15: 1091–1101.

Haverty, M. L., Nutting, W. L., and La Fage, J. P. 1975. Density of colonies and spatial distribution of foraging territories of the desert subterranean termite, *Heterotermes aureus* (Snyder). *Environ. Entomol.* 4: 105–109.

Hegh, E. 1922. *Les termites.* Brussels: Imprimerie Industrielle et Financière.

Hermann, H. R. (ed.) 1979. *Social insects.* Volume 1. New York and London: Academic Press.

Hermann, H. R. (ed.) 1982a. *Social insects.* Volume 2. New York and London: Academic Press.

Hermann, H. R. (ed.) 1982b. *Social insects.* Volume 3. New York and London: Academic Press.

Hermann, H. R. (ed.) 1982c. *Social insects.* Volume 4. New York and London: Academic Press.

Hesse, P. R. 1955. A chemical and physical study of the soil of termite mounds in East Africa. *J. Ecol.* 43: 449–461.

Hewitt, P. H. and Nel, J. J. C. 1969. Toxicity and repellency of *Chrysocoma tenuifolia* (Berg) (Compositae) to the harvester termite *Hodotermes mossambicus* (Hagen) (Hodotermitidae). *J. Entomol. Soc. S. Afr.* 32: 133–136.

Holt, J. A. and Coventry, R. J. 1982. Occurrence of termites (Isoptera) on cracking clay soils in northeastern Queensland. *J. Aust. Entomol. Soc.* 21: 135–136.

Honigberg, B. M. 1970. Protozoa associated with termites and their role in digestion. In, *Biology of termites* (K. Krishna and F. M. Weesner, eds.), Vol. 2, pp. 1–36. New York and London: Academic Press.

Howard, R. W. 1980. Effects of methoprene on colony foundation by alates of *Reticulitermes flavipes* (Kollar). *J. Georgia Entomol. Soc.* 15: 281–285.

Howard, R. W. and Blomquist, G. J. 1982. Chemical ecology and biochemistry of insect hydrocarbons. *Ann. Rev. Entomol.* 27: 149–172.

Howard, R. W. and Haverty, M. I. 1978. Defaunation, mortality and soldier differentiation: Concentration effects of methroprene in a termite. *Sociobiology* 3: 73–77.

Howard, R. W. and Haverty, M. I. 1979. Termites and juvenile hormone analogues: A review of methodology and observed effects. *Sociobiology* 4: 269–278.

Howard, R. and Haverty, M. I. 1981. Seasonal variation in caste proportions of field colonies of *Reticulitermes flavipes*. *Environ. Entomol.* 10: 546–549.

Howard, R. W., Mallette, E. J., Haverty, M. I. and Smythe, R. V. 1981. Laboratory evaluation of within-species, between-species, and parthenogenetic reproduction in *Reticulitermes flavipes* and *Reticulitermes virginicus*. *Psyche* 88: 75–88.

Howse, P. E. 1970. *Termites, a study in social behavior.* London: Hutchinson.

Howse, P. E. and Clement, J. L. (eds.) 1981. *Biosystematics of social insects.* London: Academic Press.

Hrdy, I. 1981. The potential of juvenoids in social insect regulation: Termites. In, *Regulation of insect development* (F. Sehnal, A. Zabza, J. J. Menn, and B. Cymborowski, eds.), pp. 761–767. Wroclaw, Poland: Wroclaw Technical University press.

Hungate, R. E. 1938. Studies on the nutrition of *Zootermopsis*. II. The relative importance of the termite and the protozoa in wood digestion. *Ecology* 19: 1–25.

Hungate, R. E. 1943. Quantitative analyses on the cellulose fermentation by termite protozoa. *Ann. Entomol. Soc. Amer.* 36: 730–739.

Jarzembowski, E. A. 1981. An early cretaceous termite from southern England. (Isoptera, Hodotermitidae). *Syst. Entomol.* 6: 91–96.

Johnson, R. A. 1981. Colony development and establishment of the fungus comb in *Microtermes* sp. nr. *usambaricus* (Sjost) (Isoptera, Macrotermitinae) from Nigeria. *Insectes Soc.* 28: 3–12.

Johnson, R. A. 1981. Termites, damage and control reviewed. *Span* 24: 108–110.

Johnson, R. A., Lamb, R. W., Sands, W. A., Shittu, M. O., Williams, R. M. C., and Wood, T. G. 1980. A check list of Nigerian termites (Isoptera) with brief notes on their biology and distribution. *Nigerian Field* 45: 50–64.

Johnson, R. A., Lamb, R. W., and Wood, T. G. 1981a. Termite damage and crop loss studies in Nigeria—A survey of termite damage to groundnuts. *Trop. Pest Manag.* 27: 325–342.

Johnson, R. A., Thomas, R. J., Wood, T. G., and Swift, M. J. 1981b. The innoculation of the fungus comb in newly founded colonies of some species of the Macrotermitinae (Isoptera) from Nigeria. *J. Nat. Hist.* (London) 15: 751–756.

Johnson, R. A. and Wood, T. G. 1979. Termites: Friend or foe? *W. Afr. Farm. Food Proc.* (November–December): 26–27.

Johnson, R. A. and Wood, T. G. 1980a. Termites of the arid zones of Africa and the Arabian Peninsula. *Sociobiology* 5: 279–293.

Johnson, R. A. and Wood, T. G. 1980b. Combatting termites. *W. Afr. Farm. Food Proc.* (March–April): 41–42.

Josens, G. 1971. Le renouvellement des meules a champignons construites par quatre Macrotermitinae (Isoptères) des savanes de Lamto-Pacobo (Côte d'Ivoire). *C. R. Hebd. Séanc. Acad. Sci.* (Paris) Ser. D 272: 3329–3332.

Josens, G. 1972. Etudes biologiques et ecologiques des termites (Isoptera) de la savane de Lamto-Pakobo (Côte d'Ivoire). Doctoral thesis, University of Brussels, Brussels.

Josens, G. 1973. Observations sur les bilans energetiques dans deux populations de termites a Lamto (Côte d'Ivoire). *Ann. Soc. R. Zool. Belg.* 103: 169–176.

Josens, G. 1977. Recherchez sur la structure et fonctionnement des nids hypoges de quatre especes de Macrotermitinae (Termitidae) communes dans les savanes de Lamto (Côte d'Ivoire). *Mém. Acad. R. Belg. Cl. Sci.* 2nd Serie. 42: 1–123.

Kalshoven, L. G. E. 1930. De biologie van de Djatermiet (*Kalotermes tectonae* Damm) in verband met Zijn bestrijding. *Mededelingen van het Instituut voor plantenziekten, Buitenzorg* 76: 1–154.

Kalshoven, L. G. E. 1958. Observations on the black termites, *Hospitalitermes* spp. of Java and Sumatra. *Insectes Soc.* 5: 9–30.

Karlson, P. and Lüscher, M. 1959. Pheromones: A new term for a class of biologically active substances. *Nature (London)* 183: 55–56.

Kayani, S. A., Sheikh, K. H., and Ahmad, M. 1979. Altitudinal distribution of termites in relation to vegetation and soil conditions. *Pakis. J. Zool.* 11: 123–137.

Kemner, N. A. 1934. Systematische und Biologische studien uber die Termiten Javas and Celebes. *K. Sven. Vet. Ak. Handl.* 13: 4–241.

Kemp, P. B. 1955. The termites of north-eastern Tanganyika: Their distribution and biology. *Bull. Entomol. Res.* 46: 113–135.

King, E. J. and Spink, W. T. 1974. Laboratory studies on the biology of the Formosan subterranean termite with primary emphasis on young colony development. *Ann. Entomol. Soc. Amer.* 68: 355–358.

Kofoid, C. A. (ed.) 1934. *Termites and termite control.* Berkeley: Univ. of California Press.

Kovoor, J. 1964. Modifications chimiques provoquees par un termitide (*Microcerotermes edentatus* Was.) dans du bois de peuplier sain ou partiellement degrade par des champignons. *Bull Biol. France Belg.* 98: 491–510.

Kovoor, J. 1970. Presence d'enzymes cellulolytiques dans l'intestin d'un termite superieur, *Microcerotermes edentatus* Was. *Ann. Sci. Nat. Zool.* 12: 65–71.

Krishna, K. 1961. The evolution of the family Kalotermitidae (Isoptera). *4th Cong. Un. Int. Etude Insectes Soc.* 1961: 129–132.

Krishna, K. 1961. A generic revision and phylogenetic study of the family Kalotermitidae (Isoptera). *Bull. Amer. Mus. Nat. Hist.* 122: 303–408.

Krishna, K. 1970. Taxonomy, Phylogeny and Distribution of Termites. In, *Biology of Termites* (K. Krishna and F. M. Weesner, eds.), Vol. 2, pp. 127–152. London and New York: Academic Press.

Krishna, K. and Weesner, F. M. (eds.) 1969. *Biology of termites,* Vol. 1. New York and London: Academic Press.

Krishna, K. and Weesner, F. M. (eds.) 1970. *Biology of termites,* Vol. 2. New York and London: Academic Press.

LaFage, J. P. and Nutting, W. L. 1978. Nutrient dynamics of termites. In, *Production ecology of ants and termites* (M. V. Brian, ed.), pp. 165–232. Cambridge: Cambridge Univ. Press.

Lafitte de Mosera, S., Talice, R. V., Sineiro de Sprechmann, A. M., and Aber de Szterman, A. 1979. Estudio poblacional annual de *Nasutitermes fulviceps* (Silvestri). *Rev. Biol. Urug.* 12: 77–87.

Lanzrein, B., Gentinetta, V., and Lüscher, M. 1978. In vitro-Juvenil-hormonsynthese durch corpora allata der Termite *Macrotermes subhyalinus*. *Rev. Suisse Zool.* 85: 10.

Leach, J. G. and Granovsky, A. A. 1938. Nitrogen in the nutrition of termites. *Science* 87: 66–67.

Lee, K. E. and Wood, T. G. 1971a. Physical and chemical effects on soils of some Australian termites and their pedological significance. *Pedobiologia* 11: 376–409.

Lee, K. E. and Wood, T. G. 1971b. *Termites and soils.* New York and London: Academic Press.

Lenz, M. 1976. The dependence of hormone effects in termite caste determination on external factors. In, *Phase and caste determination in insects: Endocrine aspects* (M. Lüscher, ed.), pp. 73–89. Oxford: Pergamon Press.

Lenz, M. and Barrett, R. A. 1982. Neotenic formation in field colonies of *Coptotermes lacteus* (Froggatt) in Australia, with comments on the roles of neotenics in the genus *Coptotermes* (Isoptera: Rhinotermitidae). *Sociobiology* 7: 47–59.

Lepage, M. G. 1981a. The impact of foraging populations of *Macrotermes michael-*

septentrional, Senegal: Influence de la secheresse sur la peuplement en termites. *Terre et la vie* 28: 76–94.

Lepage, M. 1974b. *Les termites d'une savane sahelienne (Ferlo Septentrional, Senegal): Peuplement, populations, consommation, role dans l'ecosysteme.* Thesis, University of Dijon, Dijon.

Lepage, M. G. 1981a. The impact of foraging populations of Macrotermes *michaelseni* (Sjöstedt) (Isoptera, Macrotermitinae) in a semi-arid ecosystem (Kajiado-Kenya). I. Foraging activity and factors determining it. *Insectes sociaux* 28: 297–308.

Lepage, M. G. 1981b. The impact of the foraging populations of *Macrotermes michaelseni* (Isoptera, Macrotermitinae) on a semi-arid ecosystem (Kajiado-Kenya). II. Food offtake comparison with large herbivores. *Insectes Soc.* 28: 309–320.

Lepage, M. G. 1981c. Study on the predation of *Megaponera foetens* (F.) on the foraging populations of Macrotermitinae in a semi-arid ecosystem (Kajiado-Kenya). *Insectes Soc* 28: 247–262.

Leprun, J. C. 1976. Une construction originale hypogee pour le stockage de l'eau par les termites en regions sahelo-soudaniennes de Haute-Volta. *Pedobiologia* 16: 451–456.

Leuthold, R. H. 1977. Postflight communication in two termite species, *Trinervitermes bettonianus* and *Hodotermes mossambicus. Proc. 8th int. congr. Int. Union Study Social Insects*, Wageningen, Netherlands, Sept. 5–10, 1977: 62–64.

Longhurst, C., Johnson, R. A. and Wood, T. G. 1978. Predation by *Megaponera foetens* (Fabr.) (Hymenoptera: Formicidae) on termites in the Nigerian southern Guinea Savanna. *Oecologia* 32: 101–107.

Longhurst, C., Johnson, R. A., and Wood, T. G. 1979. Foraging, recruitment and predation by *Decamorium uelense* (Santschi) (Formicidae: Myrmicinae) on termites in southern Guinea savanna, Nigeria. *Oecologia* 38: 83–91.

Lund, A. E. 1969. Termite attractants and repellants. In, *Termite symposium: proceedings of the meetings held at Cambridge, June 23–24, 1969*, pp. 107–111. London: British Wood Preserving Association.

Lüscher, M. 1952. Die produktion und Elimination von Ersatzgeschlechstieren bei der Termite *Kalotermes flavicollis* (Fabr.). *Zeitschrift für vergleichende Physiologie* 34: 123–141.

Lüscher, M. 1961a. Social control of polymorphism in termites. In, *Insect polymorphism* (J. S. Kennedy, ed.), pp. 57–67. *Symposium of the Royal Entomological Society of London*, No. 1.

Lüscher, M. 1961b. Air-conditioned termite nests. *Scientific American* 205, 138–145.

Lüscher, M. 1969. Die bedeutung des juvenilhormons für die differenzierung der soldaten bei der termite *Kalotermes flavicollis*. In, *Proc. 6th int. congr. Union Int. Etude Insectes Soc. Berne*: 165–170.

Lüscher, M. 1972. Environmental control of juvenile hormone (JH) secretion and caste differentiation in termites. *General and Comparative Endocrinology*, Supplement 3: 509–514.

Lüscher, M. 1973. The influence of the composition of experimental groups on caste development in *Zootermopsis* (Isoptera). *Proceedings, VIIth International*

Congress of the International Union for the Study of Social Insects, London, September 10–15, 1973: 253–256.

Lüscher, M. 1974. Kasten und Kastendifferenzienung bei niederen Termiten. In, Sozial polymorphismus bei Insekten: Probleme der kasten bildung im Tierreich (G. H. Schmidt, ed.), pp. 694–739. Winenschaftliche, Stuttgart.

Lüscher, M. 1976. Evidence for an endocrine control of caste determination in higher termites. In, Phase and caste determination in insects: Endocrine aspects (M. Lüscher, ed.), pp. 91–103. Oxford: Pergamon Press.

Lüscher, M. 1977. Queen dominance in termites. In, Proc. 8th int. Congr. Int. Union Study Social Insects, Wageningen, The Netherlands, Sept. 5–10, 1977, pp. 238–242. Wageningen, The Netherlands: Centre for Agricultural Publishing & Documentation.

McEwen, S. E., Slaytor, M., and O'Brien, R. W. 1980. Cellobiase activity in three species of Australian termites. Insect Biochem. 10: 563–567.

McKittrick, F. A. 1965. A contribution to the understanding of Cockroach-Termite affinities. Ann. Entomol. Soc. Amer. 58: 18–22.

McMahan, E. A. 1966. Studies of termite wood-feeding preferences. Proc. Hawaiian Entomol. Soc. 19: 239–250.

McMahan, E. A. 1979. Temporal polyethism in termites. Sociobiology 4: 153–168.

Martin, M. M. and Martin, J. S. 1978. Cellulose digestion in the midgut of the fungus growing termite Macrotermes natalensis: The role of acquired digestive enzymes. Science 199: 1453–1455.

Martin, M. M. and Martin, J. S. 1979. The distribution and origins of the celluylolytic enzymes of the higher termite, Macrotermes natalensis. Physiol. Zool. 52: 11–21.

Maschwitz, U., Jander, R., and Burkhardt, D. 1972. Wehrsubstanzen und wehruerhalten der termite Macrotermes carbonarius. J. Insect Physiol. 18: 1715–1720.

Mathews, A. G. A. 1977. Studies on termites from the Mato Grosso State, Brazil. Rio de Janeiro: Academia Brasiliera de Ciencias.

Matsumoto, T. and Abe, T. 1979. The role of termites in an equatorial rain forest ecosystem of West Malaysia. II. Leaf litter consumption on the forest floor. Oecologia 38: 261–274.

Mauldin, J. K. and Rich, N. M. 1980. Effect of chlortetracycline and other antibiotics on protozoan numbers in the eastern subterranean termite. J. Econ. Entomol. 73: 123–128.

Mauldin, J. K. and Smythe, R. G. 1973. Protein-bound amino acid content of normally and abnormally faunated Formosan termites. J. Insect Physiol. 19: 1955–1960.

Meyer, D. R., Lanzrein, B., Lüscher, M. and Nakanishi, K. 1976. Isolation and identification of a juvenile hormone (JH) in termites. Experientia 32: 773.

Meyer, J. A. 1960. Resultats agronomiques d'un essai de nivellement des termitieres realise dans la Cuvette centrale Conglaise. Bull Agr. de Congo Belge 51: 1047–1059.

Mielke, H. W. 1978. Termitaria and shifting cultivation: The dynamic role of the termite in soils of tropical wet-dry Africa. Trop. Ecol. 19: 117–122.

Miller, E. M. 1969. Caste differentiation in the lower termites. In, Biology of ter-

mites (K. Krishna and F. M. Weesner, eds.), Vol. 1, pp. 283–310. New York and London: Academic Press.

Milne, G. 1947. A soil reconnaissance journey through parts of Tanganyika Territory. December 1935 to February 1936. *J. Ecol.* 35: 192–264.

Moore, B. P. 1969. Biochemical studies in termites. In, *Biology of termites* (K. Krishna and F. M. Weesner, eds.), Vol. 1: 407–432. New York and London: Academic Press.

Moore, B. P. 1974. Pheromones in the termite societies. In, *Pheromones* (M. C. Birch, ed.), pp. 250–266. Amsterdam: North Holland.

Nagin, R. 1972. Caste determination in *Neotermes jouteli* (Banks). *Insectes Soc.* 19: 39–61.

Nel, J. J. C. 1968. Aggressive behaviour of the harvester termites *Hodotermes mossambicus* (Hagen) and *Trinervitermes trivervoides* (Sjöstedt). *Insectes Soc.* 15: 145–56.

Nel, J. J. C. 1969. Effect of solar radiation on the harvester termite, *Hodotermes mossambicus* (Hagen). *Nature* 223: 862–863.

Nel, J. J. C. 1970. Aspekte van die gedrag van die werkers van die grasdraertiermiet, *Hodotermes mossambicus* (Hagen) in die veld. *J. Entomol. Soc. South. Afr.* 33: 23–34.

Nel, J. J. C. and Hewitt, P. H. 1969. A study of the food eaten by a field population of the harvester termite, *Hodotermes mossambicus* (Hagen) (Isoptera, Hodotermitidae), and its relation to population density. *J. Entomol. Soc. South. Afr.* 32: 123–131.

Nel, J. J. C. and Hewitt, P. H. 1978. Swarming in the harvester termite *Hodotermes mossambicus* Hagen. *J. Entomol. Soc. South. Afr.* 41: 195–198.

Nel, J. J. C., Hewitt, P. H., and Joubert, L. 1970. The collection and utilization of redgrass, *Themeda triandra* (Forsk.) by laboratory colonies of the harvester termite, *Hodotermes mossambicus* (Hagen) and its relation to population density. *J. Entomol. Soc. South. Afr.* 33: 331–340.

Nel, J. J. C., Hewitt, P. H., Smith, L. J., and Smith, W. T. 1969. The behaviour of the harvester termite, *Hodotermes mossambicus* (Hagen) in a laboratory colony. *J. of the Entomol. Soc. of South. Afr.* 32: 9–24.

Nielsen, M. G. and Josens, G. 1978. Production by ants and termites. In, *Production ecology of ants and termites* (M. V. Brian, ed.), pp. 45–53. Cambridge: Cambridge Univ. Press.

Noirot, C. 1955. Recherches sur le polymorphisme des termites superieurs. *Ann. Sci. Nat. (Zool.) Ser. 11*, 17: 400–595.

Noirot, C. 1956. Les sexues de remplacement chez les termites superieurs (Termitidae). *Insectes Soc* 3: 145–158.

Noirot, C. 1959. Remarques sur l'écologie des termites. *Ann. Soc. R. Zool. Belg.* 89: 151–168.

Noirot, C. 1969a. Glands and secretions. In, *Biology of termites* (K. Krishna and F. M. Weesner, eds.), Vol. 1: pp. 89–123. New York & London: Academic Press.

Noirot, C. 1969b. Formation of castes in the higher termites. In, *Biology of termites* (K. Krishna and F. M. Weesner, eds.), Vol. 1, pp. 311–350. New York and London: Academic Press.

Noirot, C. 1970. The nests of termites. In, *Biology of termites* (K. Krishna and F. M. Weesner, eds.), Vol. 2, pp. 73–125. New York and London: Academic Press.

Noirot, C. 1974. Polymorphismus bei höheren Termiten. In, *Sozialpolymorphismus bei Insekten: Probleme der kastenbildungim Tierreich* (G. H. Schmidt, ed.), pp. 740–765. Stuttgart: Wissenschaftliche.

Noirot, C. 1977. Various aspects of hormone action in social insects. In, *Proc. 8th int. Congr. Int. Union Study Social Insects, Wageningen, Netherlands, Sept. 5–10, 1977:* 12–16.

Noirot, C. and Noirot-Timothee, C. 1969. The digestive system. In, *Biology of termites* (K. Krishna and F. M. Weesner, eds.), Vol. 1, pp. 49–88. New York and London: Academic Press.

Nour, H. O. A. 1979. Biodeterioration of building timber in the Sudan. A survey of the most important natural destructive agencies of wood, their interaction and possible preventive measures against them. In, *The use of home grown timber in building* (O. M. E. Fageiri and H. O. Abdel Nour, eds.), pp. 1–9. Building and Road Research Institute, University of Khartoum, Sudan.

Nour, H. O. A. 1980. The natural durability of building wood and the use of wood preservatives in the Sudan. *Sociobiology* 5: 175–182.

Nutting, W. L. 1969. Flight and colony foundation. In, *Biology of termites* (K. Krishna and F. M. Weesner, ed.), Vol. 1: 233–282. New York and London: Academic Press.

Nutting, W. L. 1979a. Termite flight periods: Strategies for predator avoidance? *Sociobiology* 4: 141–151.

Nutting, W. L. 1979b. Biological notes on a rare dry-wood termite in the southwest, *Incisitermes banksi* (Kalotermitidae). *Southw. Entomol.* 4: 308–310.

Nutting, W. L. and Haverty, M. I. 1976. Seasonal production of alates by five species of termites in an Arizona grassland. *Sociobiology* 2: 145–153.

O'Brien, R. W. and Slaytor, M. 1982. Role of microorganisms in the metabolism of termites. *Aust. J. Biol. Sci.* 35: 239–262.

O'Brien, G. W., Veivers, P. C., McEwen, S. E., Slaytor, M., and O'Brien, R. W. 1979. The origin and distribution of cellulase in the termites, *Nasutitermes exitiosus* and *Coptotermes lacteus. Insect Biochem.* 9: 619–625.

Ohiagu, C. E. 1979a. Nest and soil populations of *Trinervitermes* spp. with particular reference to *T. geminatus* (Wasmann), (Isoptera), in the Southern Guinea savanna near Mokwa, Nigeria. *Oecologia* 40: 167–178.

Ohiagu, C. E. 1979b. A quantitative study of seasonal foraging by the grass harvesting termite, *Trinervitermes geminatus* (Wasmann), (Isoptera, Nasutitermitinae) in Southern Guinea savanna, Mokwa, Nigeria. *Oecologia* 40: 179–188.

Ohiagu, C. E. and Wood, T. G. 1976. A method for measuring rate of grass-harvesting by *Trinervitermes geminatus* (Wasmann) (Isoptera, Nasutitermitinae) and observation on its foraging behaviour in southern Guinea Savanna, Nigeria. *J. Appl. Ecol.* 13: 705–713.

Okot-Kotber, B. M. 1980a. The influence of JH analogue on soldier differentiation in the higher termite *Macrotermes michaelseni. Physiological Entomol.* 5: 407–416.

Okot-Kotber, B. M. 1980b. Histological and size changes in corpora allata and prothoracic glands during development of *Macrotermes michaelseni* (Isoptera). *Insectes Soc.* 27: 361–376.

Okot-Kotber, B. M. 1981. Polymorphism and the development of the first progeny in incipient colonies of *Macrotermes michaelseni* (Isoptera, Macrotermitinae). *Insect. Sci. Appl.* 1: 147–150.

Oloo, G. W. and Leuthold, R. H. 1979. The influence of food on trail-laying and recruitment behaviour in *Trinervitermes bettonianus* (Termitidae: Nasutitermitinae). *Entomol. Exp. Appl.* 26: 267–278.

Pendleton, R. L. 1941. Some results of termite activity in Thailand soils. *Thai Sci. Bull.* 3: 29–53.

Pochon, J., de Barjac, H., and Roche, A. 1959. Recherches sur la digestion de la cellulose chez le termite *Sphaerotermes sphaerothorax*. *Ann. Inst. Pasteur* 96: 352–355.

Potrikus, C. J. and Breznak, J. A. 1981. Gut bacteria recyle uric acid nitrogen in termites: a strategy for nutrient conservation. *Proc. Nat. Acad. Sci.* (USA) 78: 4601–4605.

Potts, R. C. and Hewitt, P. H. 1973. The distribution of intestinal bacteria and cellulose activity in the harvester termite *Trinervitermes trinervoides* (Nasutitermitinae). *Insectes Soc.* 20: 215–220.

Prestwich, G. D. 1979. Termite chemical defense: New natural products and chemosystematics. *Sociobiology* 4: 127–138.

Rajagopal, D. and Veeresh, G. K. 1981. Foraging activity of the mound building termite *Odontotermes wallonensis* (Wasmann), Isoptera, Termitidae. *J. Soil Biol. Ecol.* 1: 56–64.

Ratcliffe, F. N., Gay, F. J., and Greaves, T. 1952. *Australian termites, The biology, recognition and economic importance of the common species.* Melbourne: Commonwealth Scientific and Industrial Research Organization.

Ratcliffe, F. N. and Greaves, T. 1940. The subterranean foraging galleries of *Coptotermes lacteus* (Frogg.). *Journal of the Council for Scientific and Industrial Research, Australia* 13: 150–161.

Richard, G. 1969. Nervous system and sense organs. In, *Biology of termites* (K. Krishna and F. M. Weesner, eds.), Vol. 1, pp. 161–192. New York and London: Academic Press.

Rohrmann, G. F. and Rossmann, A. Y. 1980. Nutrient strategies of *Macrotermes ukuzii*, (Isoptera: Termitidae). *Pedobiologia* 20: 61–73.

Roonwal, M. L. 1960. Biology and ecology of oriental termites No. 5. Mound structures, nest and moisture content of fungus combs in *Odontotermes obesus*, with a discussion on the association of fungi with termites. *Rec. Indian Mus.* 58: 131–150.

Roonwall, M. L. 1979. *Termite life and termite control in tropical south Asia.* Jodhpur, India: Scientific Publishers.

Roy-Noel, J. 1978. Influence de l'homme sur le peuplement en termites dans la presque ile du Cap-Vert (Senegal occidental). *Memorabilia Zool.* 29: 157–172.

Roy-Noel, J. 1979. Termites and soil properties. In, *Soils research in agroforestry* (H. O. Mongi and B. A. Huxley, eds.), pp. 271–295. Nairobi: International Council for Research in Agroforestry.

Rudman, P. and Gay, F. J. 1967. The causes of natural durability in timber. XX.

The cause of variation in the termite resistance of jarrah (*Eucalyptus marginata* Sm.). *Holzforschung* 21: 21–23.

Ruppli, E. 1969. Die Elimination überzähliger Ersatzgeschlechtstiere bei der Termite *Kalotermes flavicollis* (Fabr.). *Insectes Soc.* 16: 235–248.

Ruyooka, D. B. A. and Edwards, C. B. H. 1980. Variations in natural resistance of timber. III. Effect of fungal-termite associations on the natural resistance of selected Eucalypt timbers under laboratory and field conditions. *Mater. Org.* 15: 263–285.

Saeki, I., Sumimoto, M., and Kondo, T. 1971. The role of essential oil in resistance of coniferous woods to termite attack (*Coptotermes formosanus*). *Holzforschung* 25: 57–60.

Sands, W. A. 1961a. Nest structure and size distribution in the genus *Trinervitermes* (Isoptera, Termitidae, Nasutitermitinae), in West Africa. *Insectes Soc.* 8: 177–186.

Sands, W. A. 1961b. Foraging behaviour and feeding habits of five species of *Trinervitermes* in West Africa. *Entomol. Exp. Appl.* 4: 277–288.

Sands, W. A. 1965a. Mound population movements and fluctuation in *Trinervitermes ebenerianus* Sjöstedt (Isoptera, Termitidae, Nasutitermitinae). *Insectes Soc.* 12: 49–58.

Sands, W. A. 1965b. Termite distribution in man-modified habitats in West Africa, with special reference to species segregation in the genus *Trinervitermes* (Isoptera, Termitidae, Nasutitermitinae). *J. Anim. Ecol.* 34: 557–571.

Sands, W. A. 1965c. A revision of the Nasutitermitinae (Isoptera, Termitidae) of the Ethiopian zoogeographical region. *Bull. Brit. Mus. Nat. Hist., Entomol.*, Suppl. 4: 1–172.

Sands, W. A. 1965d. Alate development and colony foundation in five species of *Trinervitermes* (Isoptera, Nasutitermitinae) in Nigeria, West Africa. *Insectes Soc.* 12: 117–130.

Sands, W. A. 1969. The association of termites and fungi. In, *Biology of termites* (K. Krishna and F. M. Weesner, eds.), Vol. 1, 495–524. New York and London: Academic Press.

Sands, W. A. 1972a. The soldierless termites of Africa (Isoptera: Termitidae). *Bull. Brit. Mus. Nat. Hist., Entomol.*, Suppl. 18: 1–244.

Sands, W. A. 1972b. Problems in attempting to sample tropical subterranean termite populations. *Ekol. Polska* 20: 23–31.

Sands, W. A. 1982. Agonistic behavior of African soldierless Apicotermitinae (Isoptera: Termitidae) *Sociobiology* 7: 61–72.

Schaefer, D. A. and Whitford, W. G. 1981. Nutrient cycling by the subterranean termite *Gnathamitermes tubiformans* in a Chihuahuan desert ecosystem. *Oecologia* 48: 277–283.

Sen-Sarma, P. K. and Gupta, B. K. 1978. Natural resistance of six Indian timbers to termites under laboratory conditions. *Holzforsch. Holzverwert* 30: 88–91.

Sewell, J. J. and Watson, J. A. L. 1981. Developmental pathways in Australian species of *Kalotermes* Hagen (Isoptera). *Sociobiology* 6: 243–323.

Sheikh, K. H. and Kayani, S. A. 1982. Termite affected soils in Pakistan. *Soil Biol. Biochem.* 14: 359–364.

Sieber, R. 1982. The role of juvenile hormone in the development of physogastry in *Macrotermes michaelseni*, Isoptera: Macrotermitinae. *Gen. Comp. Endocrinol.* 46: 405–406.

Sieber, R. 1983. Establishment of fungus comb in laboratory colonies of *Macrotermes michaelseni* and *Odontotermes montanus* (Isoptera, Macrotermitinae). *Insectes Soc.* 30: 204–209.

Sieber, R. and Darlington, J. P. E. C. 1982. Replacement of the royal pair in *Macrotermes michaelseni*. *Insect Sci. Appl.* 3: 39–42.

Singh, U. R. and Singh, J. S. 1981. Temperature and humidity relations of termites. *Pedobiologia* 21: 211–216.

Skaife, S. H. 1955. *Dwellers in darkness: An introduction to the study of termites.* London: Longman Green & Co.

Snyder, T. E. 1949. Catalog of the termites (Isoptera) of the World. *Smithsonian Misc. Coll.* 112: 490 pp.

Spears, B. M., Ueckert, D. N. and Whigham, T. L. 1976. Desert termite control in a shortgrass prairie: Effects on soil physical properties. *Environ. Entomol.* 4: 899–904.

Spragg, W. T. and Paton, R. 1980. Tracing trophallaxis and population measurement of colonies of subterranean termites (Isoptera) using a radioactive tracer. *Ann. Entomol. Soc. Amer.* 73: 708–714.

Steward, R. C. 1983. The effects of humidity, temperature and acclimation on the feeding, water balance and reproduction of dry wood termites. *Entomol. Exp. Appl.* 33: 135–144.

Strickland, A. H. 1944. The arthropod fauna of some tropical soils. *Trop. Agr. (Trinidad)* 21: 107–114.

Stuart, A. M. 1969. Social behaviour and communication. In, *Biology of termites* (K. Krishna and F. M. Weesner, eds.), Vol. 1, pp. 193–232. New York and London: Academic Press.

Stuart, A. M. 1979. The determination and regulation of the neotenic reproductive caste in the lower termites (Isoptera): With special reference to the genus *Zootermopsis* (Hagen). *Sociobiology* 4: 223–237.

Stumper, R. 1923. Sur la composition chimique des nids de l'*Apicotermes occultus* Silv. *C. R. Hebd. Séanc. Acad. Sci.* (Paris) 177: 409–411.

Thomas, R. J. 1981. *Ecological studies on the symbiosis of* Termitomyces Heim *with Nigerian Macrotermitinae.* Thesis, University of London.

Thorne, B. L. and Noirot, C. 1982. Ergatoid reproductives in *Nasutitermes corniger*, Isoptera, Termitidae. *Int. J. Insect Morphol. Embryol.* 11: 213–226.

Tinley, K. L. 1977. *Framework of the Gorongosa Ecosystem.* Thesis, University of Pretoria.

Trapnell, C. G., Friend, M. T., Chamberlain, G. T. and Birch, H. F. 1976. The effects of fire and termites on a Zambian woodland soil. *J. Ecol.* 64: 577–588.

Ueckert, D. N., Bodine, M. C., and Spears, B. M. 1976. Population density and biomass of the desert termite *Gnathamitermes tubiformans* (Isoptera: Termitidae) in a short-grass prairie: Relationship to temperature and moisture. *Ecology* 57: 1273–1280.

UNESCO (1962). *Termites in the humid tropics.* Proceedings of the New Delhi Symposium, 4–12 October, 1960. Paris, Unesco.

Usher, M. B. and Ocloo, J. K. 1974. An investigation of stake size and shape in "graveyard" field tests for termite resistance. *J. Inst. Wood Sci.* 9: 32–36.

Usher, M. B. and Ocloo, J. K. 1979. The natural resistance of 85 West African hardwood timbers to attack by termites and micro-organisms. *Trop. Pest Bull.* 6: 47 pp.

Verma, A. N., Verma, N. D., and Tiwari, C. B. 1976. Effect of BHC and Aldrin on the termite damage in irrigated wheat crop, when insecticides were applied by different methods. *Indian J. Entomol.* 36: 221–225.

Wanyoni, K. 1974. The influence of the juvenile hormone analogue ZR 512 (Zoecon) on caste development in *Zootermopsis nevadensis* (Hagen) (Isoptera). *Insectes Soc.* 21: 35–44.

Ward, P. 1965. Feeding ecology of the black-faced dioch *Quelea quelea* in Nigeria. *Ibis* 107: 173–214.

Watson, J. A. L. 1973. The worker caste of the hodotermitid harvester termite. *Insectes Soc.* 20: 1–20.

Watson, J. A. L. 1974. Caste development and its seasonal cycle in the Australian harvester termite, *Drepanotermes perniger* (Froggatt) (Isoptera: Termitinae). *Aust. J. Zool.* 22: 471–487.

Watson, J. A. L. and Gay, F. J. 1970. The role of grain-eating termites in the degradation of a mulga ecosystem. *Search* 1: 43.

Watson, J. A. L., Lendon, C., and Low, B. S. 1973. Termites in mulga lands. *Trop. Grasslands* 7: 121–126.

Watson, J. A. L., Metcalf, E. C., and Sewell, J. J. 1977. A re-examination of the development of castes in *Mastotermes darwiniensis* Froggatt. *Aust. J. Zool.* 25: 25–42.

Watson, J. A. L. and Sewell, J. J. 1981. The origin and evolution of caste systems in termites. *Sociobiology* 6: 101–118.

Watson, J. P. 1962. The soil below a termite mound. *J. Soil Sci.* 13: 46–51.

Watson, J. P. 1974. Calcium carbonate in termite mounds. *Nature (London)* 247: 74.

Weesner, F. M. 1969. The reproductive system. In, *Biology of Termites* (K. Krishna and F. M. Weesner, eds.), Vol. 1, pp. 125–160. New York and London: Academic Press.

Weir, J. S. 1973. Air flow, evaporation and mineral accumulation in mounds of *Macrotermes subhyalinus* (Rambur). *J. Anim. Ecol.* 42: 509–520.

Wilkinson, W. 1962. Dispersal of alates and establishment of new colonies in *Cryptotermes haviland* (Sjöstedt) (Isoptera, Kalotermitidae). *Bull. Entomol. Res.* 53: 265–286.

Williams, R. M. C. 1966. The East African termites of the genus *Cubitermes* (Isoptera: Termitidae). *Trans. Roy. Entomol. Soc. Lond.* 118: 73–216.

Williams, R. M. C. 1973. *Evaluation of field and laboratory methods for testing termite resistance of timber and building materials in Ghana, with relevant biological studies.* Centre for Overseas Pest Research, Tropical Pest Bulletin, 3. London: Her Majesty's Stationery Office.

Williams, R. M. C. 1976. Factors limiting the distribution of building-damaging dry-wood termites (Isoptera: *Cryptotermes* spp.) in Africa. *Mater. Org.* (Suppl 3): 393–406.

Williams, R. M. C. 1977. The ecology and physiology of structural wood destroying Isoptera. *Mat. Org.* 12: 111–140.

Wilson, E. O. 1971. *The insect societies.* Cambridge, Mass.: The Belknap Press of Harvard Univ. Press.

Wood, T. G. 1976. The role of termites (Isoptera) in decomposition processes. In, *The role of terrestrial and aquatic organisms in decomposition processes* (J. M. Anderson and A. Macfadyen, eds.), pp. 145–168. Oxford: Blackwell Scientific.

Wood, T. G. 1978. Food and feeding habits of termites. In, *Production ecology of ants and termites* (M. V. Brian, ed.), pp. 55–80. Cambridge: Cambridge Univ. Press.

Wood, T. G. 1979. The termite (Isoptera) fauna of Malesian and other tropical rain forests. In, *The abundance of animals in Malesian rain forests* (A. G. Marshall, ed.), *Hull Univ. Dept. Geog. Misc. Ser.* 22: pp. 113–130.

Wood, T. G. 1981. Reproductive Isolating Mechanisms among species of *Microtermes* (Isoptera, Termitidae) in the Southern Guinea Savanna near Mokwa, Nigeria. In, *Biosystematics of social insects* (P. E. Howse and J. L. Clement, eds.), pp. 309–325. New York and London: Academic Press.

Wood, T. G. and Johnson, R. A. 1978. Abundance and vertical distribution in soil of *Microtermes* (Isoptera: Termitidae) in savanna woodland and agricultural ecosystems at Mokwa, Nigeria. *Memorabilia Zool.* 29: 203–213.

Wood, T. G., Johnson, R. A. and Anderson, J. M. 1983. Modification of soils in Nigerian savanna by soil-feeding *Cubitermes* (Isoptera, Termitidae). *Soil Biol. Biochem.* 15: 575–579.

Wood, T. G., Johnson, R. A., Bacchus, S., Shittu, M. O., and Anderson, J. M. 1982. Abundance and distribution of termites in riparian forest near Rabba in the Southern Guinea savanna vegetation zone of Nigeria. *Biotropica* 14: 25–39.

Wood, T. G., Johnson, R. A., and Ohiagu, C. E. 1977. Populations of termites (Isoptera) in natural and agricultural ecosystems in Southern Guinea savanna near Mokwa, Nigeria. *Geo-Eco-Trop.* 1: 139–148.

Wood, T. G., Johnson, R. A., and Ohiagu, C. E. 1980a. A review of termite (Isoptera) damage to maize and estimation of damage, loss in yield and termite (*Microtermes*) abundance at Mokwa, Nigeria. *Trop. Pest Mgmt.* 26: 355–370.

Wood, T. G. and Lee, K. E. 1971. Abundance of mounds and competition among colonies of some Australian termite species. *Pedobiologia* 11: 341–366.

Wood, T. G. and Sands, W. A. 1978. The role of termites in ecosystems. In, *Production ecology of ants and termites* (M. V. Brian, ed.), pp. 245–292. Cambridge: Cambridge Univ. Press.

Wood, T. G., Smith, R. W., Johnson, R. A., and Komolafe, P. O. 1980b. Termite damage and crop loss studies in Nigeria—pre-harvest losses to yams due to termites and other soil pests. *Trop. Pest Manag.* 26: 355–370.

Yamin, M. A. 1980. Cellulose metabolism by the termite flagellate *Trichomitopsis termopsidis*. *Appl. Environ. Biol.* 39: 859–863.

Yin, C. M. and Gillot, C. 1975. Endocrine control of caste differentiation in *Zootermopsis angusticollis* Hagen (Isoptera). *Can. J. Zool.* 53: 1701–1708.

Zhuzhikov, D. P., Zoltarev, E. K., and Mednikova, T. K. 1972. Postembryonic development of *Anacanthotermes ahngerianus* Jacobs. In, *Termites (collected articles)* (E. K. Zotolarev, ed.), pp. 46–62. Moscow: University Publishing House.

Chapter 2

ECONOMICALLY IMPORTANT TERMITES AND THEIR MANAGEMENT IN THE ORIENTAL REGION

P. K. Sen-Sarma

INTRODUCTION

Termites are an integral part of the fauna in the tropical and subtropical world. In the Oriental Region, the number and variety of termites are indeed staggering. Of approximately 2,500 species known, about 550 species have been recorded from this region and many more await discovery. About 15% are pests of living plants, wooden construction, and other materials of economic importance (Sands 1977).

There are only a few estimates of economic losses caused by termites in the world and practically none from the Oriental Region. According to Fletcher (1914) the annual loss of wheat crops in British India was about $32 million US (all dollar values U.S.). In Malaysia, timber depreciates at the rate of 0.04 to .02% in value per annum due to termite damage (Dhanarajan 1969), while annual expenditures for repairing government buildings were $150,000 in 1953 (Harris 1961). A recent survey by Rentokil (1980) showed that in Jakarta (Indonesia) an average 54% of the domestic and 43% of the commercial buildings had termite infestations, which might be due to increasing use of air conditioning. A small township in Punjab, Sri Hargobindpur, was completely destroyed by *Heterotermes indicola* (Wasm.) (Roonwal 1955). All wooden structures of houses in Nokundi village in Bulichistan were seriously damaged by *Anacanthotermes vagans* (Hagen) (Chaudhry and Ahmad 1972). Pest control operators in urban India carry out an annual business worth $5 million for termite control (personal observation). Thus, the total loss due to termite attack must run into millions of US dollars annually in the Oriental Region.

The Oriental Region discussed in this chapter is depicted in Fig. 2.1.

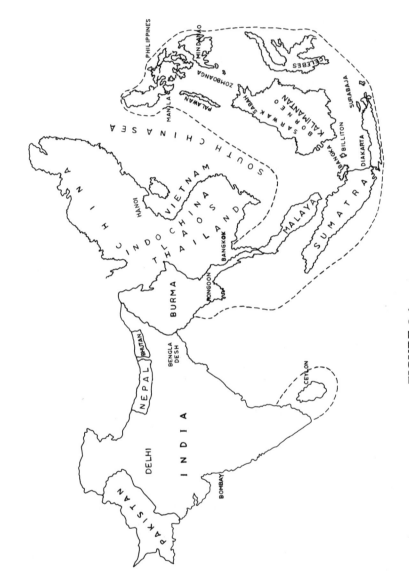

FIGURE 2.1
Map of the Oriental Region.

70

Few exclusive accounts of the economic damage caused by termites are available in the literature except those by Roonwal (1979) and Chhotani (1980). Roonwal's (1970) summary of the biology of termites of the Oriental Region also includes some information on economic damage. Sen-Sarma (1974) has given some information on plant damage caused by termites.

The distribution and ecology of termites are largely determined by the geological history, monsoon rainfall pattern, temperature, relative humidity, vegetation, altitude, and soil type (Sen-Sarma 1974). In India, termites are most active during the postmonsoon period (Chatterjee and Sen-Sarma 1962). The depth to which subterranean termites penetrate has an important bearing on preventive measures involving chemically treated soil. The only observation is that of Hoon (1962), who observed that termites could penetrate as deep as 30 m in areas having a deep alluvial deposit.

Notwithstanding their destructive propensities, termites play an important role in the breakdown of cellulose material left in the forest, thus adding nutrients to the soil. The ecosystem, therefore, is greatly dependent on the beneficial activities of termites.

In this chapter the economic damage caused by termites and the methods of termite management are briefly discussed.

DAMAGE CAUSED BY TERMITES

A crop can be susceptible to termite attack regardless of population density. While pest status depends on whether or not an economic threshold has been exceeded (Pradhan 1964), susceptibility of crops to termite attack is rather complex. Termites prefer to attack crops nearing maturity because of the higher cellulose content (Bigger 1966).

Agricultural crops commonly damaged by termites in southern Asia, Indonesia, and the Philippines include wheat, rice, maize, pearl millet, sorghum, groundnut, legumes, oil palm, castor, vegetables, sugarcane, cotton, jute, sunn hemp, and sisal (Fig. 2.2).

WHEAT (*Triticum aestivum* Linn.)—Wheat is a major cereal crop in northern India and Pakistan and recently, in suitable areas of Bangladesh. *Microtermes obesi* Holmg. causes extensive damage to wheat in Punjab, Haryana, Rajasthan, Delhi, Uttar Pradesh, Madhya Pradesh, Andrea Pradesh, and Bihar states in India, feeding on seedling roots and causing leaf yellowing and wilting. Other termite species causing major damage to wheat in India and Pakistan are *Odontotermes obesus* (Ramb.), *Microtermes* sp., *Nasutitermes* sp., and *Trinervitermes biformis* (Wasm.). Wheat is damaged by *O. obesus* in Haryana, Punjab, Rajasthan, Uttar

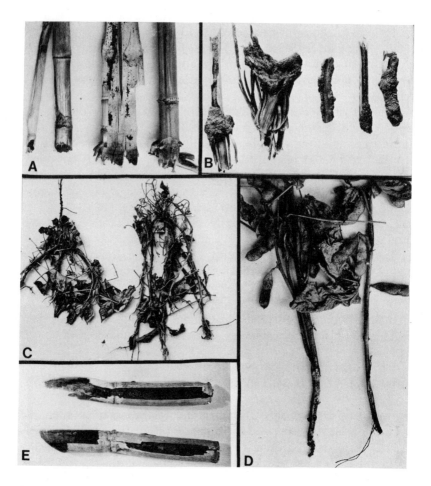

FIGURE 2.2

A. Damage to maize (*Zea mays*) by *Odontotermes wallonensis*
(*Source:* Chhotani 1980).

B. Damage to pearl millet (*Pennisetum typhoides*) by *Microtermes myco-phagus*
(*Source:* Chhotani 1980).

C. Damage to groundnut by *Odontotermes wallonensis*
(*Source:* Chhotani 1980).

D. Damage to soybean plants by *Odontotermes wallonensis*
(*Source:* Chhotani 1980).

E. Damage to sugarcane by *Odontotermes* sp. Note the soil fill in the eaten away portion.

Pradesh, Madhya Pradesh, and Gujarat in India, and by *O. gurdaspurensis* Holmg. and Holmg. in Rajasthan where damage is more pronounced in unirrigated fields (Chhotani 1980). *Microtermes* sp. has been implicated in West Bengal (Ghosh 1964). Probably the same species causes wheat damage in northwestern Bangladesh. *Trinervitermes biformis* is responsible for damage in Gujarat and Maharastra (India). In Pakistan, *O. obesus* and *M. obesi* are the major wheat pests (Janjua and Kahn 1955). In Punjab and Sind in Pakistan, *Anacanthotermes macrocephalus* (Desn.) consumes wheat in storage. *Microcerotermes tenuignathus* Holmg. attacks wheat seedlings in arid and semiarid areas in Rajasthan in India (Parihar 1978).

RICE (*Oryza sativa* Linn.)—Damage to rice (paddy) is rather uncommon. In dry areas of Tamil Nadu, *A. viarum* (Konig) causes damage to rice roots. In Rajasthan and West Bengal (also probably adjoining areas in Bangladesh), Maharastra and Delhi, *Microcerotermes* sp. cause damage to rice and *Coptotermes formosanus* (Shiraki) attacks rice in Taiwan (Menon et al. 1970, Roonwal 1979).

MAIZE (*Zea mays* Linn.)—Damage to maize is caused by *O. obesus* and *M. obesi* in Rajasthan (Srivastava 1959), *M. obesi* in Bihar (Agarwala and Sharma 1954), *Odontotermes* sp. in Assam, and *Odontotermes wallonensis* (Wasm.) in Karnataka, India. Plants are susceptible to attack from germination to harvest. Damaged plants exhibit yellowing of leaves and wilting and death occurs when rootlets are damaged.

PEARL MILLET (*Pennisetum typhoides* Stapf. and Hubb.)—Pearl millet is attacked by *M. obesi* and *Microtermes mycophagus* (Desn.) in Rajasthan. Damage to the roots and stem results in wilting and death.

SORGHUM (*Sorghum vulgare* Pers.)—Sorghum at the two- to three-leaf stage is attacked by *O. obesus*, *Odontotermes* sp., and *Microtermes* sp. in arid India and Pakistan; damage is sustained by plants less than 15 cm high (Gahukar and Jotwani 1980).

GROUNDNUT (*Arachis hypogaea* Linn.)—Subterranean termites are serious pests of groundnuts (peanuts) in the tropics but little information is available on monetary losses. Termites attack seeds, seedlings, roots, stems, and pods, as well as harvested pods drying in the field, and stored pods and seeds. Groundnut pods and seeds are attacked by *C. formosanus* in southern China, and by *O. obesus*, *O. wallonensis*, *M. obesi*, and *T. biformis* in India. In addition to direct yield loss, considerable deterioration in quality also occurs (Harris 1969a, Roonwal 1979, Amin and McDonald 1981). Although mortality of seedlings does not exceed 2%, damage by *O. obesus* has been reported from Madhya Pradesh in India up to 35%, coupled with severe damage to pods and seeds (Kaushal and Deshpande 1967). The damage to the pods increases rapidly if left in the soil beyond maturity, reducing the quality of seeds (Amin and McDonald 1981).

Scarification by termites renders pods susceptible to infection by soil fungi, of which *Aspergillus flavus* Link. is the most important (McDonald 1970). Scarification is particularly serious in mature stands (Johnson and Gumel 1981) where fungal infection produces aflatoxins in the seeds (McDonald 1970). Seeds infected by *Macrophomina phaseoli* (Mauble.) Ashby tend to increase the free fatty acids level above the acceptable limit, making the crop unsaleable. Sellschops (1965) confirmed that *Odontotermes latericus* (Hav.) are often loaded with spores of *A. flavus*. Although intensive work on this type has not been carried out in the Oriental Region, the same mechanism may operate.

LEGUMES—Legumes (pulses) are seasonal dicotyledonous crops providing inexpensive protein. Termites cause considerable damage to legume crops in southern Asia. *Odontotermes* sp. damages red gram or Pigeon pea (*Cajanus cajan* Mill sp.) in northeastern India. In Punjab damage to legumes such as black gram (*Vigna mungo* Hepper), green gram (*V. radiata* Wilczek), cow pea (*V. catiang* Walb), chickpea (*Cicer arietinum* Linn.), and lentil (*Lens culinaris* Medic.) is as high as 10% in sandy soil and rain-fed areas (Chhabra 1981). In Maharastra and Tamil Nadu, *Odontotermes* sp. are injurious to various legume crops. Soybean plants are damaged in Karnataka by *O. wallonensis* (Chhotani 1980).

OIL PALM (*Elaeis guinensis* Jacq.)—Oil palm is indigenous to tropical Africa, but is increasingly planted in Malaysia and Indonesia and to a lesser extent in India (Andaman Islands). *Coptotermes curvignathus* Holmg., *Schedorhinotermes longirostris* (Brauer), and *S. malaccensis* (Holmg.) severely damage oil palm in Malaysia and Indonesia (Hickin 1971).

CASTOR (*Ricinus communis* Linn.)—Castor is widely cultivated in many parts of southern Asia for nonedible oil. In Rajasthan, *M. mycophagus* damages roots of the seedling stage by hollowing out the stem (Parihar 1978).

SUGARCANE (*Saccharum officinarum* Linn.)—Sugarcane is an important cash crop in South and Southeast Asia. Damage takes place primarily at two growth stages of the crop: first, during the pre-monsoon period when eye buds of newly planted sets are destroyed, which results in germination failure, and second, termite penetration into the pith during the postmonsoon growth period, which leaves only the rind, often leading to death. The damage causes both loss in cane tonnage and sugar output. The damage is very serious under conditions of delayed rains, lack of irrigation, and drought (Avasthy 1967, Harris 1969b). The following species have been recorded to damage sugarcane: *C. heimi* (Wasm.), *Ermotermes paradoalis* Holmg., *O. obesus, O. wallonensis, M. obesi* and *T. biformis* in South Asia, *O. formosances* in Taiwan, South China, Thailand and Vietnam and *Heterotermes philippinensis* (Light) and *C. vastator* (Light) in the Philippines.

COTTON (*Gossypium* spp.)—In southern Asia, cotton is grown under both irrigated and unirrigated conditions. Though termites were hitherto regarded as minor cotton pests, the recent increase in cotton has resulted in termites causing major losses. On the Indian subcontinent termites attack the roots and also tunnel into the stem. The plants wilt and die when severely attacked (Butani 1973a). The American and Egyptian varieties are more susceptible than indigenous ones. Termite species recorded damaging cotton are *O. obesus, M. obesi,* and *T. biformis.*

JUTE (*Corchorus capsularis* Linn. and *C. olitorius* Linn.)—Jute plants, which are an important crop for the farmers of eastern India and Bangladesh, are infested by termites, particularly mature plants during the dry season. *Microtermes obesi* has been implicated, and their attack starts at the roots and extends to the pith, rendering the plant unsuitable for fiber production (Dutt 1962).

SUNN HEMP (*Crotolaria juncea* Linn.)—Fibers of sunn hemp are used for rope making. *Odontotermes obesus* damages sunn hemp in sandy-loam soil in India. The nature of the damage is similar to that in jute (Tripathi 1972).

SISAL (*Agave sisalana* Pers.)—Sisal is a good soil binder, which is used to control erosion and its leaves yield a hard fiber used in rope making. *Odontotermes* spp. damage leaves of sisal in India where the soil is laterite and gravelly. Plants, after yielding fiber for four to five years, reach a senescent stage when they are especially susceptible. The affected plants gradually wither (Tripathi 1972).

VEGETABLE, FLOWERING, ORNAMENTAL, AND MEDICINAL PLANTS

A number of vegetable crops such as cabbage, cauliflower, potato, eggplant, and chili are injured by termites, especially in southern Asia (Chhotani 1980). Chili (*Capsicum annum* Linn.) crop is particularly damaged in Pakistan and Gujarat by *O. obesus* and *T. biformis,* which attack the root system (Patel and Patel 1953, Chaudhry and Ahmad 1972). Termite damage is often a secondary phenomenon in chili and eggplant injury, root-rot fungi being the primary predisposing factor (Patel and Patel 1953). In India, potatoes are attacked by *O. obesus* in Bihar, by *Eremotermes* sp., and by an unidentified species of termite in Karnataka. Termites often eat away the seed potato, disrupting germination (Rataul and Mishra 1979). Termites also attack potatoes at the tuber stage, and reduce the production by 4.7 to 6.5% (Kumar 1965). Flowering plants such as rose and chrysanthemum, ornamental plants such as *Amaranthus gangeticus* Linn., and some cacti are attacked by *M. obesi* in Rajasthan and Punjab in India (Kushwaha 1960, Arora 1962). Common Japanese mint is attacked by *O. obesus* in Jammu in India, damaging the roots and stems (Gupta and Agarwal 1963).

TERMITE DAMAGE TO PLANTATION CROPS

TEA (*Camelia sinensis* O. Ktz.)—Termites cause serious injury to tea bushes in India, Bangladesh, Sri Lanka, Malaysia, Indonesia, southern China, and Taiwan (Figure 2.3A-B). A loss of 15% annually in certain districts of northeastern India is caused by *Microcerotermes* sp. and may range from 50% to 100% in acute cases (Das 1962).

In Sri Lanka, tea bush damage caused by *Glyptotermes dilatatus* (Bugn. and Popof.) has been appreciable during the last 20 years, particularly to the high-yielding soft-wooded clonal tea bushes. Faulty pruning, deterioration of plantation hygiene, application of high levels of nitrogenous fertilizers, and removal of shade trees seem to have accentuated the problem (Sivapalan et al. 1980). This termite attacks about 15 plant species including coffee, rubber, cacao, and the shade tree, *Gliricidia sepium* Walp. Colony foundation by dealated reproductive pairs primarily takes place in dried and rotten pruning snags.

Termites species damaging tea bushes in the Oriental Region are as follows: *Postelectrotermes militaris* (Desn.), *Neotermes greeni* (Desn.), and *G. dilatatus* in Sri Lanka, *P. bhimi* Roonwal and Maiti in south India, *C. ceylonicus* Holmg. in Sri Lanka and southern India, *Microcerotermes* sp. in northeastern India, *O. javanicus* Holmg. and *O. longinathus* Holmg. in Malaysia, and *O. formosanus* (Shiraki) in southern China and Taiwan, *O. obesus* in northeastern India and *Odontotermes* sp. in Bangladesh, *Microtermes pakistanicus* Ahmad in Malaysia, *M. obesi* in northeastern India and southern India. In Sri Lanka, the wood-inhabiting termites either hollow out the stem (*P. militaris*) or excavate in a honeycomb fashion (*N. greeni* and *G. dilatatus*). Subterranean termites damage the root system and hollow out the stem, often resulting in death.

COFFEE (*Coffea arabica* Linn.) and COCOA (*Theobroma cocao* Linn.)—Coffee and cocoa are damaged by *G. dilatatus* in Sri Lanka (Pinto 1941, Roonwal 1970). In southern India *Grallatotermes grallatoriformis* (Holmg. and Holmg.) and *Nasutitermes indicola* (Holmg. and Holmg.) have been implicated in the damage (Roonwal 1979).

COCONUT PALM (*Cocos nucifera* Linn.)—Coconut palm is cultivated all along the coastal regions of India, and throughout Bangladesh, Sri Lanka, and other islands and coastal areas in the tropics. Termites are often serious pests of this palm and cause considerable financial losses (Fig. 2.3C). Damage is most severe in nursery seedlings and young plantations. In the western coast of India, about 20% of the palm seedlings are killed by *O. obesus* (Nirula and Menon 1960). Seedlings are

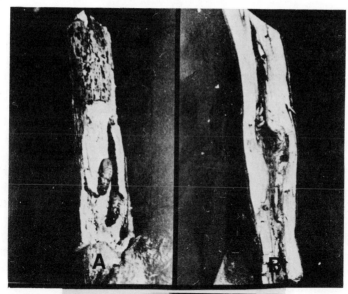

FIGURE 2.3

A. Attack to tea stem by dealated imagoes of *Glyptotermes dilatatus*: initiating new gallery in pruning snag.

B. Young colony of *G. dilatatus* in a gallery that has progressed into the sound wood
(*Source:* Sivapalam et al. 1980).

C: A termite nest in the hollow of a coconut palm.

damaged by *Macrotermes gilvus* (Hagen) in Indonesia, *C. ceylonicus, O. redemanni* (Wasm.), and *O. horni* (Wasm.) in Sri Lanka, and *Coptotermes sinabangensis* Oshima, *C. curvignathus, Globitermes sulphureus* (Hav.), and *M. carbonarius* (Hagen) in Malaysia and Indonesia. *Microcerotermes los-banosensis* Oshima and *N. luzonicus* (Oshima) attack coconut palm stems in the Philippines. Other termites causing casual damage are *C. ambroinensis* (Kemner), *N. havilandi* (Desn.), and *N. javanicus* (Holmg.) in Indonesia. Termites work through the base of the nuts or may attack the collar region. Wilting of the central shoot is the first symptom of termite infestation. In mature palms, termites may attack roots, stem, and leaf rachis, leading to reduction in nut yields. In older palms, termites seem to aggravate initial damage by borers and wounds (Harris 1961).

RUBBER (Hevea brasiliensis Muel.-Arg.)—Rubber is an important plantation crop in south and southeast Asia. *Coptotermes ceylonicus* and an unnamed species of *Glyptotermes* causes damage to rubber plants in south India. In Sri Lanka, *P. militaris, N. greeni, G. ceylonicus* (Holmg.), *G. dilatatus,* and *C. ceylonicus* infest rubber trees. *Glypotermes dilatatus* and *C. ceylonicus* are the two most injurious species to rubber plants in Sri Lanka (Fernando 1962).

Coptotermes curvignathus damages rubber plantations in Thailand, Malaysia, Vietnam, Kampuchea, and Indonesia (Caresche 1937, Ahmad 1965, Roonwal 1979). This species prevents successful plantations of rubber in these areas without resorting to preventive measures. *Coptotermes curvignathus* attacks bark, cambium, sapwood and heartwood of both young and mature rubber trees, killing them. Usual entry is at a fork of the taproot or near malformation of the taproot about 20 cm below the soil. Once entry is gained, termites penetrate into the wood causing numerous longitudinal and radial galleries. With time, the affected tree is so weakened as to be blown over. *Coptotermes curvignathus* also attacks many species of trees in adjoining forested areas. One colony may have several satellite nests and when the forest is cleared the colonies may remain in an endemic form in stumps or trees conserved for shade, and spread into the rubber plantation (Aziz bin Sheikh 1980). This species also attacks kapok trees (*Ceiba pentandra* Gaestin), which are often planted prior to rubber.

TERMITE DAMAGE TO HORTICULTURAL CROPS

Several species of fruit trees and vines during the seedling stage are severely attacked by termites. The attack starts at the roots and may spread to the upper plant parts. In the older trees, the bark and cambium are eaten. Termites may bore into the pith, hollowing it out and causing subsequent death of plants. Young grafts frequently are destroyed resulting in the disruption of plant propagation.

MANGO (*Mangifera indica* Linn.)—Mango is widely grown in southern and some parts of southeastern Asia. Termites damaging mango trees are *Neotermes assmuthi* (Holmg.), *N. bosei* Snyder, *N. fletcheri* (Holmg. and Holmg.), *N. mangiferae* (Roonwal & Sen-Sarma), *Glyptotermes almorensis* (Gardner), *Stylotermes fletcheri* (Holmg. & Holmg.), *C. heimi*, *C. kishori* (Roonwal & Chhotani), *H. indicola*, *O. assmuthi*, *O. latigula* Snyder, *O. indicus* Thakur, *O. obesus*, *O. redemanni*, *M. obesi*, and *T. biformis* in India and Bangladesh (Chhotani 1980), *Neotermes tectonae* (Damm.) in Indonesia, *C. curvignathus* in Indonesia and Malaysia, and *Globitermes sulphureus* and *Macrotermes gilvus* in Indonesia and Thailand (Roonwal 1979).

APPLE (*Malus pumila* Mill.)—The tap root and basal stem as well as dead or dry parts of plants are damaged by *Bifiditermes beesoni* (Gardner) and *C. heimi* in South Asia.

CASHEW (*Anacardium accidentale* Linn.)—In India *Neotermes greeni* attacks the dead and dry parts of cashew, often excavating the living portion of the plant. In Thailand, *Microtermes pallidus* (Hav.) damages cashew (Rajpreecha 1980).

MULBERRY (*Morus* sp.)—The seedlings are attacked by *C. heimi*, *H. indicola*, and *M. obesi* in South Asia. *Coptotermes heimi* hollows out the stem, resulting in death.

GUAVA (*Psidium guajava* Linn.)—*Odontotermes obesus* and *T. biformis* are common termite pests of guava trees in south Asia.

BER (*Zizyphus mauritiana* Linn.)—Ber is a commonly eaten fruit, providing much-needed vitamin C, and orchards are common in south Asia. The tree is excavated by the wood-inhabiting termite, *B. beesoni* in northwestern India and Pakistan. *Psammotermes rajasthanicus* (Roonwal and Bose) attacks newly planted saplings in sandy soil in Pakistan (Chaudhry and Ahmad 1972).

JACK-FRUIT (*Artocarpus integrifolia* Linn. f.)—Jack-fruit is an important fruit tree in south Asia. Unripe fruit is used as a vegetable and the ripe fruit is consumed uncooked. The dead and decayed branches are infested by *Neotermes adampurensis* Ahmad and *N. bosei* in India and Bangladesh. Infestation results in loss of tree vigor and affects fruit formation (Chaudhry and Ahmad 1972).

CITRUS sp.—Seedlings and young plants are damaged by *O. obesus* and *T. biformis* in India. *Coptotermes curvignathus* attacks citrus in Malaysia (Butani 1973b).

GRAPE (*Vitis vinifera* Linn.)—Newly planted grapevines are attacked in India by *O. obesus* and *O. redemanni*, which may hollow out the entire vine or may damage the tender sprouting shoots (Sen-Sarma 1974). The damage is caused from April to June in northern India and January to May in southern India.

Other fruit trees damaged by termites are pomegranate (*Punica granatum* Linn.) by *T. biformis* in India; apricot (*Prunus armeniacasica* Linn.) by *Microcerotermes baluchistanicus* Ahmad and *M. longignathus* Ahmad in Pakistan; and peach (*Prunus persica* Stakes) by *M. micophagus* in northwestern India. However, many more termite species probably injure fruit trees, particularly nursery stocks and seedlings in the Oriental Region.

TERMITES AS FORESTRY PESTS

During the last two decades, slow-growing indigenous forest trees have increasingly been replaced with fast-growing exotics in many parts of the Oriental Region. The monoculture of exotic trees faces considerable termite problems (Thakur and Sen-Sarma 1980).

With the introduction of community, urban, and recreational forestry, termite problems are serious in production nurseries and in plantations. Termite attack is severest in the dry season. Roots of 1- to 3-year-old transplants are damaged (Thakur and Sen-Sarma 1980). Species of *Odontotermes* and *Macrotermes* adversely affect the natural regeneration of forests and vegetation patterns in savannah, woodlands, and grasslands in Asia (Harris 1965).

Eucalptus suffer heavy damage from several species of subterranean termites in southern Asia and southeastern Asia. Termite species damaging *Eucalyptus* in India are: *A. macrocephalus*, *Microcerotermes minor* Holmg., *O. distans* (Holmg. and Holmg.), *O. gurdaspurensis* (Holmg. and Holmg.), *O. microdentatus* Roonwal and Sen-Sarma, *O. indicus* and *O. obesus* (Thakur and Sen-Sarma 1980). Incidence of damage by *M. gilvus* varies from 15% to 60% in Malaysia and Indonesia (Anonymous 1980, Harris 1969a).

In peninsular Malaysia severe damage to exotic pine plantations occurs by *C. curvignathus* leading to high rates of mortality. Conifers such as *Araucaria cunninghamii* D. Don, *A. hunsteinii* K. Schum, *Pinus caribaea* (Morelet), *P. elioti* Engelm., *P. patula* Schiede and Deppe, *P. pseudostrobus* (Lindl.), *P. strobus* Linn., are all attacked (Benedict 1971).

Coptotermes curvignathus also attacks indigenous forest trees like *Albizia lebeck* (Benth.), *A. procera* Benth., *Artocarpus elasticus* Bl., *Bombax ceiba* Linn., *Oroxylum indicum* Vent, *Sterculia campanulata* Wall., *Tetramerista glabra* Mig., and *Agathis borneensis* Warb. in Indonesia (Kalshoven 1963).

The genus *Populus* is an important fast-growing tree species because of its ease of cultivation, rapid growth, and suitability for pulp, paper, and other wood-based industries. Many exotic poplars (e.g., *Populus deltoides* Marsh., *Populus xeuramericana* (Dode) Guinier, *Populus* "G" clone, *Populus laevigata* Ait and *Populus* "casale" clone) are planted extensively as a part of agrosilviculture in many parts of southern Asia. In northeastern India, *P. deltoides* Marsh. is severely damaged by *C. heimi, C. kishori*, and *Microtermes unicolor* (Fig. 2.4). In Punjab and Haryana, India and perhaps in adjoining areas in Pakistan, *O. distans, O. gurdaspurensis*, and *O. obesus* cause large-scale mortality (P. K. Sen-Sarma, unpublished observation). In the Tarai region of India, *C. heimi* nests inside the wood of living poplar trees (Thakur 1978).

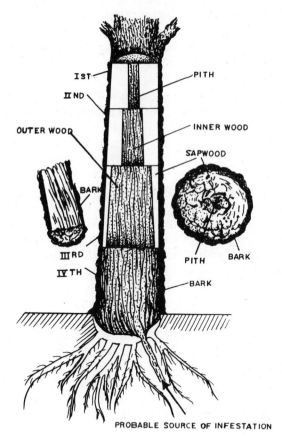

FIGURE 2.4
Diagrammatic sketch of an attacked tree of *Populus deltoides* I.C. clone showing different stages of attack by *Coptotermes heimi*.
(*Source:* Thakur and Sen-Sarma 1980).

Neotermes tectonae causes the most serious damage to teak (Tectona grandis Linn. f.) trees in Indonesia (Java, Sumatra, and Muna of South Celebes). About 50–70% of the 20- to 30-year-old growing teak are infested. Alates are attracted to rotten snags or may invade bore holes or similar wounds caused by cerambycid beetles such as Xyleborus sp. or carpenter bees. Dealates of N. tectonae also are capable of entry into the soft and rotten parts of the tree. In vigorous trees, damage to the cambium causes a partial interruption in the flow of sap, which results in the formation of callus tissues and widening of annual rings. The damage makes the valuable teak wood unsuitable for commercial purposes, causing heavy monetary losses.

The Indian Oaks are silviculturally important as companions to conifers in the Himalayas (Troup 1921). These oaks, particularly Quercus leucotrichophora A. Camus ex Bahadur, are attacked by S. bengalensis, which riddles the wood so extensively that the wood becomes virtually useless for its primary use in charcoal manufacture (Sen-Sarma et al. 1975).

The scale insects produce lac on a number of host trees in India, Thailand, and Vietnam. Earthen termite encrustations on the host trees often cover the lac insect, causing mortality of the insect by blocking its pores (Glover 1937).

TERMITES INJURIOUS TO PASTURE

Anacanthotermes macrocephalus (Desn.), which is a true harvester termite, is a serious pest of grasses in the arid regions of northwestern India and Pakistan. The foraging workers cut grass stems, seeds and tender twigs of plants and store them in underground galleries (Roonwal 1979). In Baluchistan (Pakistan), A. vagans (Hagen) causes damage to already denuded grass cover. Anacanthotermes peshwarensis Ahmad feeds on dry grasses in arid areas of Pakistan (Chaudhry and Ahmad 1972). In Peninsular India, A. viarum (Konig) and T. biformis destroy grasses in arid and semiarid areas. The damage deprives domestic animals of the already scarce food and caused deterioration of the pasture land.

TERMITES ATTACKING HUMAN DWELLINGS
AND THEIR CONTENTS

Termites cause immense damage to wood or bamboo in buildings and their contents. Damage has promoted the development of the pest control industries in almost all countries in the Oriental Region. Hickin (1971) has surveyed the rates charged for these jobs in several Oriental countries. Every year millions of dollars are spent solely to repair or replace badly damaged dwellings. Dwellings in the region are attacked by both drywood and subterranean termites, but the major problems arise from the latter.

Damage by Subteranean Termites

Subterranean termites maintain a soil connection for moisture. Among the subterranean termites attacking buildings, the most important are H. indicola and C. heimi.

Distribution of H. indicola extends to East Afghanistan (Weidner 1960) in the west and Nepal in the north (Becker 1975). The species readily develops neotenic reproductives in isolated and orphaned colonies (Sen-Sarma and Chatterjee 1965) and is relatively tolerant to toxic agents (Becker 1979). In Katmandu in Nepal (altitude 1,300 m), it seriously infests buildings, and farm houses (Becker 1975). In southern Asia, large numbers of buildings in urban and rural areas are attacked. This species also damages building contents such as cabinets, books, records, carpets, linoleum, furniture, wooden battens for supporting electric cables, basketwares, etc. This species is capable of negotiating wide gaps by means of sheltered tubes (Sen-Sarma et al. 1975). Heterotermes malabaricus Snyder causes similar damage in peninsular India.

Coptotermes heimi invades all types of niches from the hot humid tropics to temperate zones up to an altitude of 2,000 m in the Himalayas. At times this species, which has probably been introduced in some localities in southeastern Asia, has established flourishing colonies in wood without maintaining any soil connection (Thakur and Sen-Sarma 1979). This has enabled Coptotermes to become established through commerce in various tropical and subtropical regions (Gay 1969). Serious and extensive damage to building contents such as plywood paneling, carpets, office records etc., have been recorded (Sen-Sarma et al. 1975).

Anacanthotermes vagans causes damage to rafters used in houses made of either unbaked brick or mud in Pakistan (Chaudhry and Ahmad 1972) while structural timbers are attacked by P. rajasthanicus in western Rajasthan and Pakistan, Heterotermes gertrudae Roonwal in Sub-Himalayas, C. travians (Hav.) in northeastern India, O. obesus in southern Asia, O. indicus in various parts of India, and O. feae in eastern India and Bangladesh.

In Sri Lanka, major subterranean termites that damage buildings are H. ceylonicus, C. ceylonicus, and O. feae. The latter species is also a serious pest of woodwork in Burma and Thailand. Hypotermes obscuriceps (Wasm.) attacks buildings in Sri Lanka and Vietnam. In southeastern

Asia, *C. havilandi* attacks timbers in buildings in Thailand and Indonesia, while *C. travians* is a serious household pest in Malaysia, and also attacks buildings in Indonesia. In the Philippines, *C. vastator* is responsible for 70% of the damage to woodwork. In Taiwan and southern China, *C. formosanus* causes major damage to buildings (Oshima 1919). Other species of termites damaging buildings in southeastern Asia are: *G. sulphureus* in Thailand, Malaysia, Laos, Kampuchea, and Vietnam; *Schedorhinotermes malaccensis* (Holmg.) in Malaysia, Vietnam, Kampuchea, and Indonesia; *M. gilvus* (occasional pest) in Thailand, Malaysia, Indonesia, and the Philippines; *Microtermes insperatus* Kemner in Indonesia and *Nasutitermes luzonicus* (Oshima) in the Philippines.

Damage By Drywood Termites

Drywood or powder-post termites (*Cryptotermes* spp.) have been transported by man to various parts of the tropics and subtropics (Gay 1969). They attack dry wood, nest there without maintaining any soil connection, and pose serious problems in coastal areas and on islands in the Orient. They also attack bamboo rafters and poles used in huts (Sen-Sarma and Mathur 1957). New colonies are founded when alates alight on timber or bamboo having cracks or crevices or when a part of a colony becomes orphaned (Sen-Sarma and Thakur 1974). The pseudoworkers or larvae excavate galleries that may follow the grain of the wood or cut across it and ramify into an irregular network. The strength of the affected wood diminishes with time.

Cryptotermes bengalensis Snyder causes considerable damage to posts, rafters, and door and window sills in Assam, West Bengal, Tripura India, and in Bangladesh. *Cryptotermes domesticus* (Hav.) is a serious domestic pest attacking woodwork in India, Bangladesh, Sri Lanka, Thailand, Malaysia, Laos, Kampuchea, Vietnam, Indonesia, southern China, and Taiwan. *Cryptotermes cynocephalus* Light is a serious pest of dwellings in Malaysia, Indonesia, and the Philippines. *Cryptotermes dudleyi* Banks attacks wood and bamboo structures in many parts of the Orient.

TERMITES DAMAGING STORED WOOD AND WOOD IN SERVICE

Termites attack harvested timber during storage and exposed in service (Fig. 2.5). Subterranean termites cause the most damage. In southern Asia and Sri Lanka the most important of these termite species are: *A. vagans* in Baluchistan in Pakistan and southeastern Afghanistan; *A. macrocephalus* in Rajasthan, India; *P. rajasthanicus* in arid areas in northwestern India and Pakistan; *C. heimi* and *H. indicola* in Pakistan; and *Microcerotermes beesoni, M. championi, M. crassus,* and *M. tenuignathus* in India.

In southeastern Asia, the following termite species are responsible for the most damage: *Coptotermes havilandi* Holmg., *C. curvignathus, C. travians, G. sulphureus, Macrotermes carbonarius* (Hagen), *M. gilvus, O. feae, M. pakistanicus* Ahmad, *M. obesi* in Burma, Thailand, Malaysia, the former Indochina Region, and Indonesia. *Odontotermes formosanus* attacks harvested wood in Burma, Thailand, Vietnam, Taiwan, and southern China as far as latitude 35°N, including Hainan Island and Hong Kong (Roonwal 1970). *Schedorhinotermes malaccensis* and *S. medioobscurus* (Holmg.) attack wood in forests and timber yards in Thailand, Malaysia, Kampuchea, and Vietnam. *Macrotermis Coptotermes crassus* attacks timber either stored or in service in Burma, Thailand, Vietnam and Hainan in China and probably also in the Malay Peninsula, Kampuchea and Laos. *C. formosanus* is a serious destroyer of wood in Taiwan and southern China.

TERMITES CAUSING DAMAGE TO DAMS OF WATER RESERVOIRS

Odontotermes redemanni in India, and *O. formosanus* and *Macrotermes barneyi* Light in southern China cause leakage of water from reservoirs within earthen embankments. They penetrate the soil, making numerous passageways through the dam (Anonymous 1976, Hoon 1962).

TERMITE DAMAGE TO NONCELLULOSE MATERIAL

Termites often cause damage to noncellulose material such as plastics, leather goods, lead covering of cables, etc. (Snyder 1954) (Fig. 2.5B). These materials, however, do not form their diet. The only explanation for damage is that these materials are obstacles between the termites and their food. In former Indochina, telegraphic cables were damaged by *C. curvignathus* (Roonwal 1979), which also causes damage to polyethylene tubes and electric cables in Indonesia (Kalshoven 1962). Termites accelerate corrosion by moist earthen tunnels built over metal objects (Roonwal 1979). Damage to underground electric cables in an airport in India caused landing difficulties (Roonwal 1979). Damage leads to disruption of communication systems and electric and water supplies. A detailed bibliography on termite damage to electric cables has been given by Colwill (1958). Compactness and solidity of airfield runways in India are also

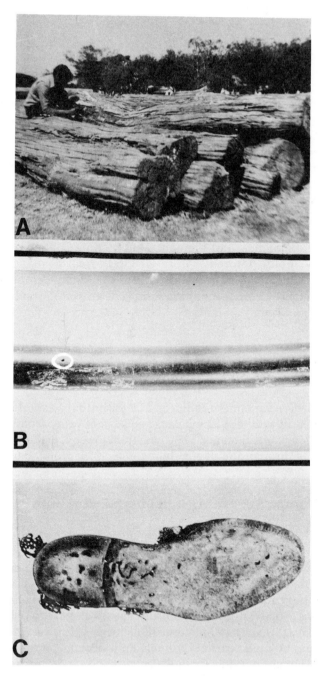

FIGURE 2.5

A. Termite (*Odontotermes wallonensis*) attack on logs of different timber species stored on land.

B. Damage to plastic pipe by *Microcerotermes beesoni* under laboratory conditions. The entrance hole is circled.

weakened due to tunneling underneath by termites (Roonwal 1979). Ammunition boots (Fig. 2.5C), and leather saddles are damaged in India by *C. heimi* (Chatterjee and Thakur 1966). Leather footballs and cricket balls were damaged by *C. curvignathus* in a warehouse in Singapore (Harris 1961). Methods of tanning largely govern termite susceptibility of leather; for example, "Chrome ratan," "semi-chrome," and "full-vegetable tan" being susceptible, while "full chrome" is relatively immune to attack (Colwill 1958).

MANAGEMENT OF TERMITES AND TERMITE PROOFING OF AGRICULTURAL CROPS

In general, vigorous plants can withstand damage by subterranean termites. But when plant vigor is lowered because of stress, the weakened plants are prone to termite attack. The common control practice is insecticidal application to the soil. The management of termites in important crops is discussed below:

WHEAT

(a) Seed Treatment: Seed treatment with 6.75% benzene hexachloride (BHC) 1 kg per 60 kg seed (Patel 1962) or spraying thinly spread seeds with water emulsion of aldrin 30 emulsifiable concentrate (EC) at 400 ml per 100 kg of seed (Verma et al. 1978) proved highly effective in controlling the damage. The yield increased by 31.1–55.9% over the untreated control.

(b) Soil Treatment. Preplanting soil treatment by means of 5% aldrin dust or hand broadcasting 0.4 kg aldrin 35 E.C. mixed with 50 g of sand per ha or 7.5% phorate granules or 0.25 gk per ha lindane or with 1.25% DDT emulsion at 1200 liter per acre, affords varying degrees of protection (Roonwal 1979).

(c) Standing crops. Dusting of termite-infested fields with 5% aldrin or 5% BHC or 6% heptachlor at 10 kg per acre (Srivastava et al. 1962) or broadcasting a 0.625% aldrin 35 EC mixed with sand (Deol, in press) affords 71 to 90% control up to harvesting.

RICE, MILLET AND LEGUMES—Little research on termite control in rice, millet, and legume crops has been conducted because of their low value. However, preplanting or postplanting soil treatment with as little as 50 g of an organochlorine insecticide has been found effective (Roon-

---

C. Ventral view of an ammunition boot damaged by *Heterotermes indicola*

(*Source:* Chatterjee and Thakur 1966).

wal 1979). Preplanting seed dressing of maize was effective in Africa (Sands 1973).

GROUNDNUT—Seed treatment using one of the organochlorine insecticides such as aldrin or lindane protects seedlings. However, where termite incidence is high, preplanting soil poisoning either by treating planting furrows with 6% heptachlor dust at 25 kg per ha (Rawat et al. 1970) or incorporating a soil insecticide (lindane, 5 kg per ha or aldrin 500 g per ha) in drills at planting protects seedlings (Feakin 1973). Repeated mechanical cultivations also reduced termite incidence (Amin and McDonald 1981). Harvesting the groundnut soon after maturity prevents scarification. The quick removal of the harvested crop from the fields prevents termite damage. ICRISAT (International Crops Research Institute for the Semi-Arid Tropics) at Hyderabad in India has recently developed cultivar NC ACC 2738 which is claimed to resist scarification (Amin and McDonald 1981).

SUGARCANE—Sugarcane is protected from attack by subterranean termites with a persistent insecticide that is applied to the sets before planting and by soil treatment after planting. Dipping the sets in 2.5% DDT emulsion suspension or other organochlorine insecticides gives good protection. Soil treatment with 5% chlordane at 20 kg per ha, heptachlor at 3 kg per ha, or aldrin at 1 kg per ha before final harrowing and furrowing for planting protects and also stimulates growth of the plants, resulting in 23 to 52% increase in yield (Roonwal 1979).

VEGETABLES CROPS AND MEDICINAL PLANTS

To protect the underground parts of vegetable crops, soil has been treated with an organochlorine insecticide. Flooding fields of Japanese mint with water reduces termite incidence. Application of 3% heptachlor at the rate of 20–25 kg per acre, however, is more effective (Gupta and Agarwal 1963). Chlorpyrifos has recently replaced the organochlorine insecticides (Khoo and Sherman 1979).

FIBER CROPS—Control of termites in cotton by preplanting soil treatment with technical carbaryl (Sevin®) or technical heptachlor (5–7.5 kg per acre in unirrigated areas and lower doses in irrigated areas) gives good protection and the yield increases considerably (Srivastava and Bhatnagar 1960). In Gujarat, a 33.3% increase in cotton yield was obtained by applying 5% BHC (0.675 gamma isomer) at the rate of 28 kg per acre (Patel 1962). For control of termites in other fiber crops such as jute, sunn hemp, or sisal, soil treatment with a persistent chlorinated hydrocarbon insecticide is effective (Dutt 1962, Tripathi 1972).

TERMITE CONTROL IN PLANTATION CROPS

TEA — Management of termites in tea can be achieved by cultural practices and application of insecticides. Pruning and hoeing cause wounds through which termites can gain access into the wood (Hickin 1971). These wounds should be treated with a wood preservative to prevent infection by decay fungi and subsequent termite attack. Sivapalan et al. (1980) observed that the incidence of attack by G. dilatatus in Sri Lanka is low in fields with shade trees (Gliricidia sepium), which act as alternate host for the termites. Furthermore, hard-wooded tea clones (TRI 777, EN 31 and NL 31) are less susceptible to termite attack when grown along with G. sepium. The soft-wooded high-yielding clones are highly susceptible when planted without G. sepium. Sivapalan et al. (1980) recommend that in termite-prone areas high-yielding resistant clones such as GMT 9, KEN 16/3 and DG 39 should be used. Other cultural practices to reduce termite infestations include isolation of the healthy fields from infected ones, systematic pruning, and removal of infested branches.

Colonies of G. dilatatus and other wood-inhabiting termites inside tea bushes are exterminated by inserting one or more 0.6 g pellets of Phostoxin through drilled holes and then sealing the hole with putty or clay (Danthanarayana and Fernando 1970).

For control of subterranean termites, treatment of soil around tea bushes with an organochlorine insecticide (e.g., aldrin at 2.3 kg active ingredient (a.i.) per ha), particularly during the dry season and early monsoon, is highly effective (Das 1962).

COCONUT PALM — Before organochlorine insecticides were marketed, the usual procedure for protecting coconut seedlings against termite attack was to envelop the seed nut with a thick layer of washed sand before filling the planting hole with soil (Harris 1961). At present, poisoning the soil with a persistent soil insecticide has been found very effective (Krishnamoorthy and Ramasubbiah 1962). Child (1964) recommended dieldrin dust, but dieldrin is banned in many countries so aldrin can be substituted. Spraying of a persistent insecticide on the stems of palms can prevent termites from building galleries on them.

RUBBER TREES — Newly planted rubber cuttings can be protected from damage by C. curvignathus by treating planting pits with 0.05% chlordane or 0.025% aldrin or heptachlor. An emulsifiable concentrate affords better protection than any other formulation. The cost is not prohibitive even to treat individual trees (Newsam and Rao 1958). For mature trees, the same insecticide can be applied by excavating a shallow channel around the tree and pouring the insecticide emulsion into it (Harris 1961). The main termitarium existing in stumps or logs in felled forest areas may be destroyed by using an explosive mixture of ammonium nitrate and fuel oil (Gray and Butcher 1969), thus reducing the termite incidence on rubber trees.

Attack by dry-wood termites can be managed by inserting pellets of aluminium phosphide into the galleries through drilled holes and then sealing the hole with putty, cement, or clay.

TERMITE CONTROL IN HORTICULTURAL CROPS

Young seedlings can be protected by soil treatment with a persistent organochlorine insecticide. Treating the top soil up to a depth of 10 cm with 5% aldrin dust at 25 kg per ha affords protection of Citrus (Butani 1973b). Termite damage in grapes can be controlled by treating the soil up to a depth of 50 cm with 5% aldrin or 6% BHC dust 22–25 kg per ha before planting. Gallery formation on the stem or grafts of fruit trees can be checked by spraying with 0.25% water emulsion of aldrin or lindane.

TERMITE CONTROL IN FORESTRY CROPS

Depredation of termites primarily occurs in production nurseries, and young plantations of afforested and reforested areas (Thakur and Sen-Sarma 1980). Most problems stem from raising plantations in termite in-fested degraded forest areas or in grasslands and savannas and is greatly accelerated by inadequate or faulty cultural management practices such as poor transplanting techniques, irregular watering (where available), and removal of shade trees (Brown 1965). Good cultural management practices such as clean-tending, mulching, and fertilizing increase vigor, and reduce termite incidence (Parry 1959, Lowe 1961). However, the best method of protecting seedlings in nurseries and plantations is to use in-secticides in soil before planting. The insecticides aldrin, γ-BHC, chlor-dane, and heptachlor (0.1–0.2% in water) have been found very effective for protection of Eucalyptus saplings (Chatterjee and Sen-Sarma 1968, Chatterjee 1972, Rajagopal et al. 1980, Nair and Varma 1981). Spot treat-ment of planting holes is costly, laborious, and time consuming, and the most suitable method is to raise seedlings in nurseries in poly-pots filled with treated soil (Fig. 2.6). Tropical pines in Malaysia can be similarly protected. Destruction of primary nests of C. curvignathus may be attempt-ed by using an explosive mixture of ammonium nitrate and fuel oil (Gray and Butcher 1969). Damage to standing trees of Tectona grandis Linn. f. by N. tectonae has defied control measures. However, removal of in-fested trees during thinning operations and splitting and exposing the split logs to sun accelerate the destruction of colonies (Kalshoven 1963).

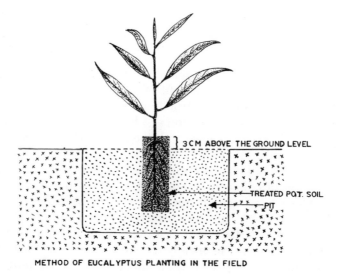

3 CM ABOVE THE GROUND LEVEL

TREATED POT SOIL

PIT

METHOD OF EUCALYPTUS PLANTING IN THE FIELD

FIGURE 2.6
Correct method of transplanting seedlings of *Eucalyptus* grown in poly-pots having insecticide-treated soil (*after* Rajagopal et al. 1980).

CONTROL OF TERMITES DAMAGING PASTURES

Control of harvester termites, *Anacanthotermes* spp., has not been done in the Oriental Region. However, baiting with sodium fluorosilicate (1.25% in water), BHC, or aldrin mixed with grain husk, chopped grass, or pellets of attractant food such as lucerne is likely to be effective in reducing damage in this region as it is in South Africa (Coaton 1958, Findlay 1971).

TERMITE MANAGEMENT IN DWELLINGS

Management of termites in buildings and mud huts falls under two categories: (1) preconstruction methods and (2) postconstruction methods. The various constructional designs that prevent access of termites to buildings and methods of wood preservation are not discussed.

    PRECONSTRUCTION MEASURES—Site hygiene is of prime importance. As a general practice, all logs, tree stumps and roots, and wooden scraps should be removed from building sites. Termite mounds in the vicinity should be destroyed. Preconstruction chemical treatment aims at creating a barrier between the building and subterranean termites. Treatment is applied step by step as the construction proceeds. In the Orient, buildings generally are constructed of baked bricks and mortar with slab floor construction. First, excavated foundation trenches are treated with an insecticidal emulsion at 5 liters per m² of surface area. Insecticides and dosage commonly used are aldrin 0.5%, heptachlor 0.5%, or chlordane 1%. Soil for backfill is also treated in a similar manner (Figs 2.7, 2.8, 2.9). After the construction is raised to plinth level, the top surface is treated with the emulsion at 5 liters per m² before laying the rubble soil or subgrade. As a second line of defense, it is advisable to use insecticidal water in mixing cement mortar used in construction at least up to windowsill level. Special attention is given to critical areas such as openings in soil made for conduits, etc. After the building is completed, its external perimeter is treated by making rod holes 15 cm apart and 30–45 cm deep and injecting insecticide emulsion at the rate of 6 liters per linear m of surface. Entrance steps, porch, attached garage and other additions are also chemically treated (Allen et al. 1961, Ebeling 1968, Anonymous 1972a). The soil treatment is effective only against subterranean termites and not against dry-wood termites.

Although long residual effects of soil poisoning have been reported (Chatterjee 1963), there are reports of detoxification of these chemicals by soil microorganisms whose activity can be suppressed by adding inexpensive neem (*Azadirachta indica* A. Juss.) seed cakes or neem oil extract in the soil (Attri and Prasad 1981).

POSTCONSTRUCTION TERMITE CONTROL—Control of termites in existing buildings is known as postconstruction, remedial, or *in situ* treatment. Treatment is more difficult and complex than precon-

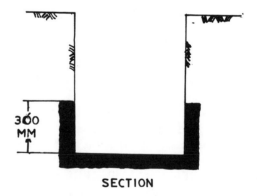

SECTION

FIGURE 2.7
Insecticide treatment of bottom and side of trench as preconstruction measures (*after* Anonymous Indian Stand. Inst. 1972a).

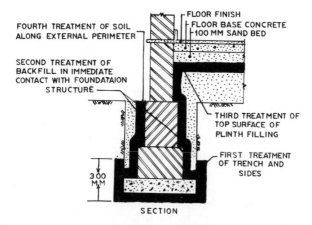

FOURTH TREATMENT OF SOIL
ALONG EXTERNAL PERIMETER

SECOND TREATMENT OF
BACKFILL IN IMMEDIATE
CONTACT WITH FOUNDATAION
STRUCTURE

FLOOR FINISH
FLOOR BASE CONCRETE
100 MM SAND BED

THIRD TREATMENT OF
TOP SURFACE OF
PLINTH FILLING

FIRST TREATMENT
OF TRENCH AND
SIDES

300 M.M

SECTION

FIGURE 2.8
Insecticide treatment for load bearing walled structure as preconstructions measures (*after* Anonymous Indian Stand. Inst. 1972a).

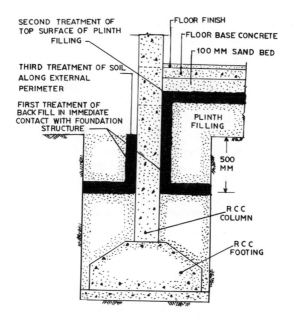

SECOND TREATMENT OF
TOP SURFACE OF PLINTH
FILLING

THIRD TREATMENT OF SOIL
ALONG EXTERNAL
PERIMETER

FIRST TREATMENT OF
BACKFILL IN IMMEDIATE
CONTACT WITH FOUNDATION
STRUCTURE

FLOOR FINISH
FLOOR BASE CONCRETE
100 MM SAND BED

PLINTH
FILLING

500
MM

R C C
COLUMN

R C C
FOOTING

FIGURE 2.9
Insecticide treatment for reinforced framed structure with columns and plinth beams as preconstruction measures (*after* Anonymous Indian Stand. Inst. 1972a).

struction treatment. The methods employed depend on the type of construction and termite species involved. The first prerequisite is to carry out a thorough inspection with regard to the nature and the extent of damage, access points to buildings, and presence of satellite nests. A termite detector is useful for this purpose (Mori 1973). Postconstruction ant/termite treatment is basically similar to preconstruction treatments. This is achieved by digging a trench, 30 cm wide and 45 cm deep, around the entire perimeter of the building and then making holes 45 cm deep at 15 cm apart by means of a crowbar or drilling and pouring in water emulsions of 0.5% aldrin, 0.5% heptachlor, or 1% chlordane at the rates 10 liters per m² surface area. The trench is subsequently filled with treated soil. After the perimeter is treated, the soil beneath the floor is treated by drilling holes through the cement, 30 cm apart, and injecting an insecticide emulsion (0.5 to 1.0 liters per hole). Special attention is given to the junction of the floor and the wall through which termites may emerge. If hollow walls are observed, that area is treated by drilling holes and the walls supporting door and window frames are also similarly treated. As most, termite infestation to a building comes through the steps, porch, and attached garage, the soil underneath these structures should be treated thoroughly (Hickin 1971, Anonymous 1972b, Roonwal 1979). The application of insecticidal water emulsion through a perforated steel pipe that is forced beneath the concrete slab (rodding) is not advocated because imperfect workmanship may cause considerable damage to the concrete floor slab.

In some countries, especially in Malaysia and the Philippines, powder treatment is claimed to be highly effective (Hickin 1971, Gonzales 1981). A fine 300-mesh toxic dust, primarily arsenic trioxide (As_2O_3), is blown into the termite galleries as a thin film at a number of points in the building by means of a small rubber bulb. Graphite or talc is commonly mixed to ensure free flow of the dust. Great care, attention, and skill are needed. The toxic dust is claimed to circulate among the members of the colony through grooming. The dust gun used by Ebeling (1968) in the United States is especially suitable for *in situ* treatment of wooden structures of a building for dry-wood termite control.

PROTECTION OF MUD HUTS

In many countries in Asia, mud huts often are built. These are composed of clay mixed with binders such as cow dung and small stubble pieces. Cheap perishable timber or bamboo is used for door and window frames and rafters and beams, and the thatching is leaves of grass or bamboo, palmyra, or nipa palm (*Nypa fruticans* Wurmb.). Bamboo has a low natural resistance compared to wood, and split bamboo is more rapidly destroyed than unsplit (Liese 1980). Occasionally, inexpensive buildings are constructed using unbaked, sun dried, bricks with clay as mortar. These houses are seriously affected by both subterranean and dry-wood termites. For protection against subterranean termites, incorporation of 1% chlordane or 0.5% aldrin to the mud or clay at 55 g of 5% BHC per 100 kg of dry soil has been found effective against subterranean termites (Roonwal 1979). Walls and floors thus treated should be thickly plastered (5–6 cm) with untreated mud or clay to prevent direct contact with poisoned surface. Pits for wooden or bamboo poles should be treated with insecticidal water emulsion at 0.5 to 1.1 liters per pit (30 cm × 30 cm). These treatments do not prevent attack by decay fungi to nondurable timber or bamboo. If the roof of these houses is projected out so as to prevent direct splash of rain, infection by decay fungi can be reduced. Infestation by dry-wood termites can be prevented only by preservative-treated wood or bamboo. However, due to the nonavailability of wood preservatives and their high cost, replacement is often more convenient and cheaper to the villagers (Liese 1980).

CONTROL OF DRY-WOOD TERMITES

Use of preservative-treated wood or bamboo is generally advocated for the prevention of dry-wood termite attack (Francia 1957). However, facilities for preservative treatment are not available in many areas. The common practice, therefore, is to treat in situ either with a toxic dust or insecticidal emulsion in water. Turning off lights during swarming prevents colony foundation in wood. Use of fine mesh wire netting outside window frames (screens) and ventilators can physically prevent alates from gaining access (Sen-Sarma 1980). Desiccant dusts, such as fluoridated silica aerogel (Dri-die® 67), when properly applied, have potential for in situ dry-wood termite control (Ebeling 1968, Gonzales 1981). A thin, uniform film of dust is blown on all exposed surfaces of wooden or bamboo components of houses (Ebeling 1968). The drill and treat method is also followed in several countries (Hickin 1971). Brushing a persistent organochlorine insecticide in oil solvents at the rate 1 liter per 3 m^2 of wooden surface has been claimed successful (Hickin 1971). Fumigation of the entire building under a sealed tarpaulin has not gained favor because of the high cost and lack of trained personnel.

RESISTANCE OF POLYMERIC CABLE COVERINGS

Various types of polymeric cable coverings have been tested under laboratory and field conditions against termites such as H. indicola (Wasm.), and C. formosanus (Becker 1964, Nigam et al. 1970, Beal and Bultman

1978). However, results of short-term exposure often are inconclusive, as long-term exposure of several apparently resistant products failed (Beal and Bultman 1978). Beal and Bultman (1978) report that, in general, either increasing hardness by reduction in plasticizer content or addition of a toxic material such as 2.6% aldrin affords protection. Addition of toxins is not favored due to hazards during manufacture. Non-polyvinyl chloride (PVC) polymers such as ethylene propylene rubber and chlorosulfonated polyethlene are almost immune to termite attack and are recommended for use in preference to the PVC's (Beal and Bultman 1978).

TERMITE PROOFING OF PANEL PRODUCTS

The main panel products are plywood, laminated wood, fiber boards, chip boards, wood-wool boards, and bamboo board, which are highly susceptible to termites in the tropics. The risks of these products are greatest when they are in contact with the soil. Earlier attempts to make various panel products, particularly composite wood, were directed at using either naturally durable wood or veneers treated with a preservative that was compatible with the adhesive used. Several discussions of the control of termites of composite wood are provided by Howick et al. (1975), Narayanamurti (1962), Sen-Sarma et al. (1982) and Shukla et al. (1981).

REFERENCES

Agarwala, S. B. D. and C. Sarma. 1954. Aldrin and dieldrin as outstanding agents in the control of *Microtermes obesi* Holmgren in maize in Bihar. *Indian J. Entomol.* 16: 78-79.

Ahmad, M. 1965. Termites (Isoptera) of Thailand. *Bull. Amer. Mus. Nat. Hist.* 131: 1–133.

Allen, T. C., G. R. Esenther, and R. S. Shenefelt. 1961. Concrete insecticide mixture toxic to termites. *J. Econ. Entomol.* 54: 1055–1056.

Amin, P. W. and D. McDonald. 1981. Termites as pests of groundnuts. Pp. 273–277. *In:* "Progress in Soil Biology and Ecology in India" (G. K. Veeresh, ed.), Univ. Agri. Sci., Bangalore.

Anonymous 1972a. Code of practice for antitermite measures in buildings. II. Preconstructional chemical treatment measures. IS: 6313: 1011. (Indian Stand. Inst., New Delhi).

Anonymous 1972b. Code of practice for antitermite measures in buildings. II. Treatment for existing buildings. IS: 6313: 1–9 (India Stand. Inst., New Delhi).

Anonymous 1976. Studies on the control of termites in the dams of water reservoirs. *Acta Entomol. Sinica* 19: 18–24 (in chinese).

Anonymous 1980. 5 years of agricultural research and development for Indonesia 1976–1980. (Ministry of Agriculture, Indonesia, Jalacarta), 128 pp.

Arora, G. L. 1962. Biological observations of some termites from Hoshiarpur, Punjab. Pp. 111–113. In: "Termites in the Humid Tropics" (Proc. New Delhi Symp. 1960), UNESCO, Paris.

Attri, B. S. and R. Prasad. 1981. Preliminary studies on the inhibition of carbofuran degradation in soil by neem cake or neem oil extractives. Pesticides 15(7): 6–7 & 12.

Avasthy, P. N. 1967. Sugarcane pests of India and their control. Gr. Brit. Trop. Prod. Inst. Manual 13:111–117.

Aziz bin Sheikh, A. K. A. 1980. Pests of rubber and covers. Pp. 83–101. In: "RRIM Training manual on crop protection and weed control in rubber plantation". Rubber Research Institute, Kuala Lampur.

Beal, R. H. and J. D. Bultman. 1978. Resistance of polymeric cable covering to subterranean termites attack after eight years of field testing in the tropics. Int. Biodeterior. Bull. 14: 123–127.

Becker, G. 1964. Testing results on termite resistance of plastics. Bull. Reilm. 25: 93–97.

Becker, G. 1975. Heterotermes indicola damaging wood in buildings at Kathmandu (Nepal). Mater. Org. 10: 275–280.

Becker, G. 1979. Testing and research results with an Indian termite species in Berling-Dahlem. J. Timber Develop Assoc. India. 25: 16–23.

Beeley, F. 1934. Experiments in control of the termite pest of young rubber tree. J. Rubber Res. Inst. 5: 160–175.

Benedict, W. V. 1971. Pilot plantations of quick growing industrial species. Protecting plantations of long fiber tree species from loss by insects and diseases. FO: SF/MAL IL Tech. Report (4) UNDP/FAO.

Bigger, M. 1966. The biology and control of termites damaging field crops in Tanganyika. Bull. Entomol. Res. 56: 417–444.

Bindra, O. S. 1961. Termite damage in North-western Madhya pradesh with results of experiments on their chemical control. Indian J. Entomol. 22: 277–282.

Brown, K. W. 1965. Termite control research in Uganda, with special reference to the control of attacks in Eucalyptus plantations. East Afr. Agri. Forest J. 32: 218–223.

Butani, D. K. 1973a. Les insectes raveugeurs du cottonnier. XVI. Les principaux problemes parasitaires du cotonnier en Inde. Coton Fibres Trop. 28: 259–268.

Butani, D. K. 1973b. A les ravageurs et les maladies de Citrus en Inde. Fruits (Paris) 28: 851–856.

Caresche, L. 1937. Le termite destructeur, de l'hevea et du kapokier. C. R. Inst. Res. Agron. For. Indochine 2 (1935–36): 195–212.

Chatterjee, P. N. 1963. Residual soil poisoning of a mixture of enzyme hexachloride and dieldrin against subterranean termites. Sci. Cult. 29: 148.

Chatterjee, P. N. 1972. Role of termites in forestry. Pp. 4–7. In: "Termite Problems in India" (M. L. Roonwal, ed.), (CSIR) New Delhi.

Chatterjee, P. N. and P. K. Sen-Sarma. 1962. Seasonal incidence of wood-destroying subterranean termites. Indian Forest. 88: 139–141.

Chatterjee, P. N. and P. K. Sen-Sarma. 1968. Important current problems of forest entomology in India. *Indian Forest.* 94: 12–117.

Chatterjee, P. N. and M. L. Thakur. 1966. Biology and ecology of Oriental termites (Isoptera). Observations on the habits and biology of some termites of the Doon Valley. *Indian Forest.* 92: 139–142.

Chaudhry, M. L. and M. Ahmad. 1972. "Termites of Pakistan. Identity, distribution and ecological relationships" (Final Tech. Rep.), Pakistan For. Res. Ins., Peshwar.

Chhabra, K. S. 1981. Incidence of termites on pulses in Punjab-A survey report. Pp. 278–280. *In:* "Progress in Soil Biology and Ecology in India" (G. K. Veeresh, ed.), Univ. Agri. Sci., Bangalore.

Chhotani, O. B. 1980. Termite pests of agriculture in the Indian Region and their control. *Zool. Surv. India Tech. Monog.* (No. 4): 94 p.

Child, R. 1964. "Coconuts." Longman Pub., London.

Coaton, W. G. H. 1958. The Hodotermitid harvester termites of South Africa. *Un. So. Afr. Dept. Agric. Sci. Bull.* 175: 1–112.

Colwill, D. J. 1958. Notes on the preservation of materials. *Pest. Abstr.* 4: 237–256.

Danthanarayana, W. and S. N. Fernando. 1970. A method of controlling termite colonies that live within plants. *Int. Pest Control* 12: 10–14.

Das, G. M. 1962. Termites in tea. Pp. 229–231. *In:* "Termites in the Humid Tropics" (Proc. New Delhi Symp.), UNESCO, Paris.

Deol, G. S., G. S. Sandhu, and A. S. Sohi. In press. *Pestology.*

Dhanarajan, G. 1969. The termite fauna of Malaysia and its economic significance. *Malay. Forest.* 17: 276–278.

Dutt, N. 1962. Preliminary observations on the incidence of termites attacking jute. Pp. 217–218. *In:* "Termite in the Humid Tropics" (Proc. New Delhi Symp. 1960) UNESCO, Paris.

Ebeling, W. 1968. Termites: their identification, biology and control of termites attacking buildings. 38, Univ. Calif., Agric. Exp. Sta. Ext. Serv. Man. 38, 74p.

Feakin, S. D. 1973. Pest control in groundnuts. *Pest. Art. News Summary* PANS manual No. 2, CORP, London.

Fernando, H. E. 1962. Termites of economic importance in Ceylon. Pp. 205–210. *In:* "Termites in the Humid Tropics" (Proc. New Delhi Sympos. 1960), UNESCO, Paris.

Findlay, J. B. R. 1971. An investigation into the use of insecticidal baits for the control of *Hodotermes mossambicus* (Hagen). *Phytophylactica* 3: 97–102.

Fletcher, T. B. 1914. Termites or white ants. *Agric. J. India* (I): 219–239.

Francia, F. C. 1957. Drywood termites and their control. *Forest Leaves* 10: 15–17.

Gahukar, R. T. and M. G. Jotwani. 1980. Present status of field pests of sorghum and millets in India. *Trop. Pest Manag.* 26: 138–152.

Gay, F. J. 1969. Species introduced by man. Vol. 9, pp. 459–494. *In:* "Biology of Termites" (K. Krishna and F. M. Weesner, ed.) Academic Press, New York.

Ghosh, S. K. 1964. Insecticidal control of termite *Microtermes* sp. damaging wheat crop. *Indian Agri.* 2: 88–91.

Glover, P. M. 1937. Lac cultivatio in India. *In:* "A Practical Manual of Lac cultivation" (P. M. Glover, ed.), 2nd ed. June (1931) 7: 8–147.

Gonzales, J. C. 1981. Urban pest control in the Philippines. *Philippines Entomologist* 4(1980): 543–547.

Gray, B. and J. B. Butcher. 1969. Termite eradication in *Araucaria* plantations in New Guinea. *Commonw. Forest. Rev.* 48: 201–207.

Gupta, R. and M. K. Agarwal. 1963. *Odontotermes obesus* (Ramb.) as a pest of Japanese mint. *J. Bombay Nat. Hist. Soc.* 60: 285–287.

Harris, W. V. 1961. "Termites, their recognition and control." Longman, London.

Harris, W. V. 1965. The role of termites in tropical forestry. *Insectes Soc.* 13: 255–266.

Harris, W. V. 1969a. "Termites as pests of crops and trees" (Commonw. Agric. Bureau, London): 1–41.

Harris, W. V. 1969b. Termites as pests of sugarcane. Pp. 226–235. *In:* "Pests of Sugarcane." Elsevier, Amsterdam.

Hickin, N. E. 1971. "Termites-A World Problem." Hutchinson Press, London.

Hoon, R. C. 1962. The incidence of white ants (termites) in the region of the Hirakud Dam Project. pp. 141–143. *In:* "Termites in Humid Tropics" (Proc. New Delhi Symp.), UNESCO, Paris.

Howick, C. D., N. Tamblyn, and J. W. Creffield. 1975. Termiticidal effects of insecticidal glue-line additives. *Proc. 17th Forest Prod. Res. Conf. Topic* 2/25: 1–4.

Janjua, N. A. and M. A. Khan. 1955. Insect pests of wheat in West Pakistan. *Agri. Pakistan* 6: 67–74.

Johnson, R. A. and M. H. Gumel. 1981. Termite damage and crop loss studies in Nigeria. The incidence of termite-scarified groundnut pods and resulting kernel contamination in field and market samples. *Trop. Pest Manag.* 27: 343–350.

Kalshoven, L. G. E. 1962. *Coptotermes curvignathus* as a cause of trouble in electric systems. *Symp. Genet. Biol. Ital.* 11: 223–229.

Kalshoven, L. G. E. 1963. *Coptotermes curvignathus* causing death of trees in Indonesia and Malaya. *Entomol. Berichte* 23: 90–100.

Kaushal, P. K. and R. R. Deshpande. 1967. Losses of groundnut by termites. *J. N. Agri. Univ. Res. Jour., Jabalpur* 1: 92–93.

Khoo, B. K. and M. Sherman. 1979. Toxicity of chlorpyrifos of normal and defaunated Formosan subterranean termites. *J. Econ. Entomol.* 72: 298–304.

Krishnamoorthy, C. and K. Ramasubbiah. 1962. Termites affecting cultivated crops in Andhra Pradesh and their control-retrospect and prospect. Pp. 243–245. *In:* "Termites in Humid Tropics" (Proc. New Delhi Symp.), UNESCO, Paris.

Kumar, R. 1965. Termites, a new pest of potato in India. *Indian Pot. J.* 7: 49.

Liese, W. 1980. Preservation of bamboo. Pp.165–172. *In:* "Bamboo Research in Asia" (G. Lessary and A. Chouirend, eds.), Entom. Res. Dev. Centre, Ottawa.

Lowe, R. G. 1961. Control of termite attack on *Eucalyptus citriodora* Hook. *Emp. Forest. Rev.* 40: 73–78.

McDonald, D. 1970. Fungal infection of groundnut fruit before harvest. *Trans. Brit. Mycol. Soc.* 53: 453–460.

Menon, R. M. G., R. N. Katiyar, S. Kuman, and S. S. Misra. 1970. Termites attacking paddy crop. *Sci. Cult.* 36: 294–295.

Mori, H. 1973. Explanation and evaluation of two electronic termite detectors, comparing the sonic detector in Japan with hysiosonic termite detector in America. *Hiyophi Sci. Rev.* 11: 58–70.

Nair, K. S. S. and R. V. Varma. 1981. Termite control in Eucalyptus plantations. Kerala For. Res. Inst., Report 6, 48 pp.

Narayanamurti, D. 1962. Termites and composite wood products. Pp. 185–197. *In:* "Termites in Humid Tropics" (Proc. New Delhi Symp.) UNESCO, Paris.

Newsam, A. and B. S. Rao. 1958. Control of *Coptotermes curvignathus* with chlorinated hydrocarbons. *J. Rubber Res. Inst. Malaya* 15: 209–218.

Nigam, B. S., S. L. Petri, and P. N. Agarwal. 1970. Susceptibility of PVC polyurethane and polythene films to insect attack. *Lab Develop. J. Sci. Tech.* 8: 145–147.

Nirula, K. K., S. J. Antony, and K. P. V. Menon. 1953. Some investigations on the control of termites. *Indian Coconut J.* 7: 26–34.

Nirula, K. K., and K. P. V. Menon. 1960. Insect pests of coconut palm in India. *World Crops* 12: 265–266.

Oshima, M. 1919. Formosan termites and methods of preventing their damage. *Philippine J. Sci.* 15: 319–383.

Parihar, D. R. 1978. Field observations on the nature and extent of damage by Indian desert termites and their control. *Ann. Arid Zone* 17: 192–199.

Parry, M. S. 1959. Control of termites in Eucalyptus plantations. *Emp. Forest. Rev.* 38: 287–292.

Patel, G. A. and A. K. Patel. 1953. Seasonal incidence of termite injury in the northern parts of Bombay State. *Indian J. Entomol.* 15: 276–278.

Patel, R. M. 1962. Effect of BHC formulations on the control of field termites in Gujarat. pp. 219–221. *In:* "Termites in the Humid Tropics" (Proc. New Delhi Symp., 1960) UNESCO, Paris.

Pradhan, S. 1964. Assessment of losses caused by insect pests of crops and estimation of insect. *Entomol. Soc. India. (New Delhi):* 17–58.

Rajagopal, D., G. K. Veeresh, and K. A. Kushalappa. 1980. Anti-termite treatment in the establishment of Eucalyptus seedlings. *My Forest, Banglore:* 127–130.

Rajpreecha, J. 1980. Insect pests of cashew in Thailand. Nat. Biol. Cont. Res. Centre, Bangkok, Spec. Pub. No. 27, 182 pp. (in Thai).

Rataul, H. S. and S. S. Mishra. 1979. Potato pests and their control. *Pesticides* 13: 27–38, 42.

Rawat, R. R., R. R. Deshpande, and P. K. Kaushal. 1970. Comparative efficacy of different modern insecticides and their methods of applications for the control of termites *Odontotermies obesus* in groundnut, *Arachis hypogaea*. *Madras Agri. J.* 57: 83–87.

Rentokil, I. 1980. Survey reveals massive infestations of termites in Jakarta, Indonesia. *Pest Control* 48: 20–22.

Roonwal, M. L. 1955. Termites ruining a township. *Zeit. Ang. Entomol.* 38: 103–104.

Roonwal, M. L. 1970. Termites of the Oriental Region. Pp. 315–391. *In:* "Biology of Termites" Vol. II (K. Krishna and F. M. Weesner, eds.), Academic Press, New York.

Roonwal, M. L. 1979. "Termite life and termite control in tropical South Asia," Scientific Publisher, Jodhpur.

Sands, W. A. 1973. Termites as pests of tropical food crops. *Pest. Art. News Summary* 19: 167–177.

Sands, W. A. 1977. Prospects for pest management procedures in termite control. Proc. VIII Int. Congr. Inst. Union Study Social Insects, Wageningen, Netherlands, p. 80.

Sellschops, J. P. F. 1965. Field observations on condition conducive to the contamination of groundnuts with the old *Aspergillus flavus* (Link ex Frics). *S. Afr. Med. J. (Suppl. S. Afr. J. Nutr.)*: 774–776.

Sen-Sarma, P. K. 1974. Ecology and biogeography of the termites of India. pp. 421–472. *In*: "Ecology and biogeography in India" (M. S. Mani, ed.), Junk Publishers, The Hague.

Sen-Sarma, P. K. 1980. Recent advances in non-toxic chemical control of termites. *13th Ann. Conv. Indian Pest Control Assoc.*, pp. 1–2.

Sen-Sarma, P. K. and P. N. Chatterjee. 1965. Colony foundation through substitute reproductives in *Heterotermes indicola* (Wasmann) under laboratory conditions. *J. Timber Develop. Assoc. India* (Dehra Dun) 11: 9–11.

Sen-Sarma, P. K. and R. N. Mathur. 1957. Further records of occurrence of *Cryptotermesdudleyi* (Banks) in India. *Curr. Sci. (India)* 26: 399.

Sen-Sarma, P. K. and M. L. Thakur. 1974. Biology, distribution and economic significance of some termites attacking buildings. *J. Indian Plywood Industr. Res. Inst. (Bangalore)* 4: 115–125.

Sen-Sarma, P. K., M. L. Thakur, S. C. Mishra, and B. K. Gupta. 1975. Studies on wood-destroying termites in relation to natural termite resistance of timber. Final Technical Rep. 1968–1973, *For. Res. Inst.* (Dehra Dun): 4–187.

Sen-Sarma, P. K., Y. Singh, S. C. Mishra, and N. L. Goswami. 1982. Studies on the antitermite characteristics of some organic fibrous material. *Proc. Nation. Sem. Build, Mat., Sci. Techn.* (New Delhi) III(1): 103. Infesting Field Crops and Fruit Trees in Rajasthan. Rajasthan Dept. Agri., Jodhpur: 5–7.

Shukla, K. S., L. Prasad and H. K. L. Bhalla. 1981. Anti-termite characteristics of treated wood-wool boards based on laboratory tests. *Holzforsch. Holzverwert.* 33: 119–121.

Sivapalan, P., A. A. C. Karunaratne, and D. G. S. Jayatilleke. 1980. Clonal susceptibility and the influence of shade trees on the incidence of *Glyptotermes dilatus* (Bugnion & Popoff). *Bull. Entomol. Res.* 70: 145–149.

Snyder, T. E. 1954. Termite attack on plastics and fabrics. *Pest Control* 23: 48.

Srivastava, A. S., B. P. Gupta, and G. P. Avasthy. 1962. Termites and their control. pp. 241–242. *In*: "Termites in the Humid Tropics" (Proc. New Delhi Symp., 1960) UNESCO, Paris.

Srivastava, B. K. 1959. Insect pests of maize in Rajasthan. *J. Bombay Nat. Hist. Soc.* 56: 665–668.

Srivastava, B. K. and S. P. Bhatnagar. 1960. Insects injurious to cotton in Rajasthan and suggestions for control of major pests. *Indiana Agri.* (Calcutta) 4: 54–58.

Thakur, C., A. R. Pradad, and R. P. Singh. 1958. Use of aldrin and dieldrin against termites and their effect on soil fertility. *India J. Entomol.* 19: 155–163.

Thakur, M. L. and P. K. Sen-Sarma. 1979. Flourishing colony of *Coptotermes heimi* (Wasm.) in a naval boat. *J. Bombay Nat. Hist. Soc.* 76: 188–189.

Thakur, M. L. and P. K. Sen-Sarma. 1980. Current status of termites as pests of forest nurseries and plantations in India. *J. Indian Acad. Wood. Sci.*, 11: 7–15.

Tripathi, R. L. 1972. Termite problems in jute and allied fiber crops. Pp. 44–45. *In:* "Termite problem in India" (M. L. Roonwal, ed.), Coun. Sci. Indus. Res., New Delhi.

Troup, R. S. 1921. "Silviculture of Indian trees," Vol. II, pp. 785–1195. Oxford Univ. Press, London.

Chapter 3

TERMITES AND THEIR ECONOMIC IMPORTANCE IN AUSTRALIA

J. R. J. French

INTRODUCTION

Today, with a human population of 15 million, Australia's forests cover about 5% of the total land (7,682,300 km²), which is in contrast with the forest area prior to European settlement when the forest area was estimated to cover about 15% of Australia. The majority of the present population of 15 million is clustered in and around the capital cities of Sydney, Melbourne, Brisbane, Adelaide, Perth, or the coastal areas.

This chapter is concerned with the Australian termite fauna of major economic importance and their general distribution. Included will be an overview of the damage caused by these termite species, current and future control measures, and some indication of the costs of some of these termite activities. All monetary figures quoted in this chapter are in Australian dollars.

There are about equal numbers of Australian termite species north and south of the Tropic of Capricorn, but there are more species of Termopsidae north of the tropic, and more species of the primitive families south of the tropic (Gay and Calaby 1970). The Termopsidae and Kalotermitidae are largely restricted to the wetter southeastern and eastern areas. Only four species, two Termopsidae and two Kalotermitidae are known from Tasmania. Only two species of Hodotermitidae and two species of Rhinotermitidae occur in rain forest areas and some species are not found in certain soil types. For example, the black earths of inland northern Australia are virtually devoid of termites although adjacent sandy-desert steppe soils have abundant fauna. The majority of termites are found in the sclerophyll forests, woodlands, and savannahs. Arid regions have few termites, but some are apparently confined to such regions (Lee and Wood 1971). Various species of *Drepanotermes*, *Amitermes*, and *Tumulitermes* are widespread in semiarid and arid Australia. In the process of white settlement across Australia the activity of termites has encroached on many economically important facets of man's activities. The most significant of these are the damage done to timber in service, such as construction timbers in buildings, fence posts, poles, bridges, railway sleepers, wharves, and wood products (Hill 1942, Ratcliffe et al. 1952, Calaby and Gay 1949, Gay 1946, 1970, Gay and Calaby 1970, Gay and Weatherly 1971), to commercial forest fruit, and ornamental trees (Ratcliffe et al. 1952, Greaves et al. 1967), to pastures and crops (Watson et al. 1973, Watson and Perry 1981) and the underground cables and plastic protective sheathings on pipelines (Watson 1981). The occurrence and risk of termite attack to these materials in Australia is, if anything, greater in regions of highest rainfall (Ratcliffe et al. 1952).

I concur with Lee and Wood (1971) in considering that the damage to timber in service by Australian termite fauna represents the area of greatest economic loss and the choice of termite species discussed in this chapter reflects this point in view.

TERMITE FAUNA OF ECONOMIC IMPORTANCE

Australia has a rich and varied termite fauna, which is comprised of about 30 genera and about 246 species (51 undescribed). Two new genera, *Hesperotermes* (1 sp.) and *Ekphysotermes* (4 spp.), have been erected for species previously assigned to Termes (Gay 1971), and Australian species of *Procryptotermes* and *Incisitermes* have been discovered (Gay and Watson 1974). Approximately half of the named genera are in the families Mastotermitidae, Termopsidae, Kalotermitidae, and Rhinotermitidae; the remainder are in the family Termitidae. Only the Termitinae and Nastitermitinae are represented in Termitidae; there are no fungus-growing Macrotermitinae.

Gay and Calaby (1970) and Gay and Watson (1974) reviewed the Australian termite fauna and the following summarizes their main findings.

Family Mastotermitidae

Mastotermes darwiniensis Froggatt is the only surviving member of the Mastotermitidae, arguably the most primitive living termite (Watson 1971). Very large colonies, containing up to seven million individuals, are found only in the regions north of the Tropic of Capricorn. This species is considered one of the more destructive termites in Australia. Radioactive tracer techniques have proved useful in tracing this termite's nesting systems (Spragg and Fox 1974), trophallaxis, and population measurements (Spragg and Paton 1980).

Family Termopsidae

The family Termopsidae is represented by four species in the subfamilies Stolotermitinae and Porotermitinae. The species are restricted to the coastal and associated mountainous areas on the eastern side of the mainland, with some extension into Tasmania. The sole Australian representative of *Porotermes*, *Porotermes adamsoni* (Froggatt) is a damp-wood feeding species found primarily in eucalypt forests (Greaves 1959), though it has been recorded attacking an exotic softwood (Minko 1965).

Family Kalotermitidae

This family is represented by a total of 18 species assigned to six genera. With exceptions, these termites are confined to hardwood forests in coastal and adjacent mountainous margins of the mainland, with some extension into Tasmania. Recently, the developmental pathways in some *Kalotermes* have been elucidated, demonstrating the plasticity of development in Kalotermitids (Sewell and Watson 1981). The genus *Cryptotermes* is perhaps the classic example of the dry-wood termites, which live wholly within wood and require no contact with the soil. *Cryptotermes* is the largest genera, with a total of 16 endemic and exotic species from the mainland and associated offshore islands of Australia (Gay and Watson 1982). *Neotermes insularis* (Walker) is found in forests within 80 km of the coast. *Kalotermes banksiae* Hill, *Glyptotermes brevicornis* (Froggatt), and *G. tuberculatus* Froggatt cause damage to timber in service.

Family Rhinotermitidae

Members of this family are widely distributed over the whole mainland except Tasmania. This family includes representatives of the three subfamilies: (1) Rhinotermitinae (two genera, *Parrhinotermes*, with a single species and *Schedorhinotermes*, with two species); (2) Heterotermitinae (one genus, *Heterotermes* which has four species); and (3) Coptotermitinae (one genus, *Coptotermes*). All the *Coptotermes* are wood eaters and are considered the most economically important termites because of their damage in so many parts of Australia.

Family Termitidae

This is the largest family in the order, and is represented by three subfamilies, 14 genera, and 136 described species. In 1970 there were more than 20 species awaiting description.

Subfamily Amitermitinae

With 57 described species *Amitermes* is the largest and most widely distributed Australian genus. They are particularly abundant in the drier inland of northern and western Australia. Most species are completely subterranean, but five species build mounds. The best known of these mounds are those of *Amitermes meridionalis* (Froggatt) often called "magnetic" or the north/south-facing mound builders. Although some species attain considerable local abundance, *Amitermes* are mostly found in low-value grazing country, and their effect on native pastures appears to be of no economic significance. However, two species, namely *A. capito* Hill and *A. herbertensis* Mjoberg, attack and damage timber in service. All 17 *Drepanotermes* species are harvesters, and gather grass, leaf litter, or other vegetable debris but only *D. perniger* (Froggatt) is considered of economic importance (Watson and Perry 1981).

Most *Microcerotermes* (11 species) eat decayed and weathered wood, but some, *M. boreus* Hill, *M. distinctus* Silvestri, *M. implacidus* Hill, *M. nervosus* Hill, and *M. turneri* (Froggatt), damage timbers in service and paper products.

Subfamily Nasutitermitinae

With over 20 described species, the *Nasutitermes* are found in the cool temperate southeastern portion of the continent. The genus includes; subterranean mound builders, such as *Nastitermes exitiosus* (Hill) and *N. triodiae* (Froggatt), which build mounds up to 7 m high; and tree-nest builders, such as *N. graveolus* (Hill) and *N. walkeri* (Hill). *Nasutitermes* food materials include grass, vegetable debris, and decayed and sound wood.

DAMAGE CAUSED BY TERMITES IN URBAN AND RURAL ENVIRONMENTS

Damage to Timber in Service

Table 3.1 shows the geographical distribution of termite species of economic importance in Australia. The economic loss to timber in service by termites constitutes the greatest problem in both urban and rural environments. Termites can attack and damage sound and decayed timber of native hardwoods, and native and exotic softwoods (Gay 1957). Only about a dozen species of termites account for most of this damage and among them *Coptotermes* and *M. darwiniensis* are the most important. According to Gay and Calaby (1970) *M. darwiniensis* is by far the most destructive termite where it occurs in Australia. For example, they men-

TABLE 3.1

Geographical Distribution of Termite Species of Economic Importance in Australia, with Particular Emphasis on Species Damaging In-Service Timber

Family/species	Australian regions[a]							
	NSW	VIC	SA	WA	QLD	TAS	ACT	NT
Mastotermitidae								
Mastotermes darwiniensis Froggatt				+	+			+
Kalotermitidae								
Neotermes insularis (Walker)	+	+			+			+
Kalotermes banksiae Hill		+	+					
Glyptotermes brevicornis (Froggatt)	+				+			
G. tuberculatus Froggatt	+							
Cryptotermes brevis (Walker)	+				+			
C. cynocephalus Light					+			
C. domesticus (Haviland)					+		+	
C. dudleyi Banks					+			
C. primus (Hills)	+				+			
Termopsidae								
Porotermes adamsoni (Froggatt)	+	+	+		+	+	+	
Rhinotermitidae								
Schedorhinotermes intermedius intermedius (Brauer)	+				+			
S. i. actuosus (Hill)	+		+	+	+			+
S. i. breinli (Hill)					+			+
S. i. seclusus (Hill)	+				+			
S. reticulatus (Froggatt)	+	+		+	+			
Heterotermes ferox (Froggatt)	+	+	+	+			+	+
H. paradoxus (Froggatt)					+			+
H. p. paradoxus (Froggatt)					+			+
H. p. intermedius (Froggatt)				+				
H. p. validus (Froggatt)								+
H. p. venustus (Froggatt)				+				+
H. vagus (Hill)								+

(continued)

TABLE 3.1 (Continued)
Geographical Distribution of Termite Species of Economic Importance in Australia, with Particular Emphasis on Species Damaging In-Service Timber

Family/species	Australian regions[a]							
	NSW	VIC	SA	WA	QLD	TAS	ACT	NT
Coptotermitinae								
Coptotermes acinaciformis (Froggatt)	+	+	+	+	+			+
C. frenchi Hill	+	+	+		+		+	
C. lacteus (Froggatt)	+	+			+		+	
C. a. raffrayi Wasmann				+				
Termitidae								
Amitermes capito Hill				+				
A. herbertensis Mjoberg					+			
Drepanotermes perniger (Froggatt)	+	+	+	+	+			+
Microcerotermes boreus Hill				+				+
M. distinctus Silvestri	+	+	+	+	+			
M. implicadus Hill	+	+			+			
M. nervosus Hill				+				+
M. turneri (Froggatt)	+				+			
Termes cheeli (Mjoberg)					+			+
Nasutitermes centralis (Hill)				+	+			+
N. exitiosus (Hill)	+	+	+	+	+		+	
N. graveolus (Hill)					+			+
N. walkeri (Hill)	+				+			

[a]NSW = New South Wales, VIC = Victoria, SA = South Australia, WA = Western Australia, QLD = Queensland, TAS = Tasmania, ACT = Australian Capital Territory, NT = Northern Territory.

tion homesteads built without adequate precautions being practically destroyed, together with their ancillary buildings and fences, in a matter of two to three years by this termite. Apart from eating timber, *M. darwiniensis* has been recorded damaging paper, leather hides, wool, horn, ivory, vegetable fibers, hay, sugar, jam, flour, salt, bitumen, rubber, ebonite, and human and animal excrement. Even though specific infestations of *M. darwiniensis* are more destructive than *Coptotermes*, *M. darwiniensis* is distributed in the sparsely populated northern half of Australia, and so the total cost of its damage and control is less than that of *Coptotermes*.

In Australia *C. acinaciformis* is responsible for greater economic losses than all the other Australian species of termites combined (Gay and Calaby 1970). This is due to its extensive range (Table 3.1), to the severe nature of its attack, and to its extraordinary success in adapting itself to urban areas. A survey (Reynolds and Eldridge 1972) on the frequency of termite species in the Sydney area attacking timber in service clearly shows the economic importance of *C. acinaciformis* (Table 3.2). It is responsible for most of the damage to buildings in Adelaide and in the country towns of New South Wales, South Australia, southern Queensland, and inland Victoria.

In a three year survey of termite damage in Melbourne and environs Howick (1966) recorded 600 separate cases of termite attack, of which the majority were in domestic dwellings. Of these 600, some 400 were caused by two *Coptotermes* species, in 160 cases the termite species were not identified, and the remaining 40 were reported variously as caused by *Kalotermes*, *Heterotermes*, *Neotermes*, and *Porotermes*. Damage to fences, stockyards, and other farm buildings is very common in rural areas of the continent, except Tasmania, and may be attributed to species other than *C. acinaciformis*, namely: *G. brevicornis*, *S. intermedius*, *S. i. actuosus*, *S. i. breinli*, *S. i. seclusus*, *S. reticulatus*, *H. paradoxus*, *C. lacteus*, *A. capito*, *A. herbertensis*, *M. boreus*, *M. distinctus*, *M. implacidus*, *M. nervosus*, *M. turneri*, *T. cheeli*, *T. sunteri*, *N. centraliensis*, *N. exitiosus*, *N. graveolus*, and *N. walkeri*. In the Canberra region the two subterranean termite species that do the most damage and are likely to be found attacking timber in buildings are *C. frenchi* and *N. exitiosus* (Watson and Barrett 1981). In Queensland, apart from *Coptotermes* and *M. darwiniensis*, *H. paradoxus*, *G. brevicornis*, *M. turneri*, *Nasutitermes*, and *Schedorhinotermes* are considered to be economically important (R. A. Yule, Personal Communication).

In the coastal regions of Queensland and in the northeast of Australia, dry-wood termites of the genus *Cryptotermes* cause considerable damage to timbers in buildings. A recent review of the genus *Cryptotermes* in Australia describes the distribution and biology of 16 endemic and exotic species (Gay and Watson 1982). However, it was the presence of such a potentially destructive species of dry-wood termite as *C. brevis* in 1966 (Heather 1971) that stimulated interest in dry-wood termites. Yule and Watson (1976) recorded the presence, for the first time on the Australian mainland, of two more species thought to be exotic, namely *C. cynocephalus* and *C. domesticus*. These authors also gave comprehensive lists of timber attacked by these two species, as well as the endemic *C. primus*.

The damp-wood termite, *P. adamsoni*, in addition to damaging living

TABLE 3.2
Frequency of Termite Species Recorded in a Survey of Sydney, Australia

Termite species	No. of samples from transmission poles only a	No. of samples from other wood in service b	Total No. of samples from wood in service a+b	No. of samples from trees, stumps, logs & mounds c	Total No. of samples a+b+c	Species percentage of total
Glyptotermes brevicornis (Froggatt)	6	0	6	11	17	1.5
Glyptotermes iridipennis Froggatt	4	0	4	2	6	0.5
Glyptotermes tuberculatus Froggatt	2	0	2	4	6	0.5
Coptotermes acinaciformis (Froggatt)	446	187	633	62	695	62.9
Coptotermes frenchi (Hill)	8	2	10	6	16	1.5
Coptotermes lacteus (Froggatt)	2	4	6	0	6	0.5
Heterotermes ferox (Froggatt)	7	30	37	39	76	6.9
Schedorhinotermes i. intermedius (Brauer)	13	31	44	31	75	6.8
Nasutitermes exitiosus (Hill)	14	15	29	53	82	7.4
Nasutitermes fumigatus (Brauer)	9	9	18	44	62	5.6
Nasutitermes walkeri (Hill)	10	3	13	52	65	5.9
Totals	521	281	802	304	1106	100.0

(Source: Modified after Reynolds and Eldridge 1972)

trees, also damages posts, flooring, poles, house piers, bridge timbers, and stored stacks of paper (Gay and Calaby 1970). Timber in service that is damaged by P. adamsoni is invariably in contact with the ground, in moist and often poorly ventilated situations, and infected with decay fungi. Termite damage to timber in buildings is so rare in Tasmania as to have no economic importance.

Damage to Forest Trees, Fruit Trees, and Crops

Termites are relatively scarce in rain forest areas, and are not considered an economic pest (Gay and Calaby 1970). However, in the native eucalypt (hardwood) forests along the coast and foothills of eastern, northeastern, and southwestern Australia, live-tree-infesting termites attack many species of eucalypt (Greaves 1959, 1962, 1967). Coptotermes acinaciformis and C. frenchi, together with N. insularis and P. adamsoni, are responsible for virtually all termite damage to commercial eucalypt forests. Usually the termites, in following decayed zones within the heartwood of the tree, hollow out the center of the trunk, and leave the characteristic "piped" appearance. Decay probably enters the tree through fire scars, broken branches, bruises caused by falling trees and branches, damage during logging operations, compaction of the soil around tree root systems and via roots snapped or damaged due to wind sway.

In Tasmania, the incidence and effects of P. adamsoni in poor quality, frequently burned forests was investigated, and up to 60% of the trees sampled were damaged. There was no significant difference in species susceptibility to P. adamsoni but the number of trees attacked increased with increasing diameter. The main effect of P. adamsoni damage is to cause downgrading from sawlog grade to pulpwood (Elliott and Bashford 1984).

Nasutitermes exitiosus is found attacking native eucalypts at elevations above 200 m in South Australia, though there are no commercial forests of native eucalypts in this state. The major commercial forests species is Pinus radiata D. Don, an exotic softwood, which is grown extensively in plantations. Minko (1965) reported termite attack in stressed living P. radiata in Victoria, and Coptotermes and Heterotermes species were found attacking burned logs and old stumps of P. radiata in a fire-damaged plantation in New South Wales (French and Keirle 1969). Severe attacks on exotic softwood and other plantations in the Northern Territory by M. darwiniensis has resulted in the discontinuation of any future expansion or establishment of such plantations. Also, this termite causes considerable damage on a local scale to many kinds of shade and ornamental trees in urban and rural gardens throughout the territory. Fox (1974) prepared a list of tree species of known susceptibility to termite attack in the Darwin region of the Northern Territory.

Other crops subject to destruction by M. darwiniensis include citrus, pineapple, vines, cassava, banana, paw paw, melon, pumpkin, carrot, maize, sorghum, potatoes, tomatoes, and sugarcane (Ratcliffe et al. 1952). The susceptibility of cassava to this termite strongly suggests that cassava cannot be envisaged as a potential crop in the Northern Territory (J. A. L. Watson, Personal Communication). Another termite species, Microcerotermes, has been recorded attacking cassava and peanuts. Coptotermes acinaciformis frequently damages stressed fruit trees around Adelaide, and citrus and vines in Queensland. Occasionally this termite has damaged potatoes, particularly during dry conditions, as has C. frenchi in Victoria.

Generally, outside the tropical belt and of M. darwiniensis activity, the impact of termites on agricultural crops is very slight. Attacks, when they occur, amount only to slight damage or at worst to severe localized problems, and may often be attributed to crops under some form of stress. For example, in the Murrumbidgee Irrigation Area, an important citrus and grape growing area in northern Victoria, C. frenchi has been recorded damaging trees beyond their prime and to vines in a poorly drained section. In this area it is not uncommon to find recently planted citrus trees attacked and killed by termites (Ratcliffe et al. 1952). In Orange (N.S.W.) termites have damaged the roots of young (2- to 4-year-old) chestnut trees.

Damage to Pastures

Large populations of grass-eating termites in northern Australia occur on native pastures where grazing is extensive (e.g., 1 sheep to 5 ha), and it has been suggested that they occur on land of poor quality (Ratcliffe et al. 1952). Instances of denudation of pastures by grass-eating termites are rare. One instance was reported in southwest Queensland, where the denudation was caused by D. perniger and D. rubriceps (Froggatt) (Watson and Gay 1970, Watson et al. 1973). The damage arose from a pattern of land use that led to the loss of mulga and, with it, the loss of a major alternative source of food when drought ended a period of favorable years. The combined effects of sheep and termites has then resulted in the destruction of perennial tussock grasses, erosion, and the exposure of numerous nests of D. perniger, which are abundant and durable enough to impede water penetration and seed lodgement for many years (Watson and Perry 1981). It seems that although Drepanotermes can cause damage, it rarely does so if the pastures are not overgrazed.

Instances of "sick" grassland in southern Australia (Ratcliffe et al. 1952) have been associated with, and attributed to, large populations of subterranean termites (Amitermes neogermanus Hill probably being the species involved) (Lee and Wood 1971). Competition among colonies of grass and debris-eating termites (A. *laurensis* Mjoberg) is greater on native pastures heavily grazed by stock than on ungrazed or lightly grazed pastures (Lee and Wood 1971). Thus as stock rates increase, either by the use of fertilizers, irrigation, better quality grasses, or by the introduction of legumes into the pasture, competition among termites is also likely to increase. Still, it is not known whether or not the termites are selective in their choice of grasses, nor is it known what proportion of the grass that termites eat would normally be consumed by stock.

Damage to Buried Cables and Pipelines

The ability of subterranean termites to damage buried cables and pipelines is well known (Ratcliffe et al. 1952). Gay and Weatherly (1969) mentioned that government departments have reported termite damage to plastic-sheathed and insulated communications and power cables, and manufacturers and their agents have reported many cases of attack on plastic piping used for irrigation, stock watering, or domestic water supplies. The plastics involved have been polyvinyl chloride, polyethylene, and cellulose acetate butyrate, and where specific identifications have been made the termites responsible were C. *acinaciformis* and M. *darwiniensis*, although other species have caused problems (J. A. L. Watson, personal communication). These attacks have been observed in all mainland states. Most of the damage to plastic piping has been confined to the low-rainfall areas of southern and southwestern portions of the continent.

Termites damage the plastic sheathings used in underground steel pipes that carry natural gas, oil, or mineral slurries. The coatings commonly used to protect pipes from rust are coal tar enamel or low- to medium-density polyethylene, both being susceptible to termite attack. The pipes are also protected cathodically, and perforation of the protective coating by termites causes a loss of electric current. Species causing economic damage to the various types of plant materials and plastics are summarized in Table 3.3.

TERMITE CONTROL MEASURES

Control in Timber in Service

Control of termites damaging timber in service has been investigated over many years in Australia (Froggatt 1905, Hill 1930, Hill and Holdaway 1934, Holdaway and Hill 1936, Ratcliffe et al. 1952, Gay et al. 1955, 1957, Gay 1961, Watson and Barrett 1981). Information on treatment of wood

TABLE 3.3
Termite Species Causing Economic Damage to Various Types of Plant Materials and to Plastic Sheathings on Buried Cables and Pipelines (Modified after Lee and Wood 1971).

Species	Damage to					
	Wood			Fruit trees and crops	Grass	Plastics on buried cables and pipelines
	living	sound dead	decayed dead			
Mastotermitidae						
Mastotermes darwiniensis	+	+		+		+
Kalotermitidae						
Neotermes insularis	+	+	+	+		
Kalotermes banksiae	+	+	+			
Glyptotermes brevicornis	+	+				
G. tuberculatus	+	+				
Cryptotermes brevis		+				
C. cynocephalus	+	+				
C. domesticus	+	+				
C. dudleyi		+				
C. primus	+	+				
Termopsidae						
Porotermes adamsoni	+	+	+			
Rhinotermitidae						
Schedorhinotermes intermedius						
intermedius	+	+		+		
S. i. actuosus		+	+			+

114

S. i. breinli	+	+				
S. i. seclusus		+		+		
S. reticulatus	+	+	+	+		
Heterotermes ferox		+	+	+		
H. paradoxus		+	+	+		
H. p. paradoxus		+	+	+		
H. p. intermedius		+	+	+		
H. p. validus		+	+	+		
H. p. venustus		+	+	+		
H. vagus	+	+				
Coptotermitinae						
Coptotermes acinaciformis	+	+	+	+		+
C. frenchi	+	+	+	+		
C. lacteus	+	+	+	+		+
C. a. raffrayi	+	+	+	+		+
Termitidae						
Amitermes capito		+	+			
A. herbertensis	+	+		+		
Drepanotermes perniger		+			+	
Microcerotermes boreus		+				
M. distinctus		+				
M. implicadus	+	+	+			
M. nervosus		+				
M. turneri	+	+		+		
Termes cheeli		+				
Nasutitermes centraliensis		+				
N. exitiosus		+				
N. graveolus	+	+				
N. triodae		+				
N. walkeri	+				+	

against termites is available from the various State Departments of Agriculture and Forestry, Federal Government Departments, and the pest control industry. But there are only a few laboratories in Australia that work mainly on termite control. These laboratories include; the Commonwealth Scientific and Industrial Research Organization (CSIRO) laboratories in Canberra (Division of Entomology) and Melbourne (Division of Chemical and Wood Technology), and the State Forest Department in New South Wales and Queensland.

Basically, there are two main strategies in termite control: eradication of the nest, and the installation of chemical and physical barriers to prevent termites from entering a building or attacking timber in contact with the ground (Watson and Barrett 1981). Over the years standards and specifications for the protection of buildings against subterranean termites have been recommended and set out by the Standards Association of Australia in the following codes:

1. Australian Standard (AS) 1694–1974, Code of practice for physical barriers used in the protection of buildings against subterranean termites.
2. Australian Standard (AS) 2057–1981, Code of practice for soil treatment for buildings under construction for protection against subterranean termites.
3. Australian Standard (AS) 2178–1978, Code of practice for the treatment of subterranean termite infestation in existing buildings.

The chemicals recommended in the standards are arsenic trioxide and the organochlorines: aldrin, dieldrin, chlordane, and heptachlor. Arsenic trioxide may only be purchased and used by licensed and registered pest control operators. The organochlorines are applied as emulsions under and around buildings, or may be mixed with concrete in slab constructions (Gay and Weatherly 1959). Long-term studies by CSIRO (J. A. L. Watson, personal communication) indicate that the recommended dose level for the four organochlorines is still 100% effective after more than 30 years.

Substitutes for arsenic trioxide and organochlorines are actively being evaluated in the CSIRO (French et al. 1979). The organophosphate chlorpyrifos ethyl (Dursban) shows promise as a termiticide. In field tests this insecticide has withstood south Australian termites for six years, although it has failed against M. darwiniensis (Watson 1981). The result of laboratory bioassays comparing chlorpyrifos and dieldrin against C. lacteus and N. exitiosus further indicate that chlorpyrifos has considerable potential as a termiticide (Howick and Creffield 1981).

Although fumigation has not been used to control subterranean termites in Australia, the use of methyl bromide against the exotic dry-wood termite, C. brevis, infesting the Queensland State Parliament Building, was extremely effective (Watson 1981). This was the largest fumigation control operation in Australia by a private pest control company, and the entire timber-constructed State Parliament Building in Brisbane was treated at a cost of nearly $1 million. Methyl bromide and sulfuryl fluoride (Vikane®) is recommended by the Australian Quarantine Service, Commonwealth Department of Health, for use in destroying insects present in timber used in packing containers (Anonymous 1981, Paton 1982). In Australia, timber for in-ground service must be treated with wood preservatives that offer protection from mechanical degradation, fungal decay, and termite attack. Greaves (1980a) recently summarized the current wood preservation situation in Australia. Over the years several wood preservatives, which normally include termiticides and fungicides, have been developed and approved for treating timber susceptible to termites. These include boric acid and zinc chloride (Gay et al. 1958, Tamblyn et al. 1968), water-borne preservative salts containing copper-chromium-arsenic (CCA) (Greaves 1980a, b), copper-fluorine-boron diffusion treatments (Johanson and Howick 1975), incorporation of arsenic in creosote and wood tar materials (Johanson 1965), arsenical diffusion treatments in railway sleepers (Johanson 1975, Perry 1979), synthetic pyrethroids, for example permethrin and fenvalerate (C.D. Howick and J. W. Creffield unpublished data), and the incorporation of organochlorines as glue-line additives in plywood (Anonymous 1981).

Laboratory evaluations of alkyl ammonium compounds (AAC) in the protection of softwood (*P. radiata*) and hardwood (*Eucalyptus regnans* F. Meull.) against *C. acinaciformis*, *M. darwiniensis*, and *N. exitiosus* indicate a possible use for only one of the AAC compounds, benzalkonium chloride, but is restricted for the protection of *P. radiata* (Howick et al. 1982). Another termiticide showing promise in laboratory tests as a dust against *C. acinaciformis* and *M. darwiniensis* is the amidinohydrazone, American Cyanamid (AC 217,300). This compound is currently being tested in the field against *C. lacteus* and *N. exitiosus*. Reappraisal of termite baiting systems using toilet rolls and small wood blocks (French and Robinson 1980, 1981) has led to the development of rapid and selective field assessment of termite wood-feeding preferences (French et al. 1981). The incorporation of toxins to baits, and the use of AC 217, 300 to dust termites aggregated at baits are under examination. Also, at the Highett Laboratory of CSIRO Division of Chemical and Wood Technology, an accelerated field simulator (AFS), has been built that is designed to accelerate the action of wood decay fungi and termites in treated and untreated timbers (Johnson et al. 1982). Within the AFS an entire live colony of *C. lacteus*, installed for almost two years, is providing valuable information regarding termite feeding preferences on decayed and undecayed wood blocks, wood consumption rates, and temperature conditions within the mound.

Laboratory studies of the development of dry-wood colonies using X ray provides a valuable nondestructive, nondisturbing method of observations (Creffield 1979). Studies on the fate of lignin in the nutrition of several Australian termites are underway (L. Cookson unpublished data), and all these developments are intended to refine current termite control measures.

The use of insect-pathogenic fungi and nematodes as biological agents against our more economically important termites is being examined. One strain of *Metarhizium anisopliae* (Metsch.) Sorok. (fungi imperfecti) was highly toxic to *N. exitiosus* in laboratory bioassays (Hanel 1981, 1982). *Metarhizium anisopliae* has also proved toxic to *C. acinaciformis* and *M. darwiniensis* when applied as a dust or spray to workers of these species. Field tests with other strains of *M. anisopliae* isolated from termites are under test in the field against *C. frenchi* and *N. exitiosus* in buildings in Canberra (J. A. L. Watson unpublished data). Large numbers of *M. darwiniensis* in the laboratory were totally eradicated within a week of baiting with the nematodes of the *Heterorhabditis* genera obtained from Darwin soil. Each termite cadaver yielded about 10,000 nematode progeny (R. A. Bedding unpublished data). The bacterium *Serratia marcescens* has been isolated from sick and dying laboratory-maintained *M. darwiniensis* workers. However, no deleterious effects were shown when healthy laboratory groups of this termite were contacted with isolates of *S. marcescens* .

Control in Forest Trees, Fruit Trees, and Crops

Species such as *C. acinaciformis* and *C. frenchi* nesting within eucalypt trees have been killed by blowing arsenic trioxide dust into the gallery system via auger holes bored into the trunks (Greaves 1959, Greaves et al. 1967). However, in forest management operations this approach is impracticable. Within an urban situation, such as in Canberra, *C. frenchi* and *N. exitiosus* found in living eucalypt trees may be successfully controlled by arsenic dust or organochlorines (Watson and Barrett 1981).

Bates (1926) reported satisfactory control of *M. darwiniensis* in the Northern Territory in sugarcane using poisoned baits and clearing the timber around the cane fields. More recently, the insecticide mirex has been incorporated within an agar-sawdust bait, which has also proved successful against *M. darwiniensis* in plantations of *Pinus caribaea* Morelet (Paton and Miller 1980). The Northern Territory is the only state where mirex is registered for termite control. Watson (1981) mentioned the problem of growing crops susceptible to *M. darwiniensis* in the Northern Territory, and considered it useful to maintain the ground as pasture for a year or so prior to planting an orchard. The plantations may also be protected by a margin of a susceptible crop, which can indicate if the termites are penetrating from uncleared areas next to the plantations, and for M. darwiniensis, cassava is ideal.

In northern Australia, cultural and management practices coupled with the judicious use of poisoned baits are more practical in combating termites in plantations, orchards and various crops, than to chemical spray controls. Termite problems in southern Australian agriculture are not severe enough to justify treatment (Watson 1981).

Control in Pastures

Lee and Wood (1971) observed that although grass-feeding termites are particularly abundant in northern Australia, generally speaking they are most numerous on land that is of low value and thus control measures must be inexpensive (if attempted). Chemical control of Drepanotermes in southern Queensland significantly reduced termite numbers but was not an effective control (Watson and Gay 1970). Again, it appears that on such land, if control is necessary, it will involve more information and management practices (Lee and Wood 1971).

Control of Buried Cables and Pipelines

No single plastic has yet proved completely resistant against a range of termites, and the most common approach to proofing such materials is the incorporation of insecticidal compounds (such as organochlorines) during processing. Currently, the aim is to find a material that has the correct combination of hardness and flexibility to make it suitable as a cable sheath, that can be extruded, and that is cheaper than nylon (Clark and Flatau 1972, Watson 1981). The large-scale burying of cables and pipelines in trenches pretreated with organochlorines is not considered a practical option.

COSTS OF TERMITE CONTROL

All the termite species that cause substantial problems have been identified, but there may be soil dwelling and dry-wood termites still to be discovered (Watson 1981).

Costs to Timber in Service

Reliable data on the costs of termite damage to timber in service, particularly buildings, are difficult to obtain. A survey questionnaire requesting information on the economic importance of termites was sent to members of the United Pest Control Association of Australia, and some chemical companies. Also, State and Federal Agricultural and Forestry agencies were contacted for further information. The following is a synthesis of all the information provided, and an attempt at estimating some direct and indirect costs.

In 1978 a survey was conducted for a large multi-national company on the Australian pest control industry (J. Webster personal communication) and an estimate of the value of this industry is shown in Table 3.4. The 1978 monetary values have been adjusted for an annual inflation rate of 10% and the figures in Table 3.4 reflect 1982 values arrived at in this fashion.

A recent estimate indicates that the annual total value of agricultural chemicals used in Australian crop protection is about $200 million of which about $4–$5 million is used in termite control (D. Matthews personal communication). Because of the procedures used in the chemical industry in the marketing of products such as organochlorines, and given the number and location of distributors selling to retailers across the continent, it has not been possible to obtain data on the quantity of organochlorines used in termite control. Also, the amount of such chemicals used by the public in termite control is unknown, but may be assumed to be small, since the majority of the public tend to call on the services of pest control operators when they discover termite activity. Further, in areas of high termite activity, the public awareness and concern is reflected in the number of termite treatments carried out.

The number of pest control operators employed in Australia is difficult to accurately assess, as several states (Australian Capital Territory, Northern Territory) do not require licenses or registration. The number

TABLE 3.4
Estimate of the Value in Australian Dollars (Thousands) of the Australian Pest Control Industry with Reference to Termite Treatments

Category	Private home treatments	Industry treatments	Total
PCO[a] total revenue	44,660	30,380	75,040
Revenue from termite treatments	14,280	9,660	23,940
Total cost of all chemicals used by PCOs	2,940	1,960	4,900
Cost of termite control chemicals used by PCOs	112	770	1,890

[a]PCO = Pest control operator. (*Source*: J. Webster, pers. comm.)

of pest control operators (PCOs) licensed to use organochlorines and arsenic trioxide in Australia is about 2,000. However, not all pest control operators licensed to use organochlorines are practicing pest control, let alone carrying out termite treatment.

Information obtained from the questionnaire indicates the average number of costs of annual termite treatments (AS 2057 and AS 2178) carried out in several states (see Table 3.5). Assuming an average private dwelling of between 135 and 145 m², the cost range for soil pretreatments as per AS 2057 varies between $78 and $185 per private dwelling; and treatments of termites in existing dwellings costs between 75¢ and 85¢ per m².

On average, more eradication treatments (as per AS 2178) are carried out per PCO each year in New South Wales and South Australia, whereas in Western Australia more soil treatments are undertaken (as per AS 2057). This is because all metropolitan councils in Perth (Western Australia) require termite pretreatments before or during building construction, whereas this is not always the case in other states. Some councils in South Australia and lending authorities in this and other states require pretreatments of new buildings as a prerequisite when financing builders. Data for Queensland and the Northern Territory were not forthcoming, but termite treatment costs are considered to be relatively comparable (R. A. Yule personal communication).

Generally, the pest control industry offers a 12-month warranty on both types of termite treatment. Annual inspections are routinely advised, and these inspections may cost from $45 to $100 dollars per property per year. Data on the number of retreatments are not known, but such information would allow a more meaningful appraisal of current termite control measures. Difficulty of treating termites under concrete slabs leads to an increase in the overall cost of termite treatment. With such a large labor component involved in these treatments, the cost can vary considerably (see Table 3.5).

Where termite damage occurs, replacement costs can also vary enormously, and an "average cost" is not possible. There are instances in which replacement costs to individual private dwellings have exceeded $100,000. A majority of PCOs in the questionnaire-survey consider that termite attacks to timber in buildings are stable in urban areas. About 60% of the attacks occur in private dwellings in urban Australia, and about 20% in the rural areas. Most inquiries from the public to PCOs regarding termite activity were recorded in the warmer months in each state (i.e., September to March), even though termite activity may be expected to continue throughout the year in the warmer regions. Currently, in New South Wales, the Forestry Commission is undertaking a survey on termite activity in private homes in Sydney area and of 1,300 homes

TABLE 3.5
Estimates of Average Number, Range and Costs of Termite Treatments per PCO[a] per year, and per private dwelling[b] (in Australian Dollars)

State	Estimate of average number and range of termite treatments per PCO per year.				Estimate of average cost ($) for termite treatments per private dwelling (up to 145 m^2).			
	Soil pretreatments (AS 2057)		Termites in buildings (AS 2178)		Soil pretreatments (AS 2057)		Termites in buildings (AS 2178)	
	Average	Range	Average	Range	Average	Range	Average	Range
NSW	37	13–70	60	5–125	$134	$75–189	$233	79–2500
VIC	38	5–80	38	10–67	105	75–140	324	170–2500
SA	22	2–37	90	8–233	78	60–100	305	200–380
WA	157	17–500	52	10–114	81	65–95	205	130–320

[a]PCO = Pest control operator.
[b]The author gratefully acknowledges all respondents to the questionnaire, particularly members of the United Pest Control Association of Australia.

inspected, about one in five showed evidence of termite activity (P. Hadlington personal communication).

Other commodities such as utility poles and railway sleepers are also subject to termite attack. It is estimated (Greaves 1983) that there are about 5.4 million wooden poles in service, of which about 45% are preservative treated. The price of treated new poles varies between $150 and $200 dollars each, which gives an overall investment of about $300 million dollars. This sum is considerably enlarged when one considers that the total replacement costs of a treated pole may be $2,000 dollars or more, and thus the standing asset value may be three or four times the overall investment figure. In the Northern Territory and South Australia, steel utility poles are used. However, in the other mainland states, even where wood-destroying termite populations are high, one can obtain a misleading impression of termite activity if one merely counts poles that are replaced annually. For example, in parts of northern Victoria, poles are placed into holes which are treated with organochlorines. So it is difficult to accurately cost termite activity unless one includes the cost of applying the poison, and such data are unobtainable.

In a CSIRO pole stub trial in Queensland, it has been observed (Barnacle and Beesley 1981) that where M. darwiniensis is prevalent, termite attack may occur sooner. Nevertheless, subterranean termites of other genera such as Schedorhinotermes may prove in time no less destructive than *M. darwiniensis*. A total of over 68 million timber railway sleepers are in tracks throughout Australia, and the majority are not treated with a wood preservative. Estimates of annual timber sleeper demand (1981–1985) is almost 3 million (Anonymous 1981). But damage by termites in timber railway sleepers is poorly documented (Creffield et al. 1978).

Now more nondurable hardwood species are used for sleepers and are minimally treated. In the iron ore rail in the Pilbara region of Western Australia, minimally treated sleepers are failing after ten years due to termite attack (J. E. Barnacle personal communication).

Although wood preservatives protect timbers from mechanical degradation, fungal decay, and subterranean termites, their annual usage and cost reflect another aspect of the costs of termite activity, albeit, somewhat indirectly. In 1982 an estimated 12,150 tons of preservative costing $18 million dollars were used to treat 402,000 m³ of timber commodities.

Costs to Forest Trees, Fruit Trees, and Crops

According to Greaves et al. (1967) the economic loss of eucalypt forest trees to termite activity may be considerable. They estimated that in virgin *Eucalyptus pilularis* Sm., *C. acinaciformis* and *C. frenchi* were responsible for 92% of the total loss and in less mature forest the corresponding figure was 64%; the revenue from felled trees averaged $149 per ha but the loss in royalties due to defects averaged $299 per ha. But in estimating such "timber" losses in forest revenues due to termite activity, it must be understood that termites are feeding on dead and decayed heartwood tissues of living trees, and these decayed areas have no commercial value in the strict economic sense.

The economic loss to exotic pine plantations and crops such as cashews, cassava, and mangos by *M. darwiniensis* in the Northern Territory is difficult to gauge in monetary terms. Suffice to say that future plans for any expansion or further establishment of exotic softwood plantations have been discontinued due to this termite (J. A. L. Watson personal communication). *Coptotermes acinaciformis*, in addition to attacks on timber in service, is responsible for serious damage to commercial fruit trees, vines, ornamental trees, and shrubs, and to forest trees in at least six genera, including some 25 species of *Eucalyptus* (Gay and Calaby 1970). Just what this constitutes in monetary terms has not been estimated.

Costs to Pastures

The biomass of harvester termites in the semiarid mulga pastures may be comparable with that of stock; however, only when plant productivity falls to very low levels, as during drought, do stock and termites compete for food. *Drepanotermes perniger* can remove up to 100 kg per ha of material annually (Watson et al. 1973). Apart from this example there is scant information on the feeding regimes of termites in pastures sufficient to hazard a guess as to their economic importance.

Costs to Buried Cables and Pipelines

Mastotermes darwiniensis and *C. acinaciformis* are the most damaging species to plastic-sheathed and insulated cables and pipelines (Gay and Weatherly 1969). It has been estimated that the damage to timber in service, forest and fruit trees, crops, and to buried cables and plastic agricultural piping by *C. acinaciformis* amounts to almost $4 million dollars annually (Gay and Calaby 1970).

The perforation of plastic sheathings by termites allows the entry of moisture, ultimately resulting in insulation breakdowns. Such faults present problems to the utility authorities in all states. Telecom Australia estimates that in the major urban cities it costs about $250,000 dollars each year to repair cables damaged by termites, and considers that to repair cables damaged by termites in underground ducts costs on average $3,000 dollars per duct. The highest incidences of termite attacks to cables in capital cities are recorded in Perth. In rural areas, Telecom Australia estimates that termites cause almost $1 million dollars worth of damage annually (M. Halsmith personal communication). Data are not available on the costs that termite damage causes to agricultural plastic piping.

SOME SOCIAL COSTS OF TERMITE CONTROL

There are considerable costs that exist in termite control, and many are difficult to quantify, as Pimentel et al. (1980) discovered when examining the environmental and social costs of pesticides. A more complete accounting of indirect costs would include additional data such as: the total costs of accidental spillages of insecticides, particularly within private dwellings; the number of termite retreatments indicating the efficiency of such treatments; potential losses resulting from the possible destruction of soil invertebrates, microflora, and microfauna; unrecorded loss of fish, wildlife, crops, and trees as a result of misuse; the costs of human insecticide poisoning, not only to PCOs but to the public where homes have been treated.

Another indirect cost is that in some Australian states licensing and registration of PCOs are not mandatory. So, coupled with the problem of variations in the level, type of education, and training of PCOs in the various states, it is difficult to assess standards where there is no reciprocity of licensing and registration between different states. The policing of PCOs is difficult but must be attempted by state and local government officials. Such officials need to be thoroughly trained and maintain a continuous adversary role in relation to the pest control industry for the benefit of all society. By the same token, society must be educated to understand why they need to pay for these statutory bodies on a continuous basis.

Information from the pest control industry survey questionnaire suggests that many architects and builders appear to have little understanding of the behavior of termites. Buildings are still being designed, sited, and built with scant regard for termite activity in an area. Apart from the upset and financial burden the discovery and treatment of termites causes a householder, such occurrences tend to reinforce the perceived community notion of equating termites in buildings with disaster.

There are no data available on the cost relating such building practices to the incidence of termite activity, but it seems these costs might be reduced if the architects and builders complied with the recommendations of the relevant Australian Standards (ie., AS 1694, AS 2057, AS 2178).

REFERENCES

Anonymous. 1981. Cargo containers: Quarantine aspects and procedures. Commonw. Dept. Health, Aust. Govt. Pulb. Serv., Canberra, p. 57.

Barnacle, J. E. and J. Beesley. 1981. Report of pole test in Queensland after 21 years. CSIRO Div. Chem. Wood Tech., Melbourne.

Bates, G. 1926. Cane pest combat and control. Giant white ants Mastotermes darwiniensis Frogg. Queensl. Agric. J. 25: 4–5.

Calaby, J. H. and F. J. Gay. 1959. Aspects of the distribution and ecology of Australian termites. Monograph Biol. 8: 211–223.

Clark, R. A. and G. Flatau. 1972. Development of nylon jacketed telephone cable resistant to insect attack. 21st Int. Wire and Cable Symp., pp. 245–251.

Creffield, J. W. 1979. The use of x-rays in studies of drywood termites. J. Inst. Wood Sci. 8: 1–5.

Creffield, J. W., F. A. Dale, and H. W. G. Lowe. 1978. Protection of sleepers against Mastotermes darwiniensis (the giant Northern termite). Rail Track. 2: 24–28.

Elliott, H. J. and Bashford, R. 1984. Incidence and effects of the dampwood termite Porotermes adamsoni, in two Tasmanian east coast eucalypt facsti. Aust. For. 47: 11–15.

Fox, R. E. 1974. Check list of tree species of known susceptibility to termite attack in the Darwin region of the Northern Territory. For. Timb. Bur. Aust. Tech., Note No. 9.

French, J. R. J. and R. Keirle. 1969. Studies in fire-damaged radiata pine plantations. *Aust. For.* 33: 175–180.

French, J. R. J. and P. J. Robinson. 1980. Field baiting of some Australian subterranean termites. *Z. Angew. Entomol.* 90: 444–449.

French, J. R. J. and P. J. Robinson. 1981. Baits for aggregating large numbers of subterranean termites. *J. Aust. Entomol. Soc.* 20: 76–77.

French, J. R. J., P. J. Robinson, and N. R. Bartlett. 1981. A rapid and selective field assessment of termite wood feeding preference of the subterranean termite *Heterotermes ferox* (Frogg.) using toilet roll and small wood-block baits. *Sociobiology* 6: 135–151.

French, J. R. J., P. J. Robinson, Y. Yazaki and W. E. Hillis. 1979. Bioassays of extracts from white cypress pine (*Callitris columellaris* F. Muell.) against subterranean termites. *Holzforschung* 33: 144–148.

Froggatt, W. W. 1905. White ants (Termitidae), with suggestions for dealing with them in house and orchards. *Agri. Gaz. N.S.W.* 16: 632–656.

Gay, F. J. 1946. Case of house infestation by a tree dwelling colony of *Coptotermes frenchi* Hill. *J. Counc. Sci. Ind. Res.* 19: 330–334.

Gay, F. J. 1957. Termite attack on radiata pine timber. *Aust. For.* 11: 86–91.

Gay, F. J. 1961. The control of termites in Australia. *Symp. Genet. Biol. Ital.* II: 47–60.

Gay, F. J. 1970. Isoptera (termites). Pp. 275–293. In "Insects in Australia." Melbourne Univ. Press, Melbourne.

Gay, F. J. 1971. The Termitinae (Isoptera) of temperate Australia. *Aust. J. Zool.* (Suppl. Series. 3): 36.

Gay, F. J. and J. H. Calaby. 1970. Termites of the Australian region. Vol. 1, pp. 393–448. In "Biology of termites" (K. Krishna and F. M. Weesner, eds.). Academic Press, New York.

Gay, F. J. and J. A. L. Watson. 1974. Isoptera (termites). Suppl. 1974, pp. 37–39. In "Insects of Australia." Melbourne Univ. Press, Melbourne.

Gay, F. J. and J. A. L. Watson. 1982. The genus *Cryptotermes* in Australia. *Aust. J. Zool.* (Suppl. Series 88): 64.

Gay, F. J. and A. H. Weatherly. 1959. The termite-proofing of concrete. *Constr. Rev.* 32: 26–28.

Gay, F. J. and A. H. Weatherly. 1969. Laboratory studies of termite resistance. V. The termite resistance of plastics. CSIRO Div. Entomol. Tech. Paper No. 10.

Gay, F. J. and A. H. Weatherly. 1971. Laboratory studies of termite resistance. VI. The termite resistance of particleboards, fibreboards and plywoods. CSIRO Div. Entomol. Tech. Paper No. 11, p. 28.

Gay, F. J., K. M. Harrow, and A. H. Weatherly. 1958. Laboratory studies of termite resistance. III. A comparative study of the anti-termite value of boric acid, zinc chloride, and tanalith U. CSIRO Aust. Div. Entomol. Tech. Paper No. 4.

Gay, F. J., T. Greaves, F. G. Holdaway and A. H. Weatherly. 1955. Standard laboratory colonies of termites for evaluating the resistance of timber, timber preservatives, and other materials to termite attack. CSIRO, Bull. No. 277, p. 60.

Gay, F. J., T. Greaves, F. G. Holdaway, and A. H. Weatherly. 1957. The development and use of field testing techniques with termites in Australia, CSIRO, Bull. No. 280.

Greaves, H. 1980a. Current technology for wood preservation in Australia. *Commonw. For. Rev.* 59: 337–348.

Greaves, H. 1980b. Wood preservation research in Australia. *Proc. AWPA Annu. Meet.* (*Nashville*, Tenn.): 154–168.

Greaves, H. 1983. Research and development for the preservation of hardwood poles. Symp. Nat. Timber Res. Inst. (Pretoria), March 1983, p. 20.

Greaves, T. 1959. Termites as forest pests. *Aust. For.* 23: 114–120.

Greaves, T. 1962. Studies of foraging galleries and the invasion of living tress by *Coptotermes acinaciformis* and *C. brunneus* (Isoptera). *Aust. J. Zool.* 10: 630–651.

Greaves, T. 1967. Experiments to determine the populations of tree-dwelling colonies of termites (*Coptotermes acinaciformis* (Froggatt) and *C. frenchi* (Hill)). *Tech. Paper Div. Entomol. CSIRO Aust.* 7: 19–33.

Greaves, T., R. S. McInnes and J. E. Dowse. 1965. Timber losses caused by termites, decay and fire in an alpine forest in New South Wales. *Aust. For.* 29: 161–174.

Greaves, T., R. S. McInnes and J. E. Dowse. 1967. Control of termites (*Coptotermes* spp.) in blackbutt (*Eucalyptus pilularis*). *Tech. Paper Div. Entomol. CSIRO Aust.* 7: 35–43.

Hadlington, P. W. and N. G. Cooney. 1971. "A guide to pest control in Australia." pp. 105–127. New South Wales Univ. Press, Sydney.

Hanel, H. 1981. A bioassay for measuring the virulence of the insect pathogenic fungus *Metarhizium anisopliae* (Metsch.) Sorok. (fungi imperfecti) against the termite *Nasutitermes exitiosus* (Hill). *Z. Angew. Entomol.* 92: 9–18.

Hanel, H. 1982. Selection of fungus species, suitable for the biological control of the termite *Nasutitermes exitiosus* (Hill). *Z. Angew. Entomol.* 94: 237–245.

Heather, N. W. 1971. The exotic drywood termite *Cryptotermes brevis* (Walker) and endemic Australian drywood termites in Queensland. *J. Aust. Entomol. Soc.* 10: 134–141.

Hill, G. F. 1930. Some aspects of wood preservation in Australia. 3. Termites (White ants). *J. Counc. Sci. Indust. Res.* 3: 141–146.

Hill, G. F. 1942. Termites (Isoptera) from the Australian region. CSIRO, Melbourne, p. 479.

Hill, G. F. and F. G. Holdaway. 1934. Observations on the use of zinc chloride for soil treatments in the control of termites. *J. Counc. Sci. Indust. Res.* 7: 169–172.

Holdaway, F. G. and F. G. Hill. 1936. The control of mound colonies of *Eutermes exitiosus* Hill. *J. Counc. Sci. Indust. Res. Aust.* 9: 135–136.

Howick, C. D. 1966. The incidence and distribution of termite attack in Melbourne and environs. *Quant. Surv.* 13: 18–19.

Howick, C. D. and J. W. Creffield. 1981. Laboratory bioassays to compare the efficacy of chlorpyrifos and dieldrin in protecting wood from termites. *Int. Pest Control* (March/April): 8.

Howick, C. D., J. W. Creffield, J. A. Butcher, and H. Greaves. 1982. The termiticidal efficacy a quaternary ammonium compound and a tertiary amine salt in wood protection. *Mater. Org.* 17: 81–92.

Johanson, R. 1965. The incorporation of arsenic in creosote and wood tar material to increase termiticidal effectiveness. *Wood Sci. J.* 15: 36–44.

Johanson, R. 1975. Arsenical diffusion treatment of *Eucalyptus diversicolor* rail sleepers. *Holzforschung* 29: 187–191.

Johanson, R. and C. D. Howick. 1975. Copper-fluorine-boron diffusion wood preservative. III. Termiticidal effect of fluorine, boron and F-B-complex. *Holzforschung* 29: 25–29.

Lee, K. E. and T. G. Wood. 1971. "Termites and soil." p. 251. Academic Press, London.

Minko, G. 1965. Termites in living *Pinus radiata* D. Don. Vic. For. Comm. Tech. Paper No. 15, pp 1–3.

Paton, R. 1982. Australian Plant Quarantine requirements for preservative treatments. *B.W.P.A. Ann. Conv.* (London): 7.

Paton, R. and L. R. Miller. 1980. Control of *Mastotermes darwiniensis* Froggatt with mirex baits. *Aust. For. Res.* 10: 249–258.

Perry, D. H. 1979. Protection of timber in contact with the ground from the termite *Mastotermes darwiniensis* in the Pilbara region of Western Australia. Western Australia For. Dept., Res. Paper No. 56, p. 10.

Pimentel, D., D. Andow, R. Dyson-Hudson, D. Gallahan, S. Jacobson, M. Irish, S. Kroop, A. Moss, I. Schreiner, M. Shepard, T. Thompson, and B. Vinzant. 1980. Environmental and social costs of pesticides: a preliminary assessment. *Oikos* 34: 126–140.

Ratcliffe, F. N., F. J. Gay and T. Greaves. 1952. Australian termites. The biology, recognition, and economic importance of the common species. CSIRO, Melbourne, p. 124.

Reynolds, J. L. and R. H. Eldridge. 1972. The distribution and abundance of termites attacking wood in service in the Sydney area. 16th For. Prod. Res. Conf., Melbourne, Vol. I. Topic 3/39. p. 6.

Sewell, J. J. and J. A. L. Watson. 1981. Developmental pathways in Australian species of *Kalotermes hagen* (Isoptera). *Sociobiology* 6: 243–322.

Spragg, W. T. and R. E. Fox. 1974. The use of radioactive tracers to show the nesting system of *Mastotermes darwiniensis* Froggatt. *Insect. Socio.* 21: 309–316.

Spragg, W. T. and R. Paton. 1980. Tracing, trophallaxis and population measurement of colonies of subterranian termites using a radioactive tracer. *Ann. Entomol. Soc. Am.* 73: 708–714.

Tamblyn, N., S. J. Colwell, and G. M. Vickers. 1968. Preservative treatment of tropical building timbers by a dip diffusion process. *9th Br. Commonw. For. Conf.*

Watson, J. A. L. 1981. Termite studies in Australia. Baiyi Yanjiu (Termite Research) 5: 2–8 (in Chinese). English copies of this paper may be obtained on request to J. A. L. Watson, CSIRO, Div. Entomol. Canberra, ACT.).

Watson, J. A. L. and R. A. Barrett. 1981. Termites in the Canberra region. CSIRO, Canberra, p. 38.

Watson, J. A. L. and F. J. Gay. 1970. The role of grass-eating termites in the degradation of a mulga ecosystem. *Search* 1: 43.

Watson, J. A. L. and D. H. Perry. 1981. The Australian harvester termites of the genus *Drepanotermes*. *Aust. J. Zool (Suppl. Series 78)*: 1–153.

Watson, J. A. L., C. Lendon and B. S. Low. 1973. Termites in mulga lands. *Grasslands* 7: 121–126.

Yule, R. A. and J. A. L. Watson. 1976. Two further domestic species of Cryptotermes from the Australian mainland. *J. Aust. Entomol. Soc.* 15: 349–351.

Chapter 4

ECONOMIC IMPORTANCE AND CONTROL OF TERMITES IN THE UNITED STATES

Joe K. Mauldin

INTRODUCTION

Although termites are beneficial in many respects, a few species cause extensive damage, primarily to wooden structures and products. Of the 2,200+ species of termites known worldwide, Ebeling (1968) lists 69 that infest buildings. These are found among six families as follows: Rhinotermitidae, 33 species; Termitidae, 18 species; Kalotermitidae, 12 species; Mastotermitidae, one species; Termopsidae, two species; and Hodotermitidae, three species. About 15 species found in the United States damage human dwellings, commercial buildings, furniture, utility poles, marine pilings, fence posts, paper products, logs, and lumber and are particularly serious and costly pests. Termites are also minor pests of some crops in the United States.

Subterranean termites are found in all parts of the United States (Fig. 4.1) except Alaska. The predominant subterranean termites, *Reticulitermes* spp., *Coptotermes formosanus* Shiraki, and *Heterotermes* spp., account for about 95% of the termite damage to wood and wood products (Moreland 1981). Although of lesser economic importance, dry-wood termites, primarily *Incisitermes* sp., *Cryptotermes brevis* (Walker), and *Neotermes castaneus* (Burmeister), may cause considerable damage to wood and wood products. Dry-wood termites occur primarily in the southern and southeastern coast states, Arizona, California, and Mexico (Fig. 4.2). Damp-wood termites, *Zootermopsis* spp., cause damage and economic loss in Oregon, Washington, California, and in extreme southwestern Canada (Fig. 4.2).

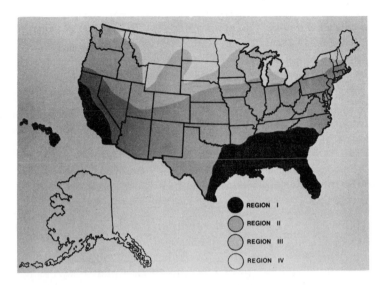

FIGURE 4.1.
Distribution of subterranean termites in the United States.

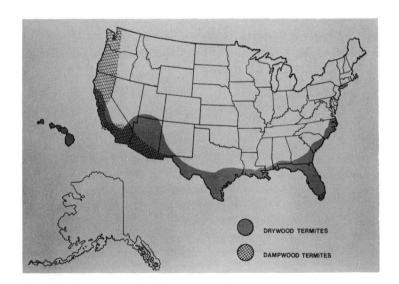

FIGURE 4.2.
Distribution of dampwood and dry-wood termites in the United States.

DAMAGE TO WOODEN BUILDINGS AND PRODUCTS

In the United States, termites are serious pests because: (1) wooden structures and products are highly valued, (2) many wooden structures are located in formerly forested areas, (3) available, effective preventive control methods are not used as widely or as effectively as they should be, and (4) construction practices that can minimize the chances for termite infestations are not consistently used. Each of these reasons will be discussed briefly.

Value of Single-Family Housing Units

For most U.S. families, the purchase of housing is their largest single financial investment. The dollar value (U.S. dollar value is used in this chapter) of all wooden structures in the United States is difficult to determine, but some estimates can be made based on the value and number of single family dwellings (Pinto 1981). In 1980, a total of 86,769,389 housing units existed in the United States (U.S. Department of Commerce 1981a) (Table 4.1). Because many of these units are part of multifamily units for which dollar values are difficult to obtain, only the 53,943,450 detached single-family units will be considered. The mean value of a housing unit in 1980 was $47,200 (U.S. Department of Commerce 1981b).

TABLE 4.1
Types and Number of Housing Units in the United States in 1980[a]

Type of housing unit	Number of units
Single-family units—detached	53,943,425
Single-family units—attached	3,530,964
Two-family units—attached	5,315,579
Three- and four-family units—attached	4,442,012
Five- or more family units—attached	15,212,089
Mobile home or trailer	4,325,320
Total year-round housing units	86,769,389

[a]Modified from the U.S. Dept. Commerce, Bur. Census. 1981. Provisional estimates of social, economic, and housing characteristics; States and elected standard metropolitan statistical areas. p. 80, Suppl Rep. PHC80-S1-1. 114 pp., plus appendices.

Thus, the estimated total value of U.S. single-family units, most of which contain significant quantities of wood, in 1980 was $2.5 trillion. This amount, plus the unknown value of multifamily units, government-owned buildings, and other wooden products, makes the total value of structures and products containing wood a significant portion of the U.S. net worth (Pinto 1981).

Formerly Forested Areas Used as Building Sites

Many houses and other structures are built on what was formerly forested land. In this situation subterranean termites are deprived of their natural food. However, a large quantity of susceptible wood is available in these buildings. If termite control methods are not used, buildings in most parts of the United States will be attacked by termites, primarily subterranean, sometime during the life of the structure. A house in Region I (Fig. 4.1) has an 80 to 100% chance of being attacked, a 50 to 80% chance in Region II, a 25 to 50% chance in Region III, and less than a 25% chance in Region IV.

Preventive Treatments Are Not Being Used

Many subterranean termite infestations in buildings could be avoided by using well-known and effective control methods (Moore 1979, Beal et al. 1983). Williams and Smythe (1978) estimated that 2.7% of the houses existing in 11 southeastern states were treated for subterranean termite prevention and remedial control in 1970. The most effective and least expensive treatment can be made during construction, but many houses are not treated at this time (Williams and Smythe 1978).

Two possible reasons why effective preventive methods are not used are: (1) architects, home builders, and the general public are not aware of potential termite problems, available control procedures, and the relatively low cost of preventive control; and (2) those who are aware of the problems, procedures, and costs are not willing to pay for treatment.

Poor Construction Practices

Poor construction practices increase the chances of subterranean and dry-wood termite infestations. Ebeling (1968), Moore (1979), and Beal et al. (1983) thoroughly reviewed good construction practices. These publications and several popularized summaries such as "You can protect your home from termites" (Haverty 1977), "Porches and planters: They can cause termite and decay damage" (Amburgey 1978), and "The best way to fight termites is before your house is built" (Smith and Beal 1978) are available. However, many new buildings are still constructed in ways that invite termite attack. Therefore, the most important procedures for minimizing the chances of a termite infestation are: (1) remove all stumps, logs, large roots, and surface vegetation from the building site; (2) prepare the building site so that water quickly drains away from the building in all directions; (3) if site drainage is poor and cannot be corrected, and if there is a crawl space or basement, place tile drain around the outside of the foundation; (4) remove all wooden materials from earth fill used in site preparation and later in landscaping; (5) remove all boards and stakes used to form the foundation and cement slab; (6) in areas where termite infestation is likely, chemically treat the soil before the foundation footings and cement slabs are poured; (7) do not disturb soil after it has been treated; (8) avoid wood-to-soil contact, such as wooden steps placed on soil or support columns that penetrate through cement slabs, except when the wood has been properly treated for this use, such as in all-wood foundations; (9) ensure that eave gutters and downspouts direct water away from the building and connect them to a storm sewer system, if possible; (10) properly install termite shields, if used, and inspect them annually, because termites can construct tunnels over them; (11) in houses with a crawl space, provide for at least an 18-in (46 cm) clearance between the bottom of floor joists and the ground and 12 in (30 cm) between support beams or girders and the soil; (12) be sure the final outside grade of houses with a crawl space is not higher than the soil underneath the house, because this situation causes water to be trapped underneath the house; (13) place a vapor barrier on the soil underneath buildings with a crawl space to prevent moisture buildup on and in structural timbers; (14) provide for adequate ventilation of crawl-space areas; and (15) provide for an 8-in (14-cm) clearance between wooden siding and the final outside grade.

Some procedures that discourage dry-wood termite attack are: (1) carefully inspect wooden articles, such as furniture, picture frames, and crates, to avoid introducing an infestation by taking infested wooden objects into the building; (2) keep all outside cracks and joints tightly caulked and all wood surfaces covered with a continuous coat of paint or varnish; (3) inspect for infestations under siding and between and underneath wood, asbestos, and fiberglass shingles; and (4) construct buildings so that they are not in the framing stage during swarming of dry-wood termites.

Estimates of Economic Impact

The total economic impact of termites in the United States is not well documented, but several estimates of control and damage costs are available. Ebeling (1968) estimated that preventing termite attack, control of infestations, and repair of damaged wood costs the people of the United

States $500 million annually. Ebeling's figure is quoted most often, but other estimates range from $100 million (Lund 1967) to $3.4 billion (USDA 1974) annually. Williams and Smythe (1978) used treatment and cost records to estimate that $168.8 million 1976 dollars were spent for prevention, control, and repair on single-family homes in 11 southeastern states. These states are in a region of the United States with the highest termite hazard. However, the estimate does not include losses to multifamily dwellings, commercial establishments, public buildings, new houses built between 1970 and 1980, and military structures.

The U.S. Environmental Protection Agency (EPA) (1981) extrapolated the Williams and Smythe data using nationwide survey data on pesticide usage (EPA 1979) and arrived at an annual loss figure of $470.8 million 1976 dollars for the entire United States. Included in this estimate are costs of prevention and remedial control, annual renewal fees to insure against reinfestations, and damage repair. Using the 1981 housing census data (U.S. Department of Commerce 1981c), the $470.8 million figure was adjusted to $753.4 million. As the EPA (1981) noted, the $753.4 million figure has an upward bias because it is extrapolated from 11 states with the highest termite infestation. The EPA also justifiably noted that the upward bias in the estimate is undoubtedly offset by the downward bias resulting from the exclusion of losses in multifamily dwellings, commercial establishments, and public buildings. Therefore, $753.4 million for annual termite damage in the United States during 1981 is probably the best estimate available (Mauldin 1982).

Subterranean Termite Control Is a Good Investment

Prevention or remedial control of subterranean termites is a good investment. Homeowners pay from $300 to $1,000 annually to insure their houses against fires and natural disasters. In 1977, the average cost for a termite control treatment was under $300 (Pinto 1981). If this $300 is inflated to $400 1980 dollars, the cost is less that 1% of the $47,200 average value of a single-family dwelling. If the soil treatment is properly done, this $300 should be a one-time cost. However, some homeowners pay an annual fee to insure against future termite damage and this figure was $30 in 1977 (Pinto 1981). The cost for protecting the family's largest monetary investment is very small, particularly considering that repair costs can reach hundreds or even thousands of dollars.

Subterranean Termite Control

The ten chemicals currently registered as soil treatment termiticides in the United States are aldrin, chlordane, chlorpyrifos, dieldrin, endosulfan, heptachlor, isofenphos, lindane, pentachlorophenol, and permethrin. Aldrin, chlordane, heptachlor, and a mixture of chlordane and heptachlor are the most commonly used termiticides. Chlorpyrifos was registered in 1980 and is now being used. Isofenphos and endosulfan were registered in 1982 and permethrin in 1984 but these are not being marketed as termiticides. Dieldrin and pentachlorophenol are not used as termiticides in the United States. Lindane is used in some areas. All of these soil treatment termiticides, except pentachlorophenol, are effective because they form a chemical barrier that separates subterranean termites from wood in the structure.

Proper subterranean termite control procedures are presented and thoroughly described by the National Pest Control Association's Approved Reference Procedures for Subterranean Termite Control (1980), Ebeling (1968, 1975), Moore (1979), and Beal et al. (1983). In the field tests conducted by Raymond H. Beal (personal communication), several insecticides have remained effective as soil treatments for as long as 34 years in Mississippi (Table 4.2). The same chemicals are also proving effective in field tests nationwide and in Panama (Table 4.3) (Beal 1981). Some termiticides, primarily aldrin, chlordane, and heptachlor, have been used successfully by the U.S. pest control industry for more than 25 years.

Some subterranean termites, particularly Coptotermes formosanus, construct nests in wall or ceiling voids and may live there indefinitely with no ground contact as long as they have a source of moisture. An insecticide barrier in the soil does not control termites in this situation.

TABLE 4.2
Number of Years Selected Insecticides have Remained Effective in Preventing Termites from Penetrating Treated Soil in Mississippi

Insecticide	Percentage concentration	Years effective[a]
Aldrin	0.5	33
Chlordane	1.0	34
Chlordane : heptachlor (1 : 1)	0.75	24
Heptachlor	0.5	30
Chlorpyrifos	1.0	17
Isofenphos	1.0	10

[a]Experiments are still in progress. Data for chlorpyrifos and isofenphos are for soils covered with concrete. Soils treated with the other insecticides were not covered with concrete.

TABLE 4.3
Long-Term Effectiveness of Termiticides as Soil Treatments at Different Test Sites

Insecticide	Location	Percentage concentration[a]	Years effective[b]
Aldrin[c]	Arizona	0.125	18
	Florida	0.125	16
	Maryland	0.125	19
	South Carolina	0.125	17
Chlordane[c]	Arizona	0.500	18
	Florida	0.500	16
	Maryland	0.250	19
	South Carolina	0.250	17
Chlorpyrifos[d,e]	Arizona	2.000	11
	Florida	0.500	12
	Louisiana	2.000	7
	Maryland	0.500	12
	Mississippi	1.000	12
	South Carolina	0.250	12
Dieldrin[c]	Arizona	0.125	18
	Florida	0.125	16
	Maryland	0.125	19
	South Carolina	0.125	17
Heptachlor[c]	Arizona	—[f]	—[f]
	Florida	0.125	16
	Maryland	0.125	19
	South Carolina	0.125	17

[a]Lowest concentration that is still 100% effective.
[b]Experiments are still in progress.
[c]Data taken from Mauldin (1982).
[d]Chlorpyrifos data taken from Beal (1980).
[e]Chlorpyrifos treatments were placed under concrete slab.
[f]All heptachlor treatments have dropped below the 100% effective mark. However, the 0.5% concentration is still 90% effective.

Termites in these aerial nests can be controlled by: (1) removing the above-ground moisture source at the same time the soil treatment is applied; (2) injecting the nest with such products as PT® 250 Baygon®[1] or PT® 270 Dursban® (Whitmire Research Laboratories, Inc., Saint Louis, Missouri), Ficam® W or Ficam® D (NOR-AM® Chemical Co., Wilmington, Delaware), and wood-X FWP (Norel Enterprises, Inc., Germantown, Tennessee).

Dry-wood Termite Control

As discussed earlier, good construction methods will decrease, but not eliminate, the chances of dry-wood termite infestations.

Prevention

Although not yet practical economically, dry-wood and subterranean termite damage can be prevented by constructing buildings with wood that has been pressure treated with an approved wood perservative. Pressure-treated wood is more expensive than untreated wood. However, in areas where dry-wood termites are common, the use of pressure-treated wood —at least for rafters, ceiling beams, and wall studs—is advisable. In crawl-space buildings, sills and floor joists should also be treated with a wood preservative. Two insecticides, Drione® (Fairfield American Corporation, Medina, New York) and wood-X FWP, are currently registered.

Based on the success of many pest control operators in southern California, Ebeling (1975) suggests that silica aerogel (silica gels of lowest bulk density and greatest porosity) should be blown into attics, wall voids, and other enclosed spaces during construction. Ebeling also gives detailed application methods for the silica aerogels during and after construction.

Remedial Control

Localized infestations can be controlled by drilling small holes into the infested wood and injecting toxic chemicals. Ebeling (1975) describes the "drill and treat" method in detail and discusses use of some insecticides that are no longer available as termiticides in the United States. More recently, Whitmire Research Laboratories, Inc., has introduced Perma-Dust PT® 240, PT® 270 Dursban, PT® 250 Baygon®, and Tri-Die PT® 230 for use in the "drill and treat" method. The "drill and treat" method pro-

---

[1]Mention of a company or trade name is for identification purposes only and does not imply endorsement by the U.S. Department of Agriculture.

vides residual protection only in the areas treated and some termites may survive because treatment of all infested wood is difficult.

Another insecticide mixture used against localized dry-wood termite infestations is Wood Treatment TC (Farmco Industries, Inc., Watertown, South Dakota). Wood Treatment TC contains pentachlorophenol and heptachlor in a heavy-bodied mayonnaise-type oil emulsion and can be applied with a paint brush or caulking gun. When applied to wood, the toxicants slowly penetrate. Ebeling (1975) and Moore (1979) state that the Wood Treatment TC is more reliable than the "drill and treat" method because it is easy to treat wood around the suspected area of infestation. Wood Treatment TC does not penetrate readily into some painted wood; therefore, painted surfaces should be roughened or scratched (sanded) before application. Special procedures are necessary before treated surfaces can be painted. Wood Treatment TC provides good residual protection but only in the treated areas.

Widespread dry-wood termite infestations can be eliminated effectively only by fumigation of the entire structure with methyl bromide or sulfuryl fluoride. Fumigation provides no residual protection, is a hazardous procedure and should be done only by experienced pest control operators, requires occupants to vacate the building for one or more days, and is expensive. Small pieces of furniture infested with dry-wood termites can be fumigated in small vaults by some pest control operators. Specific instructions for fumigation are given by Ebeling (1975) and Moore (1979).

DAMAGE TO TREES, AGRICULTURAL CROPS, AND ORNAMENTALS—INCLUDING CONTROLS

Some agricultural crops damaged by subterranean termites are sugarcane, grasses, citrus, elm, fruit and nut trees, grapevines, and shrubs and flowers. No economic estimates of termite damage to crops are available, but they are considered small in comparison to damage to wooden structures. Some specific examples are discussed below.

Sugarcane and Other Plants

Snyder (1916) reported *Reticulitermes* spp. damage to sugarcane, primarily seed cane, in Florida; pampas grass in Texas; cotton, Irish potatoes, and corn in the southern United States; and apple and peach tree sprouts in nurseries, pecan seedlings, grapevines, and shrubs and flowers in gardens and greenhouses throughout the United States. Muir and Swezel (1926) concluded that predation by the common ant *Pheidole megacephala*

(Fabricius) in Hawaiian sugarcane fields prevents C. formosanus from being a major pest. Heinz et al. (1979) tested several insecticides for control of subterranean termites in sugarcane seed pieces and growing stalks. Heptachlor at 1 or 2 lb per acre (= 1.12 and 2.25 kg per ha) gave 100% control. Lorsban® gave 95% control on seed pieces and 96.7% control in 5-month-old stalks. Lorsban® and heptachlor are not registered for this use.

Light (1937) reported that Paraneotermes simplicornis Banks attacks and sometimes kills grapefruit and tangerine trees in the Coachella Valley in California. No recent reports of this problem are available.

Ornamental Trees and Grapevines

Beal and Stauffer (1967) found that C. formosanus often make hollows inside the Chinese Elms in New Orleans, Louisiana. However, no trees were reported to have died from the damage.

The western dry-wood termite Incisitermes minor Hagen sometimes attacks dead heartwood of grapevines (Barnes 1970). Vines are weakened so that arms and trunks are broken off easily. A subterranean termite, Reticulitermes hesperus Banks, is occasionally found in dead heartwood of older vines. Barnes recommended using termite-resistant or preservative-treated wood stakes so that the subterranean termites do not have a good place to start an infestation.

Range Grasses

Watts et al. (1982) stated that Gnathamitermes tubiformans (Buckley) is a potential pest of range grasses in the southwestern United States. Significant grass defoliation is occasionally encountered. Bodine and Ueckert (1975) reported that chlordane controlled termites in range grasses, but the effect of insecticides on the ecosystem should be studied thoroughly before large-scale desert control is attempted. Chlordane is not approved by the EPA for this use.

DAMAGE TO CABLE COVERINGS—INCLUDING CONTROL

Subterranean termites sometimes attack and damage coverings for buried power transmission and communication lines (Armitt 1962, Beal and Bultman 1978). Such attacks occur during the normal foraging and tunneling by termites. Polymeric cable coverings can be made more termite-resistant by reducing the plasticizer content, thus making the cable harder, or by adding a toxic chemical to the plastic (Beal and Bultman 1978). Of the soft polymers tested, chlorosulfonated polyethylene and ethylene proplyene rubber were most resistant and are preferred over polyvinyl chloride plastics as cable coverings. Armitt (1962) suggested: (1) removing all cellulose-containing debris from the vicinity of buried cable; (2) laying the cable in well-drained soil, if possible; and (3) wrapping the cable in copper or bronze tape inside a nonmetallic jacket. Armitt also suggested several insecticides that could be used, but such uses for some of the chemicals are not permitted in the United States.

SUMMARY

Termites are important pests primarily because they cost the U.S. home-owners approximately $750 million in prevention, remedial control, and repair of damage. Economical and effective subterranean termite control methods are available to prevent much of this damage, but are not always used. Localized dry-wood termite infestations can be controlled by treating only the infested wood, but widespread infestations can be eliminated only by fumigating the entire structure. Termites can be minor pests of crops and ornamentals but, in most situations, chemical controls are not available or necessary.

This publication reports research involving pesticides. It does not contain recommendations for their use, nor does it imply that the uses discussed have been registered. All pesticides must be registered by appropriate State and/or Federal agencies before they can be used.

CAUTION: Pesticides can be injurious to humans, domestic animals, desirable plants, and fish or other wildlife—if they are not handled or applied properly. Use all pesticides selectively and carefully. Follow recommended practices for the disposal of surplus pesticides and pesticide containers.

REFERENCES

Amburgey, T. L. 1978. Porches and planters: They can cause termite and decay damage. U.S. Dept. Agric., For. Service, Southern For. Expt. Stn., and Co-operative Ext. Serv. 4 pp.

Armitt, H. T. 1962. Insect and rodent damage to cable. *Elect. Constr. Maint.* 1962: 96–98.

Barnes, M. M. 1970. Grape pests in southern California. *Calif. Agric. Exp. Stn. Ext. Serv. Circ.* 553. 8 pp.

Beal, R. H. 1980. New materials for termite control. *Pest Control* 48: 46, 48, 50, 52, 54.

Beal, R. H. 1981. Termite control studies in Panama. U.S. Dept. Agric. For. Serv. Res., Note SO-280. 6 pp.

Beal, R. H. and J. D. Bultman. 1978. Resistance of polymeric cable coverings to subterranean termite attack after eight years of field testing in the tropics. *Int. Biodeterior. Bull.* 14: 123–127.

Beal, R. H., J. K. Mauldin, and S. C. Jones. 1983. Subterranean termites—their prevention and control in buildings. U.S. Dept. Agric., For. Serv., Home and Gard. Bull. No. 64. 36 pp.

Beal, R. H. and L. S. Stauffer. 1967. How serious the Formosan termite invasion? *For. People* 17: 12–13, 28, 40–41.

Bodine, M. C. and D. N. Ueckert. 1975. Effect of desert termites on herbage and litter in a shortgrass ecosystem in West Texas. *J. Range Manag.* 28: 353–358.

Ebeling, W. 1968. Termites: Identification, biology, and control of termites attacking buildings. Univ. Calif., Calif. Agric. Exp. Stn. Ext. Serv. Man. 38, 74 p.

Ebeling, W. 1975. Urban Entomology. Univ. Calif., Div. Agric. Sci., Los Angeles, Calif. 695 pp.

Haverty, M. I. 1977. You can protect your home from termites. U.S. Dep. Agric., For. Serv., South. For. Exp. Stn. 13 pp.

Heinz, D. J., M. K. Carlson, and R. S. Tabusa. 1979. 1979 Annual Report of the Experimental Station. *Hawaiian Sug. Plant. Assoc.*: 1980. 79 pp.

Light, S. F. 1937. Contributions to the biology and taxonomy of *Kalotermes (Paraneoterms) simplicicornis* Banks (Isoptera). *Univ. Calif. Publ. Entomol.* 6: 423–464.

Lund, A. E. 1967. The study of subterranean termites: A laboratory and field approach. *17th Annu. Conv. Brit. Wood Preserv. Assoc.* 1967: 119–127.

Mauldin, J. K. 1982. The economic importance of termites in North America. In: "The Biology of Social Insects," pp. 138–141. (M. D. Breed, C. D. Michener, and H. E. Evans, eds.) Proc. 9th Congress Int. Union Study Social Insects. Westview Press, Boulder, Colo. 420 pp.

Moore, H. B. 1979. Wood-inhabiting insects in houses: Their identification, biology, prevention and control. U.S. Dept. Agr., For. Serv., Dept. Housing and Urban Dev. Man. 133 pp. (Interagency Agreement IAA-25-75).

Moreland, D. 1981. Subterranean termites. *Pest Control Technol.* 9: 30–34.

Muir, F. and O. H. Swezel. 1926. Entomologists' report on termite problem. *Hawaiian Plant. Rec.* 30: 331–335.

National Pest Control Association. 1980. Approved reference procedures for subterranean termite control. (George W. Rambo, ed.) Nat. Pest Control Assoc., Vienna, VA. 157 pp., plus Appendices I-IV.

Pinto, L. J. 1981. The structural pest control industry: Description and impact on the nation. *Nat. Pest Control Assoc. Rep.* 36 pp.

Smith, V. K., and R. H. Beal. 1978. The best way to fight termites is before your house is built. U.S. Dept. Agric., For. Serv., South. For. Exp. Stn., and Cooperative Ext. Serv. 4 pp.

Snyder, T. E. 1916. Termites, or "White ants," in the United States: Their damage, and methods of prevention. *U.S. Dept. Agric. Bull.* No. 333. 32 pp.

U.S. Dept. of Agriculture. 1974. Insects affecting man and his possessions. Research needs in the southern region. Joint Task Force Rep. of the South. Region Agric. Exp. Stns. and U.S. Dept. Agric. 34 pp.

U.S. Dept. Commerce, Bur. Census. 1981a. Provisional estimates of social, economic, and housing characteristics: States and selected standard metropolitan statistical areas. P. 80, Suppl. Rep. PHC80-S1-1. 114 pp., plus appendices.

U.S. Dept. Commerce, Bur. Census. 1981b. 1980 Census of housing. Selected housing characteristics by states and countries. P. 10, Suppl. Rep. HC80-S1-1. 36 pp.

U.S. Dept. Commerce, Bur. Ind. Econ. 1981c. Building materials outlook for 1981. *Const. Rev.* 27: 49.

U.S. Environmental Protection Agency. 1979. National Household Pesticide Usage Study, 1976–1977. Office of Pesticide Programs. EPA-540/9-80-002.

U.S. Environmental Protection Agency. 1981. Comparative benefit analysis of the seven chemicals registered for use on subterranean termites. Draft of part of the Benefits and Field Stud. Div., Off. Pestic. Programs, Environ. Protection Agency's cluster analysis. 185 pp.

Watts, J. G., E. W. Huddleston, and J. C. Owens. 1982. Rangeland entomology. *Annu. Rev. Entomol.* 27: 283–311.

Williams, L. H. and R. V. Smythe. 1978. Estimated losses caused by wood products insects during 1970 for single-family dwellings in 11 southern states. U.S. Dept. Agric. For. Serv. Res. Pap. SO-145. 10 pp.

Chapter 5

BENEFICIAL ASPECTS OF TERMITES
Elizabeth A. McMahan

INTRODUCTION

The harm that termites inflict on the human economy gives them a very negative image and tends to overshadow any beneficial roles they may play. Because of their habit of feeding on cellulosic material they menace buildings and other constructions, ruin libraries and stored goods, damage agricultural crops, and compete for grazing lands. Their mound-building activities produce obstructions to mechanical agriculture, landing fields, railroads, and other engineering projects. Nevertheless, only 10% of the 2,000 + known species have been recorded as pests, and the same feeding and soil-working propensities that lead to destructive results may lead to results with potential benefit. Hickin (1971) says:

> It is perhaps not unreal to suggest that the continuation of the biosphere as we now comprehend it, depends more on maintaining termites alive and in their present numbers in the tropical tree-growing areas of the world, than on destroying termites where man utilizes wood in his own social economy.

"Beneficial" and "detrimental" are relative terms. Usually they are defined from the human point of view and with regard to immediate and direct effects. But even in this restrictive sense some termite activities have direct and positive effects for man. The soil in termite mounds, for example, is used in some parts of the world for making bricks. In most cases, however, the beneficial effects for man are more indirect.

In tropical regions termites comprise a relatively large component of the ecosystem and, through their activities leading to wood-cellulose breakdown, play an important role in nutrient recycling. Assessment of the human gain resulting from such termite activity is difficult, for few studies have measured their effects in ways that permit quantitative evaluation. In the paragraphs that follow we will consider ways in which termites have been credited with playing beneficial roles in the human economy or in the ecosystem.

Although the average layman sees little reason to feel kindly toward termites, most people seem to find accounts of life in a termite colony (indeed of any social insect colony) entrancing. Anyone who has seen a documentary movie on termites, such as the Roots' "Mysterious Castles of Clay" (1978) realize the potential for entertainment. From brood care to foraging activities, from colony defense to new nest construction, from alate emergence to the founding of incipient colonies, termites exhibit coordinated activity that cannot help but excite our profound interest.

MODELING AND TEACHING OF BIOLOGICAL PRINCIPLES

A termite mound is a microcosm, a model ecosystem, a community of organisms, which can be used to illustrate evolutionary, ecological, behavioral, and physiological principles. The following are a few examples of the use of termite colonies as models for facilitating an understanding of the processes by which animals carry out their various activities.

Alfred E. Emerson (1939, 1952b), who spent his life in close contemplation of termites and their societies, enlarged on the ideas of William Morton Wheeler (1911) and others in applying the supraorganismic concept (as he called it) to the insect society. Emerson (1939) pointed out ontogenetic and phylogenetic correlations between such a society and the organism. His emphasis was on biological mechanisms, likening the castes of social insects to cellular division of labor, the control of caste development to hormonal control of tissue differentiation, etc. He was not intimidated by critics who pointed out that the analogies were limited. He replied (Emerson 1952b) that he was quite aware of these limitations but believed that the concept of the supraorganism could serve a useful purpose in elucidating integrating principles. He used the supraorganismic concept chiefly in the analysis of evolutionary trends in termites, finding it useful to consider the colony and its adaptations as a single evolving unit. Using the termite colony as a model, he analyzed and illustrated a variety of evolutionary concepts and processes: homology, convergence, divergence, parallelism; and progressive, regressive, and adaptive evolution (Krishna 1969).

Termite nest architecture has been cited in behavior theory modeling to illustrate the view that genetic controls extend not only to bodily structures but also to nervous structure and thence to behavior (Emerson 1952b, Howarth 1966, Noirot 1970). African termites of the genus *Apicotermes* build extremely complex underground nests whose architecture does not appear to vary much within a species. It differs considerably between species especially with regard to the pattern and arrangement of perforations in the external walls that serve to regulate the gaseous exchange between the nest interior and exterior. Nest differences are more distinct than are those between the termites themselves. Emerson (1956) used the term "ethospecies" to designate termite species that are more readily identified by nest characteristics than by bodily structure. Noirot (1970) regards the study of the nest of *Apicotermes* as providing "some very important results in the field of phylogeny of behavior."

Termites, no less than other organisms, have been used to generate data for modeling ecosystems and elucidating ecological principles. For example, Usher (1979) extrapolated from an analysis of termite succession in baitwood blocks in Ghana to conclude that ecological succession is a non-stationary Markovian (non-random) process. Studies of termite phylogeny and distribution have been used also in defining zoogeographic regions and in reconstructing former continental relationships (Emerson 1952a).

A special case of the usefulness of termites—at least their symbionts—is found in the field of cell biology. Certain termite protozoans have aided in understanding the plasma membrane. Kirby (1941–1949) first reported the fact that certain devescovinid flagellates are able to rotate one part of the cell independently of another part. He suggested that the clockwise rotation of the anterior cytoplasm relative to the posterior part of the cell demonstrated the "fluidity and lability of the surface layers." This phenomenon was independently rediscovered by Tamm and Tamm (1974) in a devescovinid found in the hindgut of *Cryptotermes cavifrons*. They proceeded to study it, using modern methods of electron microscopy and the freeze fracture technique of specimen preparation (Tamm and Tamm 1974, Tamm 1979). They found that the lipid bilayer of the plasma membrane is continuous across the shear zone and that the distribution and density of intramembranous particles within this zone are similar to those of other regions of the membrane. The termite's devescovinid has turned out to be a useful system with obvious advantages for the study of the fluid properties of cell membranes.

As classroom teaching tools termites have the advantage of being easily obtainable in warmer climates and easily reared. In introductory biology laboratories they are useful in demonstrating a number of biological phenomena. For example, the semitransparent exoskeleton of the sterile castes of most species makes the microscopic observation of heart contractions easy. Termite hindguts team with flagellates and bacteria, permitting not only a close study of symbiotic relationships, but providing protozoans for demonstrations of unicellular organisms. Maceration of a *Reticulitermes* hindgut in a droplet of Insect Ringer releases thousands of jostling *Trichonympha* and other symbionts, providing the young microscopist with a sight he or she will long remember.

BENEFICIAL EFFECTS OF TERMITES IN THE ECOSYSTEM

Soils

As components of the ecosystem termites play their major roles. They occupy the position of primary consumers (herbivores and detritivores), causing the comminution, decomposition, humification, and mineralization of the materials they attack (see LaFage and Nutting 1978). They play a role in the turnover of complex carbohydrates by decomposition to simpler organic compounds. In addition, they serve in the dissemination of fungi that also help to break down cellulose (Bakshi 1960). They tunnel extensively, bring up the subsoil, and (in case of many tropical species) make mounds and other constructions into which they incorporate soil, fecal, and salivary particles. Their relative effects vary according to species and habits, but the resulting soil modifications are sometimes great, involving changes in soil profiles, soil textures, and ratios of organic and inorganic matter. Holt et al. (1980) estimate that in one generation of mound building, termites in the Charters Towers areas of North Queensland, Australia, could rework up to 20 tons per hectare of soil. Most of the literature dealing with the effects of termites on soils has been summarized by Snyder (1956–1968), Harris (1961), Kevan (1962), Lee and Wood (1971), and Wood and Sands (1978). These authors judge some of the effects to be detrimental to soil fertility and plant growth, and some to be beneficial. They believe that no clear and definite assessment of the overall effects of termites on soil is possible at the present time. In this chapter we note possible benefits.

Lee and Wood (1971) attribute to Drummond the view that the fertility of the Nile Valley owes much to "the humble termites in the forest slopes about Victoria Nyanza." Presumably the disintegrated nest material was inferred to compose a significant part of the Nile's burden of silt. Wood and Sands (1978), however, conclude that erosion of termite mounds contributes very little to the nutrient status of soils. Most claims of fertility enhancement are based on the observation of agricultural practices of native farmers in regions where termite mounds abound. In tropical Africa and Asia, crops such as cotton, sisal, and tobacco are said to be grown preferentially on mounds (Pendleton 1942, Harris 1961). Harris reports that where these crops are grown on a large scale, the presence of mounds in the fields are indicated by increased growth and intensified green color. Sheppe (1970b) reports that *Odontotermes* mounds are preferred sites for maize cultivation, and Mielke (1978) suggests not only that termites in southeastern Tanzania and northern Malawi may enhance soil quality (as evidenced by the coexistence of termitaria and cultivated fields) but that attempts at "disrupting the termite may eliminate one of the creatures which can best assist humans in survival when the high energy and material demands of technological agriculture meet harder times." Preliminary tests with *Reticulitermes tibialis* have indicated that termites can modify coal strip mine spoils toward greater porosity and friability (Ettershank et al. 1978). On the other hand, species that utilize only subsoil in construction may be expected to produce mounds less fertile than the surrounding soil (Kirkpatrick 1957, Lee and Wood 1971).

Relatively recently the microbial symbionts of termites have shown the ability to fix atmospheric nitrogen (Benemann 1973, Breznak et al. 1973, Sylvester-Bradley et al. 1978). Breznak et al. (1973) suggest that in tropical regions termites "may play a hitherto unrecognized role in soil fertility by replenishing combined nitrogen compounds in the soil." Kozlova reports (1951) that *Anacanthotermes* mounds of the deserts of Central Asia are a source of nitrate fertilizer.

Several authors have discussed the role of termites as agents of dung removal in Africa and Australia. At certain seasons of the year they replace dung beetles in importance as major removers of elephant (Weir 1972, Coe 1977) and bovine dung (Ferrar and Watson 1970). Coe estimates, for example, that the 19,900 elephants in the Tsavo National Park in Kenya can be expected to produce $8.7 \times 10^3/km^2$ (dry weight) of feces per km^2 per year. However, instead of burying the dung where it enriches the soil immediately, termites transport it to their nests where it is used as food or incorporated into the nests. There it may be immobilized for the life of the mound, which may range from less than ten years for *Cubitermes fungifaber* to more than 80 years for certain *Macrotermes* (Wood 1976). Other organic nutrients that may be immobilized in a mound include termite feces, salivary secretions, and termite corpses (Gillman et al. 1972, Wood 1976, Sen-Sarma 1974, Agarwal 1975). Eventually the mount disintegrates and the nutrients are released.

Since termites make extensive tunnels and galleries in the soil (going to depths of 70 m to reach the water table) (Lee and Wood 1971), they have a potentially large effect on soil porosity. This is correlated with improved soil aeration, increased water infiltration, and greater penetration and proliferation of plant roots (Ghilarov 1960). The suggestion (Drummond 1886) that the termite is the "tropical analogue" of the earthworm in promoting soil porosity and preserving soil fertility is probably overstated, however, and no quantitative data are available for comparison. Furthermore, some species such as *Drepanotermes rubriceps* and *D. perniger* produce a hardpan sort of surface above their subterranean nests that actually retards the percolation of water (Lee and Wood 1971). Nevertheless, termite tunneling probably does serve to increase the rate of rain infiltration and drainage and to aerate both the topsoil and the subsoil (Robinson 1958, Kevan 1962, Lee and Wood 1971, Sen-Sarma 1974). The presence or absence of termite mounds in some areas may also serve as useful indicators of the relative depths of the water table.

Vegetation

Termite activity may initiate reforestation and affect vegetation patterns. Thomas (1941) has asserted that termite mounds in the Sese Islands in Lake Victoria, Uganda, greatly accelerate the regeneration of forest land from savannah. Islands of vegetation arise in conjunction with *Macrotermes* mounds and eventually coalesce to form a continuous forest. Similar descriptions of the termite role in the colonization of grassland by forest are given by Troll (1936), Burtt (1942), and Grassé and Noirot (1961). Griffith (1938) suggested that mounds stimulate plant growth through affording them space out of the range of fires. Recently, Salick et al. (1983) studied termite populations in nutrient-poor habitats along the Rio Negro in Venezuela. Their data indicate that "termites exaggerate resource patchiness by concentrating nutrients in their termitaria." Abandoned termitaria serve as fertile microsites, encouraging seedling establishment.

White (1970) studied striped patterns of vegetation of the laterite plateau surfaces in southern Niger. Here the woodland is separated by bare ground, the orientation of the lines corresponding to the direction of runoff and at right angles to the direction of sheet flow. He suggests that these linear patterns are the result of intense root competition, a response to a shallow soil condition in a zone of limited rainfall. He hypothesizes that the differentiation of the vegetation into strips enables the available moisture to be utilized most efficiently, allowing a more luxurient growth than would be possible if the plants were distributed evenly. He believes that the bare areas are probably initiated by termite activity and that the patterns are probably self-perpetuating. Similarly, Glover et al. (1964) suggest that vegetation patterns on the Loita plains of Kenya coincide with soil zones formed by colloidal material washed from termitaria. In desert areas, the clay tunnels and chamber walls of *Anacanthotermes ahngerianus* become important agents of sand binding, forming a sort of carapace, according to Ghilarov (1960), and preventing wind erosion.

Mineral Concentration and Geochemical Prospecting

The transport of subsoil particles to the surface has already been mentioned as an activity characteristic of termites. Harris (1961) suggests that this activity may have significance for geochemical prospecting, especially in parts of Africa, where the land is covered by a featureless blanket of red laterite earths and there are few outcrops. He believes that the soils of the larger termite mounds are good places to search for evidence of such underlying metals as tungsten, molybdenum, and nickel. Watson (1972) put this idea to the test by examining mounds of *Odontotermes latericius*, *Macrotermes mossambicus*, and *Rhadinotermes coarctatus* at a gold anomaly in Kalahari sand in Rhodesia, Africa. He found that the termites probably did carry gold particles up from the subsoil, but the mean concentration of gold in their mounds did not differ significantly from that in the surrounding soil. He concluded that most of the gold in the mound had been obtained by termites from a depth of less than 3 m. d'Orey (1975) sampled Macrotermitid mounds and found anomalous concentrations of copper and nickel. The usefulness of termite mounds in geochemical prospecting has yet to be realized.

The concentration of minerals in termite mounds is the basis for reports (Howse 1970) that mound material is eaten by primitive tribes, presumably to supplement minerals in short supply in the diet. Weir (1969, 1972) has reported that *Macrotermes* mounds have higher concentrations of potassium, calcium, magnesium, and other minerals than the surrounding soil and that elephants sometimes exploit them as salt licks. Ghilarov (1960) has noted that nests of *Anacanthotermes* are foci of salt concentration in Central Asia.

Members of the Food Chain

As a major component of tropical ecosystems, termites are an important means of recycling nutrients. They serve as food for predators of a wide variety. Many predators are opportunistic feeders on alates during periods of swarming. Some appear to feed almost exclusively on termites, being especially adapted for tearing open mounds or carton nests and extracting the inhabitants. Sheppe (1970a) watched geoplanid turbellarians trapping termites in secreted slime and feeding on them. The Micheners (1951) observed dragonflies capturing swarming alates, biting off the abdomens and discarding the presumably less-palatable head, thorax, and wings. Sen-Sarma (1974) also mentions dragonfly predation on *Odontotermes* alates. Phasmids have been seen capturing termites (Sheppe 1970a), and Bhatnagar (1966) speculates that an earwig found in the Kashmir Valley may eat them. Several species of assassin bugs (Reduviidae) have been observed in association with termite nests. Lent and Wygod-zinsky (1947) found adults and young of *Neivacoris steini* in *Cornitermes* nests in São Paulo, Brazil. *Reduvius, Petalochirus, Lisarda, Tegea,* and *Acanthaspis* have been seen carrying termites and eating them (Miller 1956, Odhiambo 1958, Casimer 1960, Araujo 1970, Sheppe 1970a, Root and Root 1978). The reduviid *Salyavata variegata* uses a bait-and-capture strategy of predation that appears to be so uniquely adapted to preying on *Nasutitermes* species that it may feed almost exclusively on them (McMahan 1982). The larvae of a termitophilous neuropteran (*Lomayia latipennis*) apparently paralyze termites with a specific allomone as a capture mechanism (Johnson and Hagen 1981). Sheppe (1970a) saw staphylinids of the genus *Zyras* and a carabid, *Rhopalomelus bennigseni,* preying on termites and Araujo (1970) lists tiger beetles as predators. Ants are usually considered to be the worst enemies of termites (Wheeler 1938, Sheppe 1970a), and many species have special behavioral mechanisms for enhancing termite capture. For example, *Decamorium uelense* uses scouts to detect foraging termites. They then recruit nest mates which emerge in a column to attack the prey (Longhurst et al. 1979b). *Megaponera foetens* coordinates its termite predation by chemicals from at least two glands (Longhurst et al. 1979a). Foraging ants, on finding groups of termites, release chemicals that attract other workers to the site and then instigate digging activities into the termite galleries to expose the prey. Termites were found to compose most of the prey of the ponerine ant, *Odontomachus chelifer* in Central Paraguay (Fowler 1980). Schaefer and Whitford (1981), from their study of *Gnathamitermes tubiformans* in a Chihuahuan desert ecosystem, concluded that subterranean termites are important as regulators in desert nutrient cycles primarily through predation losses to ants. Other invertebrates that reportedly feed on termites are onychophorans, spiders, scorpions, solpugids, centipedes, thysanurans, cockroaches, mantids, various beetles, robber flies, calliphorid and tachinid larvae, and wasps (Snyder 1948, 1956–1968, Mathur 1960, Harris 1961, Nutting 1969, Araujo 1970, Sheppe 1970a, Sen-Sarma 1974, Wood and Sands 1978, Scarborough and Sraver 1979, Deligne et al. 1981).

Examination of the stomach contents of a great variety of vertebrates has indicated that for them also termites are a major food item. In three species of Australian frogs of the genus *Pseudophryne* the diet was shown to consist mainly of ants and termites (Pengiliey 1971). The simultaneous emergence of termites and the spadefoot desert toad, *Schaphiopus couchi* after spring rains is thought to be essential to the survival of the toad, whose diet was found to consist mainly of *Gnathamitermes pexplexus* (Dimmitt and Ruibal 1980). Snyder (1948) lists three species of toads, *Bufo cognatus, B. compactilis,* and *B. punctatus,* as preying on termites in North America. Sen-Sarma (1974) lists *Bufo melanostictus,* as well as the frogs

Rana tigrina and *R. breviceps*, as predators on Indian termites. Termites ranked fourth in prey frequency in the stomachs of the limbless lizard *Ophisaurus apodus* in the Surkham-Darya River area of the Soviet Union (Yadgarov 1974). They are also the major food of the lizards *Cnemidophorus tigris, C. scarlaris,* and *C. inornatus* in the southern part of their range through southern Arizona and northern Sonora (Pianka 1970, Barbault and Maury 1981). The blind snakes *Leptotyphlops d. dulcis* and *Typhlops pusillus* feed on termites (as well as on ants and other insects) and have been shown to follow readily their pheromone trails, thereby reducing the energy spent in foraging (Gehlbach et al. 1971). The stomach contents of *L. h. humilis* and *L. d. dulcis* showed that ants and termites composed 54–64% of the total diet for the animals examined (Punzo 1974). Snyder (1948) reports that during termite swarming, alates also serve as food for fishes and crocodiles. Araujo (1970) quotes Carvalho as observing small fishes in flooded forests and savannas of Central Brazil acutally leaping from the water to seize termites emerging from nests partially submerged.

The importance of termites as food for birds has been mentioned frequently. Snyder (1948) lists 52 species of Nearctic birds known to feed on termites, and Thiollay (1970) lists 150 tropical species that prey on alate swarms. Sen-Sarma (1974) reports that termites are an important food item for a number of common Indian birds. Wood and Sands (1978) have summarized studies that attempted to quantify such bird predation. They quoted Rowan as stating that termites are used as food by nearly 43% of all birds, nearly 39% of passerines, and about 47% of nonpasserines. DeBont (1964), Bouillon (1970), and Moreau (1972) have suggested that were it not for African termites, which are vital to the survival of European migrant (insectivorous) birds overwintering in Africa, insect pests would be a greater menace to the economy of Europe. Winged termites are also credited with influencing the timing and spread of the migrations of some intratropical migrants, such as raptors, rollers, and bee-eaters (Thiollay 1970). Ward and Jones (1977) showed that alates figure prominently in the diet of three races of the weaver bird, *Quelea quelea*, in Tanzania at the beginning of the rainy season, a time when their normal seed food is disappearing. The queleas perform a migration at this time, and the termites provide food important for premigratory fattening. Ward and Jones measured the fat content of 200 (unidentified) termites taken from the crops of these birds and found it to be 30% of the net weight or 57% of the dry weight. In addition to being a source of fat, termites are apparently also a good source of amino acids required for gonad maturation and change into breeding plumage.

Termite colonies provide poultry food in rural sections of Puerto Rico and other tropical countries where broken nest fragments are exposed to chickens. Wood and Sands (1978) mention Hegh's report that in northern Ghana the Farafara tribe attract *Odontotermes* sp. to dung-filled earthenware pots where they are cultured to feed domestic fowl. Kalshoven (1958) reports that *Gallus varius*, the peacock, and the pungklor (*Eucychla*) can be raised in captivity on *Macrotermes* and *Microtermes*. Both he and Roonwal (1970) note, however, that domestic fowl are not attracted to foraging columns of *Hospitalitermes*.

Termites are also eaten by many kinds of mammals. Most insectivorous forms take termites, but some mammals feed on them almost exclusively. Wood and Sands (1978) summarize Smithers' study of the gut contents of Botswana mammals. He found termites in the elephant shrew, hedgehog, vervet monkey, wildcat, bat-eared fox, Cape fox, black-banded jackal, side-striped jackal, small spotted genet, rusty spotted genet, suricate, seleous mongoose, yellow mongoose, white-tailed mongoose, and dwarf mongoose. The small burrowing rodent *Steatomys opimus* (Dendromuridae) of Central Africa was shown by stomach content analysis to feed mostly on termites, which are accessible to it at all seasons (Genest-Villard 1979). Rood (1975) studied a population of about 100 banded mongooses in the Rwenzori National Park in Uganda. Their dens were in termite mounds, and termites made up a sizeable proportion of their diet. Snyder (1948) reported that in North America termites have been found in the stomachs of the common mole, the hooded skunk, the white-throated wood rat, the armadillo, and the black bear. Sen-Sarma (1974) notes that insectivorous bats and *Vulpes bengalensis* have also been recorded feeding on swarming termites.

The animals usually thought of first as termite predators are the pangolins, aardvarks, and aardwolves of Africa (Pages 1970, Prince 1971, Kruuk and Sands 1972); the anteaters of Central and South America (Araujo 1970, Prince 1971); the echidnas and numbats of Australia (Hicken 1971, Wood and Sands 1978); and the scaly-anteater (pangolin) and the sloth bear of India (Sen-Sarma 1974). These predators are specialized for extracting termites from their mounds and carton nests, most of them possessing formidable claws, long mobile and adhesive tongues, and acute senses of smell and hearing. Their success at termite predation is indicated by Zur-Strassen's (1967) record of 40,000 termites in the stomach of one aardwolf, and Hicken's (1971) mention of a pangolin (*Manis gigantea*) whose stomach contained about six liters of termites.

Primates, including man, are known to eat termites. Goodall (1964) first recorded the use by chimpanzees of small twigs and stems to extract termites (*Macrotermes bellicosus*) from their mounds at Gombe Stream Reserve in Tanzania. Other investigators (Jones and Sabater-Pi 1969, Sugiyama and Koman 1979, Nishida and Uehara 1980, Uehara 1982 and McBeath and McGrew 1982) found further evidence of such tool use by chimpanzees in termite probing. Termites probably provide a relatively minor part of the chimpanzees' diet, however.

Human consumption of termites has been reported many times. Termites form a useful source of animal protein in areas where meat is scarce (Harris 1961). Although there have been reports (Snyder 1948, Howse 1970) of sickness or death from too much feeding on termites, Campbell (1968) notes that Bushmen of the Central Kalahari Game Reserve in Botswana regularly include termites in their diets. Williams (1959) mentions that while termite alates in Eastern Uganda are not collected regularly as food, at times of swarming they are eaten by children. Harris (1961) provides a photograph of a trap made of grasses for ensnaring emerging alates. He states that "in practically all countries where termites reach adequate size, the indigenous inhabitants capture the flying forms for food." According to Schlippe (1956) members of the Azande tribe in tropical Africa spend 26% of their total daily work effort in catching termites during the rainy season of March through June. This is the termite swarming season, which coincides with the period in the yearly cycle when new crops have just been planted and human food stores are at their lowest. The termites provide food at a critical time. Mathur (1954) and Sen-Sarma (1976) report that termites constitute the chief source of protein, fat, and calories among some tribes in India. Snyder (1948) mentions the fact that huge termite queens are considered a special delicacy in some parts of Africa and are reported to have "properties that will invigorate old men."

Like chickens, man seems to prefer some termite species over others. *Macrotermes bellicosus* and *M. natalensis* have been sold in market places over much of tropical Africa (Harris 1961). Howse (1970) quotes Smeathman as considering roasted termites to be superior in taste to shrimp. He also reports that termites are sometimes pressed to extract a cooking oil. Wood and Sands (1978), after reviewing estimates of net caloric values of alates, reported an average of about 31 kJ per g (dry weight) for Macrotermitinae and about 29 kJ per g for other species. Tihon (1946) analyzed the nutritional content of alates, lightly roasted and offered for sale in Leopoldville. He reported 44.4% fat and 36% protein, with 560 calories per 100 g. Castlewitz (1976) quotes Defoliant as ascribing to termites a "higher protein content than beef or fish." Moore (1969) says termites are a good source of vitamin B complex, which "accounts for their importance as an item of diet for many tropical animals, including aboriginal man."

A human food source in parts of the world where fungus-growing termites (Macrotermitinae) abound are the fruiting structures produced by the termite-associated fungi (*Termitomyces*) during the rainy season (Pegler and Pierce 1980). Sands (1969) reports Alasoadura's observation that the agaric phase structures are commonly sold in Nigerian and Ugandan markets. Cheo (1948) comments that *Termitamyces (Collybia) albumi-*

nosa is a well-known Chinese delicacy (called "chicken drumstick" in Yunnan), which is preserved dry by salting. Cheo further notes that a preparation made from another species, *Xylosphaera nigripes* (associated with abandoned fungus combs of *Odontotermes* sp.) is used in China for relieving spasms in children.

IMPORTANCE OF TERMITE MOUNDS

In Human Constructions

Termites tend to concentrate relatively fine soil particles in their mounds, reflecting the size of their mandibles. This material hardens after watering, reportedly becoming harder and stickier than normal and being especially well suited for making road surfaces, tennis courts, wall plastering, pottery, ovens, and bricks (Gunther 1955, Harris 1961, Howse 1970, Sen-Sarma 1976). Howse reports that 450,000 bricks have been made from a single termite mound. Froggatt (1903) noted that the fine earth from termite mounds has been used also by native jewelers in Sri Lanka to polish gems.

As Shelters for Other Organisms

Termite mounds are regularly inhabited by a wide range of animals that can be classified variously as predators, scavengers, commensals, termitophiles, or termitariophiles. In many cases the relationship between the termites and their guests is unknown, as is the case of the Margarodid scale insects in galleries of *Heterotermes* and *Neocapritermes* (Silvestri 1938, Jukubski 1965) and the tineid caterpillars reported by Emerson (1938) and Silvestri (1944). According to the classic definition, termitophiles are arthropods that have an obligatory relationship to the termite society.

Kistner (1969) lists major groups and discusses their biology. They include millipedes, mites, collembola, phorid flies, certain ants, and a variety of beetles. The best known are the termitophilous beetles belonging to the family Staphylinidae (see also Seevers, 1957).

Other animals associate not with the termites themselves but with termitaria, seeking a protected haven or a nesting site. These organisms are more properly called termitariophiles. For example, Araujo (1970) lists several kinds of beetles (anacanthocerids, dynastids, ptillids, etc) that associate with termite mounds in Brazil. Stingless bees (*Trigona* primarily) frequently build their nests within the confines of termite nests, lining the cavities with bitumen, presumably to exclude the termites (Wille and Michener 1973). Some ant species also nest in termitaria (Holt and Greenslade 1979). Hindwood (1959) observed two instances of honey bees building their hives in arboreal termitaria near Sydney, Australia; and Batra (1979) reported the nesting of a enumenid wasp in a senescent *Odontotermes obesus* mound in India. Snakes use termite mounds as places to deposit their eggs. Kalshoven (1958) found snake eggs in the arboreal termitaria of *Lacessitermes batavus*. Similarly, a large reptile of Natal called the Nile monitor (*Varanus niloticus*) reportedly digs into *Nasutitermes* nests and deposits its eggs there. The termite nests provide conditions of constant temperature and humidity favorable to incubation (Snyder 1948, Howse 1970).

A number of tropical birds hollow out nest cavities in arboreal termitaria. The termites plaster over the broken galleries and continue to live in the undamaged portions. Hindwood (1959) has listed about 50 species of birds that habitually lay eggs in termitaria in the tropics. He even mentions a case in which kookaburras and sacred kingfishers occupied simultaneously different cavities in the same *Nasutitermes walkeri* nest. Additional birds that nest in arboreal termite nests are reported by Berger (1961, quoting Haverschmidt) in South America and by Dorst (1974) and Skutch (1977) in Mexico and Central America.

Burrowing owls are said to utilize terrestrial termite mounds, nesting in excavations that have been made by armadillos and anteaters (Hindwood 1959) or underneath the mounds (Araujo 1970). The activity of termites in hollowing dead trees has been credited with providing the white-tailed black cockatoo in Australia with nesting sites (Saunders 1979). The pigmy parrot in New Guinea not only nests in termitaria but also uses them as roosting sites (Hindwood 1959).

Some mammals make dens in termite mounds, as in the case of the banded mongoose in Uganda (Rood 1975), or use them as scratching posts or lookout posts (Root and Root 1978). According to Lee and Wood (1971), Drummond noted their use in providing cover for hunters in Africa.

OTHER BENEFITS

A few additional, and rather exotic "benefits" derived from the activities of termites may be listed. Termitaria have been used as models for art objects. Wolcott (1948) describes a concrete arbor sold in Puerto Rico whose manufacturer incorporated into it a replica of a *Nasutitermes* arboreal nest to enhance the rustic appearance. The 56½-foot, tenth-century Jain statue of Gommateswara at Sravana Belgola (Karnataka), India includes sculptured vines entwining the sage's arms and termite nests about his feet signifying "the timeless concentration of the questing soul for ultimate illumination" (Sen 1964). Australian aborigines seek termite-hollowed saplings for making digeridoos, their rhythm-producing musical instruments (Lee and Wood 1971). Farmers in Thailand use termite mounds near farmsteads as bases on which to build straw stacks or place the rice-milling mortar and walking-beam pestle (Pendleton 1942). The fumes of burning carton are inhaled for chest ailments in Puerto Rico, and both termites and their nests are reportedly used for medicinal purposes in India (Cleghorn 1896). Members of the Azande tribe in the Sudan reportedly use termites as oracles, deciding issues on the basis of which of two pieces of wood placed in a termite mound is attacked (Howse 1970). And finally, termite mounds have been used as burial mounds by members of the Bakutu tribe in the Congo (Gunther 1955).

We end our account, not entirely facetiously, with the undeniably positive role that termites play in the lives of pest control operators and of professional biologists with specialties involving termites. The detriment to the client homeowner is the benefit to the pest control operator, who owes his or her livelihood to the tiny despoilers. In a similar vein, the consternation we "termite specialists" in academia feel on discovering that termites have somehow invaded our basements is tempered somewhat by the knowledge that they are also responsible for much of the satisfaction we derive from life.

Alfred Emerson (1947), in his vice-presidential address to the American Association for the Advancement of Science, tried to explain some of his own fascination with termites. He admitted that friends were often shocked to learn that his major interest was not in termite eradication. He considered the benefits of termites to outweigh their damaging aspects, and he never lost his bias in their favor. An understanding of termite social life and ecological adjustments, he felt, would give us a perspective "far more important for human life than the knowledge we may gain concerning their economic control" (Emerson 1949). He hoped that such understanding would aid in attaining "a wise control over the physical, biotic, and social forces which are shaping our future" (Emerson 1942). It is a point of view we should not lose.

REFERENCES

Agarwal, V. B. 1975. Studies on the biology and ecology of mound building termites *Odontotermes microdentatus* Roonwal and Sen-Sarma and *Odontotermes obesus* (Rambur). Ph.D. Thesis, U. of Meerut, Dehra Dun, India, p. 53.

Araujo, R. L. 1970. Termites of the Neotropical region. *In*: "Biology of Termites," Vol. II, pp. 527–576 (K. Krishna and F. M. Weesner, eds.) Academic Press, New York.

Bakshi, B. K. 1960. Fungi in relation to termites. In: "Termites in the Humid Tropics." (Proc. Symp. New Delhi) UNESCO. Pp. 117–119.

Barbault, R. and M. E. Maury. 1981. Ecological organization of a Chihuahuan desert lizard community. Oecologia 51:335–342.

Batra, S. W. T. 1979. Nests of the eumenid wasp, Anterhynchium abdominale bengalense, from a termite mound in India. Orient. Insects 13:163–165.

Benemann, J. R. 1973. Nitrogen fixation in termites. Science 181:164–165.

Berger, A. J. 1961. "Bird Study." John Wiley & Sons, New York.

Bhatnagar, R. D. S. 1966. Anechura himalayana Singh (Dermaptera, Forficulidae), an earwig from Kashmir probably predacious on termites. Entomol. Monthly Mag. 102:129–130.

Bouillon, A. 1970. Termites of the Ethiopian region. In: "Biology of Termites," Vol. I, pp. 153–280 (K. Krishna and F. M. Weesner, Eds.), Academic Press, New York.

Breznak, J. A., W. J. Brill, J. W. Mertins, and H. C. Coppel. 1973. Nitrogen fixation in termites. Nature (London) 244:577–580.

Burtt, B. D. 1942. Some East African vegetation communities. J. Ecol. 30:65–146.

Campbell, A. G. 1968. Central Kalahari Game Reserve. II. Afr. Wildl. 22:321–327.

Casimer, M. 1960. Tegea atropica Stal, an unusual predator. Proc. Linn. Soc. N. S. W. 85:230–232.

Castlewitz, M. 1976. Insects as human food. Insect World Dig. 3:11.

Cheo, C. C. 1948. Notes on fungus growing termites in Yunnan, China. Lloydia 2:139–147.

Cleghorn, P. R. 1896. White ants. J. Agri. Soc. India 10:527–533.

Coe, M. 1977. The role of termites in the removal of elephant dung in the Tsavo (East) National Park Kenya. East Afr. Wildl. J. 15:49–55.

DeBont, A. F. 1964. Termites et densite d'oiseaux. In: "Etudes sur les termites africains." Pp. 273–283. (A. Bouillon ed.) Masson et Cie, Paris.

Deligne, J., A. Quennedey, and M. S. Blum. 1981. The enemies and defense mechanisms of termites. In: "Social Insects," Vol. II, pp. 1–76 (H. R. Hermann, ed.), Academic Press, New York.

Dimmitt, M. A. and R. Ruibal. 1980. Exploitation of food resources by spadefoot toads (Scaphiopus). Copeia 1980:854–862.

d'Orey, F. L. C. 1975. Contribution of termite mounds to locating hidden copper deposits. Texas Instn. Min. Metall. (Sect. B.) 84:B150–151.

Dorst, J. 1974. "The Life of Birds," Vol. I. Columbia Univ. Press, New York.

Drummond, H. 1886. On the termite as the tropical analogue of the earthworm. Proc. R. Soc. Edinb. 13:137–146.

Emerson, A. E. 1938. Termite nests. A study of the philogeny of behavior. Ecol. Monogr. 8:247–284.

Emerson, A. E. 1939. Social coordination and the supraorganism. Amer. Midl. Nat. 21:182–209.

Emerson, A. E. 1942. The modern naturalist. Transyl. Coll. Bull. 15:71–77.

Emerson, A. E. 1947. Why termites? Sci. Monthly 64:337–345.

Emerson, A. E. 1949. Termite studies in the Belgian Congo. Deuxieme Rapport Annuel de l'Inst. Recherche Sci. Afrique Centrale, pp. 149–160.

Emerson, A. E. 1952a. The biogeography of termites. *In:* "The problem of land connections across the South Atlantic, with special reference to the Mesozoic". *Bull. Amer. Mus. Nat. Hist.* 99:217–225.

Emerson, A. E. 1952b. The supraorganismic aspects of the society. *In:* "Structure et physiologie des societes animales." Centre. Nat. Recher. Sci., Paris 34: 333–354.

Emerson, A. E. 1956. Ethospecies, ethotypes, taxonomy and evolution of *Apicotermes* and *Allognathotermes*. *Amer. Mus. Nov.* 1771, 1–31.

Ettershank, G., N. Z. Elkins, P. F. Santos, W. G. Whitford, and E. F. Aldon. 1978. The use of termites and other soil fauna to develop soils on strip mine spoils. US Dept. Agri., Forest Ser. Res. Note. RM-361: 4 pp.

Ferrar, P. and J. A. L. Watson. 1970. Termites associated with dung in Australia. *J. Aust. Entomol. Soc.* 9:100–102.

Fowler, H. G. 1980. Populations, prey capture and sharing, and foraging of the Paraguayan ponerine *Odontomachus chelifer* Latreille. *J. Nat. Hist.* 14:79–84.

Froggatt, W. W. 1903. The White Ant City. A Nature Study. *Agri. Gaz. N. S. W.* 14:726–730.

Gehlbach, F. R., J. F. Watkins, and J. C. Kroll. 1971. Pheromone trail-following studies of typhlopid, leptotyphlopid, and colubrid snakes. *Behaviour* 40:282–294.

Genest-Villard, H. 1979. Ecologie de *Steatomys opimus* Pousargues, 1894 (Rongeurs, Dendromurides) en Afrique centrale. *Mammalia* 43:275–294.

Ghilarov, M. S. 1960. Termites of the USSR, their distribution and importance. *In:* "Termites in the Humid Tropics." (Proc. Symp. New Delhi) UNESCO, Paris, pp. 131–135.

Gillman, L. R., M. K. Jefferies, and G. N. Richards. 1972. Non-soil constituents of termite (*Coptotermes acinaciformis*) mounds. *Aust. J. Biol. Sci.* 25:1005–1013.

Glover, P. E., E. C. Trump, and L. E. D. Wateridge. 1964. Termitaria and vegetation patterns on the Loita plains of Kenya. *J. Ecol.* 52:367–377.

Goodall, J. 1964. Tool-using and aimed throwing in a community of free-living chimpanzees. *Nature (London)* 201:1264–1266.

Grassé, P. P. and C. Noirot. 1961. Nouvelles recherches sur la systematique et l'ethologie des termites champignonnistes du genre *Bellicositermes* Emerson *Insect. Soc.* 8:311–357.

Griffith, G. 1938. A note on termite hills. *East Afr. Agri. J.* 4:70–71.

Gunther, J. 1955. "Inside Africa." Harper, New York.

Harris, W. V. 1961. "Termites, Their Recognition and Control." Longman, Green, and Co., London.

Hickin, N. E. 1971. "Termites: A World Problem." Hutchinson, London, p. 24.

Hindwood, K. A. 1959. The nesting of birds in the nests of social insects. *Emu* 59: 1–36.

Holt, J. A., R. J. Coventry and D. F. Sinclair. 1980. Some aspects of the biology and pedological significance of mound-building termites in a red and yellow earth landscape near Charters Towers, North Queensland, *Aust. J. Soil Res.* 18:97–109.

Holt, J. A. and Greenslade, P. J. M. 1979. Ants in mounds of *Amitermes laurensis*. *J. Aust. Entomol. Soc.* 18:349–361.

Howarth, E. 1966. Instincts and their vicissitudes—A discussion of ethological models. *J. Psychol. Res.* 10:110–115.

Howse, P. E. 1970. "Termites, a Study in Social Behavior." Hutchinson Univ. Lib. Pub., London.

Johnson, J. B. and K. S. Hagen. 1981. A neuropterous larva uses an allomone to attack termites. *Nature (London)* 289:506–507.

Jones, C. and Sabater-Pi. 1969. Sticks used by chimpanzees in Rio Mundi, West Africa. *Nature* (London) 223:100–101.

Jukubski, A. W. 1965. A critical revision of the families Margarodidae and Termitococcidae (Hemiptera: Coccididea). Trustees British Museum Nat. Hist. London, 1–187.

Kalshoven, L. G. E. 1958. Observations on the black termites, *Hospitalitermes* spp., of Java and Sumatra. *Insect Soc.* 5:9–30.

Kevan, D. K. McE. 1962. "Soil Animals." H. F. and G. Witherby Ltd., London.

Kirby, H. 1941–1949. Devescovinid flagellates of termites. I. The genus *Devescovina*. II. The genera *Caduceia* and *Macrotrichomonas*. III. The genera *Foaina* and *Parajoenia*. IV. The genera *Metadevescovina* and *Pseudodevescovina*. V. The genus *Hyperdevescovina*, the genus *Bullannympha*, and undescribed or unrecorded species. *Univ. Calif. Publs. Zool.* 45:1–421.

Kirkpatrick, T. W. 1957. "Insect Life in the Tropics." Longman, Green and Co. Ltd., London.

Kistner, D. H. 1969. The biology of termitophiles. In: "Biology of Termites," Vol. I, pp. 525–557 (K. Krishna and F. M. Weesner, eds.), Academic Press, New York.

Kozlova, A. V. 1951. Nitrate accumulation in Turkmenian termite mounds. *Soil Fertil.* 15:15.

Krishna, K. 1969. Introduction In: "Biology of Termites," Vol. I, p. 4 (K. Krishna and F. M. Weesner, eds.), Academic Press, New York.

Kruuk, H. and W. A. Sands. 1972. The aardwolf (*Proteles cristatus* Sparrman 1783) as predator of termites. *East African Wildlife* 10:211–227.

LaFage, J. P. and W. L. Nutting. 1978. Nutrient dynamics of termites. In: "Production Ecology of Ants and Termites" pp. 165–232. (M. V. Brian ed.) Cambridge Univ. Press, London.

Lee, K. E. and T. G. Wood. 1971. "Termites and Soils." Academic Press, London.

Lent, H. and P. Wygodzinsky. 1947. Contribuico as conhecimento dos *Reduviinae americanos* (Reduviidae, Hemiptera). *Rev. Brasil. Biol.* 7:341–368.

Longhurst, C., R. Baker, and P. E. Howse. 1979a. Termite predation by *Megaponera foetens* (FAB.): Coordination of raids by glandular secretions. *J. Chem. Ecol.* 5:703–719.

Longhurst, C., R. A. Johnson, and T. G. Wood. 1979b. Foraging, recruitment and predation by *Decamorium uelense* (Sanstchi) on termites in Southern Guinea Savanna, Nigeria. *Oecologia* 38:83–91.

Mathur, R. N. 1954. Insects and other wild animals as human food. *Indian For.* 80:429.

Mathur, F. N. 1960. Enemies of termites. *In:* "Termites of the Humid Tropics." (Proc. Symp. New Delhi) UNESCO, Paris, pp. 137–139.

McBeath, N. M. and W. C. McGrew. 1982. Tool use by wild chimpanzees to obtain termites at Mt. Assirik Senegal: The influence of habitat. *J. Human Evol.* 11:65–72.

McMahan, E. A. 1982. Bait-and-capture strategy of a termite-eating assassin bug. *Insect Soc.* 29:346–351.

Michener, C, D. and M. H. Michener. 1951. "American Social Insects." D. Van Nostrand Co., New York.

Mielke, H. W. 1978. Termitaria and shifting cultivation: The dynamic role of the termite in soils of tropical wet-dry Africa. *Trop. Ecol.* 19:117–122.

Miller, N. C. E. 1956. "The Biology of the Heteroptera." Leonard Hill Books, Ltd. London.

Moore, B. P. 1969. Biochemical studies in termites. *In:* "Biology of Termites," Vol. I, pp. 407–432 (K. Krishna and F. M. Weesner, eds.), Academic Press, New York.

Moreau, R. E. 1972. "The Palaearctic-African Bird Migration Systems." Academic Press, New York.

Nishida, T. and S. Uehara. 1980. Chimpanzees, tools, and termites: another example from Tanzania. *Curr. Anthropol.* 21:671–672.

Noirot, C. 1970. The nests of termites. *In:* "Biology of Termites," Vol. II, pp. 73–125 (K. Krishna and F. M. Weesner, eds.), Academic Press, New York.

Nutting, W. L. 1969. Flight and colony foundation. *In:* "Biology of Termites," Vol. I, pp. 233–282 (K. Krishna and F. M. Weesner, eds.), Academic Press, New York.

Odhiambo, T. R. 1958. Some observations on the natural history of *Acanthaspis petax* Stal living in termite mounds in Uganda. *Proc. R. Entomol. Soc. London (A)* 33:167–175.

Pages, E. 1970. On the ecology and adaptation of sympatric *Orycteropus* and pangolins in Gabon. *Biol. Gabonica* 6:27–92.

Pegler, D. N. and G. D. Pierce. 1980. The edible mushrooms of Zambia. *Kew Bull.* 35:475–492.

Pendleton, R. L. 1942. Importance of termites in modifying certain Thailand soils. *J. Amer. Soc. Agron.* 34:340–344.

Pengiliey, R. K. 1971. The food of some Australian anurans (Amphibia). *J. Zool. (London)* 163:93–103.

Pianka, E. R. 1970. Comparative autecology of the lizard *Cnemidophorus tigris* in different parts of its geographic range. *Ecology* 51:703–720.

Prince, J. H. 1971. "Animals in the Night." Thomas Nelson, Inc. New York.

Punzo, F. 1974. Comparative analysis of the feeding habits of two specis of Arizona blind snakes, *Leptotyphlops h. humilis* and *Leptotyphlops d. dulcis*. *J. Herpetol.* 8:153–156.

Robinson, J. B. D. 1958. Some chemical characteristics of "termite soils" in Kenya coffee fields. *J. Soil Sci.* 9:58–65.

Rood, J. P. 1975. Population dynamics and food habits of the banded mongoose.

East Afr. Wildl. J. 13:89–111.

Roonwal, M. L. 1970. Termites of the Oriental region. *In:* "Biology of Termites," Vol. II, pp. 315–356 (K. Krishna and F. M. Weesner, eds.), Academic Press, New York.

Root, C. and A. Root. 1978. Mysterious Castles of Clay. Benchmark Films Inc., New York.

Salick, J., R. Herrera and C. F. Jordan. 1983. Termitaria: Nutrient patchiness in nutrient-deficient rainforests. *Biotropica* 15:1–7.

Sands, W. A. 1969. The association of termites and fungi. *In* "Biology of Termites," pp. 495–524 (K. Krishna and F. M. Weesner, eds.), Academic Press, New York.

Saunders, D. A. 1979. The availability of tree hollows for use as nest sites by White-tailed Black Cockatoos (*Calyptorhynchus baudinii latirostris*). *Aust. Wildl. Res.* 6:205–216.

Scarborough, A. G. and B. E. Sraver. 1979. Predatory behavior and prey of *Atomosia puella. Proc. Entomol. Soc. Wash.* 81:639.

Schaefer, D. A. and W. G. Whitford. 1981. Nutrient cycling by the subterranean termite, *Gnathamitermes tubiformans* in a Chihuahuan desert ecosystem. *Oecologia* 48:277–283.

Schlippe, P. de. 1956. "Shifting cultivation in Africa: The Zande System of Agriculture." Routledge and Kegan Paul, London.

Seevers, C. H. 1957. A monograph on the termitophilous Staphylinidae (Coleoptera) Fieldiana. *Zoology* 40:1–334.

Sen, G. E. 1964. "The Story of Early Indian Civilization." Orient Longmans, New Delhi, p. 39.

Sen-Sarma, P. K. 1974. Ecology and biogeography of the termites of India. *In:* "Ecology and Biogeography in India," pp. 421–472 (M. S. Mani, ed.), Junk, The Hague.

Sen-Sarma, P. K. 1976. Termites. *In:* "The Wealth of India," p. 196. Publ. and Inform. Directorate, CSIR. New Delhi.

Sheppe, W. 1970a. Invertebrate predation on termites of the African savanna. *Insect Soc.* 17:205–218.

Sheppe, W. 1970b. Daily and seasonal patterns of construction activity by *Odontotermes latericius. Insect Soc.* 17:225–232.

Silvestri, F. 1938. Ridescrizione del genere *Termito coccus* Silv., con una specie nuova del Brasile e descrizione de un nuovo genere affine. *Bol. Lab. Zool. Portici* 30:32–40.

Silvestri, F. 1944. Nuovo genere di Lepidottero Tineide termitofilo del Brasile. *Boll. Lab. Entmol. Agr. Portici* 5:243–251.

Skutch, A. F. 1977. "A Bird Watcher's Adventures in Tropical America." Univ. of Texas Press, Austin.

Snyder, T. E. 1948. "Our Enemy the Termite." Comstock Press, New York.

Snyder, T. E. 1956–1968. Annotated Subject-Heading Bibliography of Termites, 1350 B.C. to A.D. 1954. Smithsonian Miscellaneous Collections. Wash. D.C. 1961. Supplement, 1955 to 1960. 1968 Supplement, 1961 to 1965.

Sugiyama, Y. and J. Koman. 1979. Tool-using and tool-making behavior in wild chimpanzees of Bassau, Guinea. *Primates* 20:513–524.

Sylvester-Bradley, R. A. Gomes-Bandeira and L. Antonio de Oliveira. 1978. Nitrogen fixation (reduction of acetylene) in termites of central Amazonia. *Acta Amazon. (Brazil)* 8:621–628.

Tamm, S. L. 1979. Membrane movements and fluidity during rotational motility of a termite flagellate: A freeze-fracture study. *J. Cell. Biol.* 80:141–149.

Tamm, S. L. and S. Tamm. 1974. Direct evidence for fluid membranes. *Proc. Nat. Acad. Sci. USA* 71:4589–4593.

Thiollay, J. M. 1970. L'exploitation par les oiseaux des essaimages de fourmis et termites dans une zone de contact savane-foret en Cote-d'Ivoire. *Alauda Rev. Int. Ornithol.* 38:255–273.

Thomas, A. S. 1941. The vegetation of the Sese Islands, Uganda. *J. Ecol.* 29:330–353.

Tihon, L. 1946. A propos des termites au point de vue alimentaire. *Bull. Agri. Congo Belge* 37:865–868.

Troll, C. 1936. Termitensavannen. *In:* "Landekundliche Vorschrift Festschrift fur Norbert Krebs," pp. 275–312. Engelhorn, Stuttgart.

Uehara, S. 1982. Seasonal changes in the techniques employed by wild chimpanzees in the Mahale mountains, Tanzania to feed on termites (*Pseudacanthotermes spiniger*). *Folia Primatol.* 37:44–76.

Usher, M. B. 1979. Markovian approaches to ecological succession. *J. Anim. Ecol.* 48:413–426.

Ward, P. and P. J. Jones. 1977. Premigratory fattening in three races of the red-billed quelea, *Quelea quelea* (Aves: Ploceidae) an intra-tropical migrant. *J. Zool. (London)* 181:43–56.

Watson, J. P. 1972. The distribution of gold in termite mounds and soils at a gold anomaly in Kalahari sand. *Soil Sci.* 113:317–321.

Weir, J. S. 1969. Chemical properties and occurrence on Kalahari sand of salt licks created by elephants. *J. Zool.* 158:293–310.

Weir, J. S. 1972. Spatial distribution of elephants in an African National Park in relation to environmental sodium. *Oikos* 23:1–12.

Wheeler, W. M. 1911. The ant-colony as an organism. *J. Morphol.* 22:307–325.

Wheeler, W. M. 1938. Ecological relations of Ponerine and other ants to termites. *Proc. Amer. Acad. Arts Sci.* 71:159–243.

White, L. P. 1970. Brousse tigree patterns in southern Niger. *J. Ecol.* 58:549–553.

Wille, A. and C. D. Michener. 1973. The nest architecture of stingless bees with special reference to those of Costa Rica. *Rev. Biol. Trop.* 21: (Suppl. 1).

Williams, R. M. C. 1959. Flight and colony foundation in two *Cubitermes* species. *Insect Soc.* 6:203–218.

Wolcott, G. N. 1948. The insects of Puerto Rico. *J. Agr. Univ. Puerto Rico* 32:1–224.

Wood, T. G. 1976. The role of termites in decomposition processes. *In:* "The Role of Terrestrial and Aquatic Organism in Decomposition Processes," pp. 145–168 (J. M. Anderson and A. MacFadyen, eds.), Blackwell Scientific, Oxford.

Wood, T. G. and W. A. Sands. 1978. The role of termites in ecosystems. *In:* "Production Ecology of Ants and Termites," pp. 245–292 (M. Brian, ed.), Cambridge Univ. Press, Cambridge.

Yadgarov, T. 1974. Feeding of the scheltopusik *Ophisaurus apodus* (Pallas) in the basin of the Surkhan–Darya River, UZB. *Biol. Z.* 18:68–70.

Zur-Strassen, W. H. 1967. Protection of vermin in Southwest Africa. *Afr. Wild.* 21:144–147. (Abstract).

Chapter 6

THE ECONOMIC IMPORTANCE AND CONTROL OF LEAF-CUTTING ANTS

J. M. Cherrett

INTRODUCTION

Leaf-cutting, fungus-growing ants of the genera *Atta* and *Acromyrmex* have a formidable reputation in the New World tropics and subtropics as pests. Mariconi (1970) quotes a Portuguese saying attributed to the French naturalist St. Hilaire traveling in South America [during 1816–1822] "Ou o Brasil mata a saúva ou a saúva mata o Brasil" (Either Brazil kills the sauva or the sauva will kill Brazil). He also reports Clark as saying in 1867 "O Brasil e um grande formigueiro" (Brazil is one giant ant colony). In a similar vein, Belt (1874), describing his experiences in Nicaragua, wrote ' . . . they are one of the greatest scourges of tropical America, and it has been too readily supposed that their attacks cannot be warded off."

Today, using modern methods, farmers would not take as pessimistic a view of the damage caused by leaf-cutting ants as these pioneer European settlers in South America. Cherrett (1982) attributes this change of view to the success of modern control techniques. When assessing the economic importance of leaf-cutting ants today, it is necessary to distinguish between their potential importance, as experienced by those first farmers equipped with virtually no chemical defenses, and their actual importance in, for example, a well-run citrus orchard with access to modern poison baits. To illustrate this, he quotes a series of studies on the economic loss to citrus growers in Trinidad caused by leaf-cutting ants.

Lewis and Norton (1973) estimated the average annual loss to the island from 1968 to 1971 at $27,600 U.S. dollars,* taking into account

---

*All dollar values hereafter are U.S. dollars.

only the cost of replacing the young trees which the ants killed. This is the most obvious and easily assessed form of damage although there were undoubtedly other yield losses resulting from the partial defoliation of mature bearing trees that were not assessed. Cherrett and Sims (1968), working in the same area, estimate losses at $169,700 based on a somewhat larger estimate of the citrus acreage. This figure includes the costs of the control measures being taken against the ants at the time the damage was recorded, in effect estimating the savings to be expected by citrus growers if leaf-cutting ants were eliminated. Finally, Cherrett and Jutsum (1983) were able to obtain some idea of the potential losses in Trinidad citrus orchards in the absence of any control, by following the fate of an orchard four to seven years after it had been abandoned, consequent on changes in the economics of citrus growing in the island. They report that of 108 trees abandoned in 1972, 20 (19%) had died by 1976, and well in excess of 70 (65%) by 1979. They attribute these losses to repeated defoliation by *Atta cephalotes*, whose populations increased enormously once control was stopped. They claim that without ant control, citrus cultivation would not be possible. This conclusion confirms Belt's (1874) comments "Again and again have I been told in Nicaragua, when inquiring why no fruit trees were grown at particular places, 'It is no use planting them; the ants eat them up.'" In 1967, the value of the citrus industry in Trinidad was estimated to be $3.2 million, which represents the potential loss to leaf-cutting ants in the absence of any control.

The statement can be argued that the success of a pest control scheme is best assessed by comparing present losses with potential losses. The control effectiveness index (CEI) can be determined as follows:

$$CEI = \frac{(\text{Potential loss} - \text{current damage cost} - \text{current control cost}) \times 100}{\text{Potential loss}}$$

The control effectiveness index can range from a minus percentage value, where current losses plus treatment costs actually exceed the likely losses if no control is attempted (e.g., when intensive insecticide use reduces the effectiveness of natural enemies in suppressing pest population growth, and induces resistance to the insecticide on the part of the pest), to very high values approaching 100%.

In the case of leaf-cutting ant control in Trinidad citrus in the late 1960s and early 1970s, the index would be 100 ($3,200,000 – $169,700)/ $3,200,000 = 95%. This is a very high figure, which emphasizes the gratifying degree of success that pesticides have had in reducing the ant problem to manageable proportions, and the marked distinction between actual and potential pest status. However, there are also dangers inherent in such a high index:

1. As actual pest status is often low in well-managed orchards, it is easy to forget the ants' potential pest status. This danger was not emphasized by either Cherrett and Sims (1968) or Lewis and Norton (1973).
2. Increases in potential pest status, as greater citrus acreages are planted in vulnerable areas, may not be fully appreciated. Leaf-cutting ants are no longer a key factor in determining where crops should be grown.
3. The marked difference between actual and potential pest status, which we have come to accept as the norm, is at present entirely maintained by the use of chemical pesticides. Accordingly, it is very vulnerable to changes that may occur in their cost or effectiveness, or to new perceptions of chemical use resulting from environmental or human health considerations.

These general considerations about the economics of leaf-cutting ant pest status and control have been illustrated by reference to citrus growing in Trinidad in the late 1960s and early 1970s. Since then, the ants have become less important as the oil price boom has reduced the significance of agriculture to the country's economy, undermining the citrus industry. However, the lessons to be drawn have wider application for ant problems in agriculture, forestry, and range management, and only the data are lacking as the following section demonstrates.

TYPES AND EXTENT OF DAMAGE CAUSED

Since there are 37 species of leaf-cutting ants, some of which occur throughout most of the American mainland from the southern United States to northern Argentina, a wide range of crops is affected. Because the workers only drink sap from the leaves they cut (Littledyke and Cherrett 1976) and rely on the fungus they cultivate as the major protein source, they can circumvent the chemical defense mechanisms of many plants. This permits an exceptional degree of polyphagy and Cherrett (1980) considers it to be an important reason for the evolution of the fungus-growing habit. Consequently, few crops grown by man in the neotropics and subtropics entirely escape leaf-cutting ant damage.

Agricultural and Horticultural Crops

In a questionnaire survey of 27 countries, Cherrett and Peregrine (1976) found that 47 crops are reported as being damaged by leaf-cutting ants, 14 being mentioned more than once. A survey of the literature reveals 26 crops damaged, of which 17 were mentioned more than once. If the frequency with which a crop is cited is used to indicate the significance of leaf-cutting ants in its cultivation, it will be seen from Table 6.1 that both lists agree on the six most vulnerable crops, citrus heading both lists.

TABLE 6.1

The Agricultural and Horticultural Crops Most Frequently Recorded as Being Damaged by Leaf-Cutting Ants[a]

From a questionnaire survey[b] *(1972)*		*From a literature survey*	
Citrus	(17)	Citrus	(32)
Coffee	(11)	Cocoa	(25)
Maize	(10)	Manioc	(13)
Cotton	(8)	Coffee	(11)
Manioc	(8)	Maize	(7)
Cocoa	(7)	Cotton	(6)
Mango	(5)	Rice	(6)
Roses	(5)	Peaches	(4)
Beans	(4)	Roses	(4)
Sweet potatoes	(3)	Mango	(3)
Groundnuts	(2)	Coca	(3)
Plantains	(2)	Cabbage	(3)
Rice	(2)	Coconuts	(3)
Wheat	(2)	Sugarcane	(3)
		Avocado	(2)
		Wheat	(2)
		Legumes	(2)

[a]Only crops reported more than once are recorded; the number of citations appear in parentheses.
[b]After Cherrett and Peregrine (1976).

Cherrett (1982) suggests that the predominance of trees and shrubs in the six (50%) most vulnerable crops may be because of the relatively small amount of tillage required for these semipermanent crops. The lack of tillage permits a slow buildup of leaf-cutting ant populations in undisturbed nests.

Some agricultural crops are virtually immune to attack: limes are rarely cut, possibly because of their pungent leaf oils, and sisal and pineapple foliage escape, presumably because of their toughness. Belt (1874) has remarked that man's agricultural crops seem much more susceptible to ant attack than the natural vegetation they replace. This frequently results in an increased ant population and may also be accompanied by changes in the species present (Cherrett 1968). This vulnerability of crop plants has recently been attributed to a combination of factors: less effective chemical deterrents, lack of latex, reduced toughness, and man's interference with defensive ant faunas (Cherrett 1981).

Few detailed studies appear to have been carried out on the actual losses caused by leaf-cutting ants to agriculture. The detailed studies on citrus loss in Trinidad have already been mentioned. Combining similar studies on cocoa in Trinidad suggests a national annual loss of $160,000 (4% of the crop value) for the same period (Cherrett and Sims 1968, Lewis and Norton 1973). Troppmair (1973) carried out a detailed study of eight farms in São Paulo State in 1962, and for 11 crops, he records an average loss of 4.2% due to leaf-cutting ant activities. Amante (1972) claimed that a mixed population of *Atta capiguara* and *A. laevigata* nests at an average density of 2.4 per ha will take up to 4 tons fresh weight of sugar cane per hectare per year. He estimates annual sugar cane losses in São Paulo State at $6.3 million. Schade (1973) reported that farmers in the Acahay district of Paraguay are virtually prohibited from planting cotton and watermelons and experience crop losses of 40–45%, and Robinson (1979) gives a figure for annual losses over the whole country of between $6.3 and $7.9 million.

Pastures

Cherrett and Peregrine (1976) found 13 species of range plants damaged by leaf-cutting ants in the 27 countries surveyed, although only three receive more than one mention. In the literature survey (Cherrett 1982) pastures are the third most frequently cited crop after citrus and cocoa.

The extent to which leaf-cutting ants reduce the carrying capacity of pastures for cattle by competing with them for grass is still uncertain. Casual observation of such grass-cutting species as *Acromyrmex landolti*, *Atta capiguara*, and *A. vollenweideri* reveal the impressive volume of grass being carried into the nest. However, not all vegetation taken in is necessarily the preferred food of cattle, and short-term observations during daylight in favorable seasons may give a distorted idea of the annual harvest. Cherrett et al. (1974) estimated that an average *Acromyrmex landolti* nest in Guyana harvests 5.96 g in 24 hours based on a single 24-hour count. Fowler and Robinson (1975), in a one-month study of the same species in Paraguay, estimated the daily harvest at 4.23 g. Jonkman (1980) criticized these short-term studies, and instead employed a technique first used by Autuori (1947). He weighed the accumulated spent fungus garden substrate obtained by excavating large *Atta* nests and then estimated the original amount of grass cut using a conversion factor. On this basis, he concludes that most estimates of the amount of grass taken are greatly exaggerated. Robinson and Fowler (1982) produced a revised figure of 2.33 g for a mean 24-hour harvest based on counts made periodically throughout the year. Although this confirms the validity of Jonkman's (1980) warning, Robinson and Fowler feel that his method of calculating the intake of *Atta vollenweideri* nests produces gross underestimates by at least a factor of 2. There is clearly a need for long-term (12-month) regular observations on the species, quantity, and age of pasture plants cut by ants.

As large numbers of *Acromyrmex landolti* nests per hectare are recorded in *Panicum maximum* pastures (4,400 in Paraguay, 5,930 in Venezuela, and 6,000 in Peru (see Cherrett 1981) considerable quantities of grass are undoubtedly taken. Amante (1972) records up to 64 *Atta* (mainly *A. capiguara*) nests per hectare, and quotes mean figures of 13 per ha for the *Alta sorocabana* and ten per ha for the *Alta paulista* pastures of São Paulo State. On the assumption that ten nests consume as much grass as one cow, he estimates that the carrying capacity of the pastures in these regions was being reduced by 1,071,000 head of cattle in 1970. At a value of $66 per cow, this represents a loss of $70 million in potential production. Jonkman (1979), studying *Atta vollenweideri* in the Paraguayan Chaco, obtained maximum infestation rates of 4.5 per ha. From an aerial photograph survey of 80,000 km^2, he concludes that even in the infested areas, average densities are around 0.2 per ha. Assuming 20 nests are equivalent to one cow in terms of grass consumption, he concludes that this species is not a significant pest of the Paraguayan Chaco (Jonkman 1980). This interpretation has been challenged by Robinson and Fowler (1982).

Throughout South America, the forest is being felled to increase pasture land, and the productivity of the pasture is being raised by fertilization and the cultivation of higher-yielding grass cultivars. Both will increase the potential losses from grass-cutting ants.

Forests

Cherrett and Peregrine (1976) found that 15 species of forest trees are cited as suffering leaf-cutting ant damage, but that only pines and teak are mentioned more than once. In the literature, eucalyptus, *Gmelina*, and rubber are also mentioned. Several studies have been conducted on the role of leaf-cutting ants in natural forests (Cherrett 1983, Rockwood 1975). Lugo et al. (1973) have calculated that they may take 0.2% of the gross productivity, an impact that reduces potential productivity by 1%. This loss is of no practical significance as long as forestry is based on extracting timber from natural stands, a practice in South and Central America until relatively recent times. Once plantations are established, the situation is transformed especially where exotic trees are involved. Eucalyptus and pines are now widely grown in South America, and suffer severely from *Atta* defoliation. From calculations based on average nest density and the amount of leaves taken into a nest, Amante (1972) estimates annual wood losses at 17.5 million m^3 for eucalyptus and 3.5 million m^3 for pines in the state of São Paulo. This represents an annual loss of $85 million. Jonkman (1980) rightly criticizes these estimates on the grounds that leaf loss cannot be readily translated into wood loss. However, in terms of potential loss, these figures could be an underestimate, since eucalyptus cannot be established in many areas without leaf-cutting ant control (Vaughan 1946). A eucalyptus plantation in Brazil, planted in 1936 with 143,437 trees, suffered a 26% loss from leaf-cutting ants (Troppmair 1973). Mendes Filho (1979) reported that eucalyptus dies after the third successive ant defoliation, and estimates that a plantation with four *Atta* nests per hectare loses 14% and one with 200 *Acomyrmex* nests per hectare suffers damage amounting to 30%.

In the Rio Jari scheme in Amazonia, where some 90,000 ha have been planted to *Gmelina arborea* and *Pinus caribaea*, Ribeiro and Woessner (1979) claim that leaf-cutting ants are the major entomological problem, with nest densities of two to 30 per ha. After studying six control techniques, they found mirex bait to be the cheapest, a $40 per ha. Although control is mainly needed during the early years of plantation establishment, it seems likely that the annual cost in this forestry project must have been between $0.3 and $1.0 million per annum.

Since leaf-cutting ants are responsible for so many different types of losses to so many different industries over such a wide geographical area, it is much more difficult to produce a realistic global assessment of their impact than it would be for monophagous herbivore attacking a single agricultural crop in a limited area. Even minor damage, when it is inflicted on many crops, adds up to a considerable loss. The most commonly quoted global estimate of leaf-cutting ant damage is $1 billion per annum (Cramer 1967). It was originally suggested in 1923 by Townsend for total damage in tropical America, and has been used ever since without any correction for inflation. If this figure is meant to refer to potential loss then; as Cherrett (1982) points out, it is justifiable if only because the value of the citrus industry in countries with leaf-cutting ant problems could be two or three times as much (FAO 1979).

At the present time, we simply do not have enough data to attempt any global estimate of the extent of potential or actual leaf-cutting ant losses; all we can say is that they rank among the worst animal pests in the New World subtropics. This view is substantiated by the questionnaire survey (Cherrett and Peregrine 1976) where the percentages of correspondents placing them among the 20 worst national pests are: agriculture, 76%; forestry, 36%; and range management, 30%.

THE SPECIES INVOLVED

Although restricted to the New World, there are 14 species of *Atta* and 23 of *Acromyrmex*, collectively found between the southern United States and northern Argentina (Weber 1972). Some species are more restricted geographically in their distribution (Cherrett and Peregrine 1976); some cut only dicotyledonous plants, others only monocots, while a few can exploit both (Table 6.2). As a consequence, some species are widely known as pests, reflected by the large number of common names they have, while others have only a local reputation for damage. A survey of the literature (Table 6.2) shows that of the ten species most frequently cited as being pests, six belong to the genus *Atta*, only one is an exclusive cutter of monocots, and seven cut only dicots. Accordingly, the most typical leaf-cutting ant pest species is one which is widespread, cuts dicots, and forms very large nests. *Atta cephalotes* of central and northern South America and *Atta sexdens* of South America are the most famous examples.

With the adoption of agriculture, there is commonly a shift in species composition compared with the virgin vegetation that preceded it. *Atta capiguara*, *Atta sexdens*, *Acromyrmex landolti*, and *Acromyrmex octospinosus* are all reported as followers of man-made vegetation (Cherrett 1981).

CONTROL TECHNIQUES

Only in unusual cases, such as the accidental introduction of *Acromyrmex octospinosus* into the Caribbean island of Guadeloupe in the early 1950s (Blanche 1961), is the objective of control techniques the eradication of an ant from a large area. In most cases this has been impractical even if, given conservation considerations, it was thought desirable. Even in Guadeloupe, attempts to eradicate the introduced population before it spread have not succeeded. The objective is usually to control damage, not necessarily ant populations, and this involves regulating damage so as not to exceed acceptable levels. Attempts to do this have a long history, admirably reviewed by Mariconi (1970). In this section we examine the various techniques available.

Growing Resistant Crops

If damage to desirable crops cannot be limited, then the problem can be minimized by growing only those crops with some innate resistance. This was the traditional approach, as farmers soon discovered which crops were rewarding to cultivate and which were not, and is illustrated by Belt's (1874) observation in Nicaragua that farmers did not bother to grow fruit trees in some areas because of leaf-cutting ant depredations. Pena (1974), confronted with damage by *Acromyrmex landolti* to *Panicum maximum* pastures in Peru, set up a replicated resistance trial with 26 grass

TABLE 6.2
Some Indicators of the Pest Status of Leaf-Cutting Ant Species

Species	No. countries affected	Type of vegetation attacked	No. pest literature citations	No. local names[a]
1 *Atta bisphaerica* For.	1	M	11	4
2 *A. capiguara* Gonçalves	2	M	9	3
3 *A. cephalotes* (L.)	17	D	48	22
4 *A. colombica* Guérin	5	D	1	1
5 *A. goiana* Gonçalves	1	D & M	3	0
6 *A. insularis* Guérin	1	D	11	1
7 *A. laevigata* (F. Smith)	6	D & M	21	5
8 *A. mexicana* (F. Smith)	6	D	4	4
9 *A. opaciceps* Borgm.	1	D	6	1
10 *A. robusta* Borgm.	1	D	4	1
11 *A. saltensis* For.	3	D	2	0
12 *A. sexdens* (L.)	14	D	100	18
13 *A. texana* (Buckley)	2	D	20	6
14 *A. vollenweideri* For.	5	M	8	5
1 *Acromyrmex ambiguus* (Emery)	4	D	2	3
2 *A. aspersus* (F. Smith)	4	D	2	1
3 *A. coronatus* (Fabricius)	10	D	6	1
4 *A. crassispinus* (For.)	3	D	3	2
5 *A. disciger* (Mayr)	2	D	3	2
6 *A. gallardoi* Santschi	1	?	0	0
7 *A. heyeri* (For.)	4	M	5	2
8 *A. hispidus* Santschi	6	D	5	5
9 *A. hystrix* (Latreille)	5	D	1	1
10 *A. landolti* (For.)	9	M	13	11
11 *A. laticeps* (Emery)	5	D	3	2
12 *A. lobicornis* (Emery)	5	D & M	5	3
13 *A. lundi* (Guérin)	5	D	8	3
14 *A. mesopotamicus* Gallardo	1	?	0	0
15 *A. niger* (F. Smith)	1	D	3	3
16 *A. nobilis* Santschi	1	?	0	0
17 *A. octospinosus* (Reich.)	20	D	21	6
18 *A. pulvereus* Santschi	1	?	0	0
19 *A. rugosus* (F. Smith)	7	D	4	4
20 *A. striatus* (Roger)	5	D & M	11	5
21 *A. subterraneus* (For.)	6	D	17	5
22 *A. sylvestrii* Emery	3	?	0	0
23 *A. versicolor* Pergande	2	D	1	1

No. (%) species cutting	Atta	Acromyrmex
Dicotyledons (D)	9 (64)	14 (78)
Monocotyledons (M)	3 (21)	2 (11)
Both (D & M)	2 (14)	2 (11)

[a]Cherrett, unpublished

173

cultivars: 16 (62%) proved susceptible, seven (27%) resistant, and three (12%) failed to germinate. Mariconi (1970) has even advocated reforestation as a means of controlling the advance of the grass-cutting ant *Atta capiguara*, which does not attack dicotyledonous trees.

Refraining from growing desirable crops because of leaf-cutting ant damage is an admission of defeat and does not help the people who need the crops. Breeding resistance to leaf-cutting ant damage into desirable crops is possible in principle but in practice there are major difficulties. As leaf-cutting ants are polyphagous they must be capable of dealing with the chemical defenses of a wide range of plant species. Accordingly, selective breeding that produces minor changes in the biochemistry of the crop is less likely to produce resistance to ant attack than to the attack of more specialist herbivores. Physical defenses such as leaf toughness (Cherrett 1972a) and the possession of sticky leaf latex (Stradling 1978) deter ant attack; but as the ants possess large, heavily chitinized jaws, major changes in physical defenses will usually be required. These may, however, be undesirable for crop utilization.

Preventing Access to Crops

If the ants can be prevented from reaching susceptible crops, no damage will occur. Nelson (1951) states that concrete moats filled with oil have been constructed around nests in an attempt to contain the foraging workers, and Harrison et al. (1916) report the use of water-filled trenches. In addition to being labor-intensive, these are unlikely to be effective, as *Atta* spp. dig underground tunnels that could easily pass under such a moat. Where tree crops are to be protected, the task is easier. Foliage susceptible to damage can only be reached if the ants climb up the trunk. Circular ceramic troughs called "ant pots" filled with water or oil have been placed around the base of saplings and ornamentals such as roses for many years (Sampaio de Azevedo 1894). Similarly, metal collars (Wille 1929) or grease bands (Weyrauch 1942) can be fixed around the trunks. Such time-consuming methods are only practical for protecting specimen plants in gardens, as the devices have to be inspected regularly to prevent fallen leaves and twigs from bridging the impassable zone. They are also ineffective where canopies interlock with unprotected vegetation.

There is no detailed information on how leaf-cutting ant scouts wandering over the ground locate plant stems, nor their precise reaction when they find them. Belt (1874) describes how Indians tie "grass-skirts" around tree trunks, Barreda (1922) reports the use of flax, and Wille (1943) states that fresh banana leaves, tied around trunks, repel ants provided they are renewed every two or three days. As leaf-cutting ants undoubtedly cut away grass or banana leaves when clearing trails through leaf litter or grass, these devices probably work by providing physical or chemical camouflage. They conceal the existence of the trunk from the scout ants which do not persist in exploring the plant further.

A final method of preventing leaf-cutting ants from gaining access to susceptible trees is to encourage defensive ant faunas to live in the trees and repel them. This possibility is mentioned by Eberhard and Kafury (1974) and Leston (1978). Jutsum et al. (1981) demonstrated that *Azteca* sp. largely prevent *Atta cephalotes* from defoliating the trees they inhabit. However, since this species tends scale insects and attacks farm laborers attempting to pick the fruit, it is not suitable as a control agent. Other defensive ant species may present fewer problems to the farmer, and foresters could well find this technique of value in protecting stands of a vulnerable age.

Biological Control

In nature many predators, parasites and diseases act to reduce the numbers of leaf-cutting ants. Weber (1972) lists two parasites, 39 predators, and parasitoids ranging from phorid flies to birds and mammals, and two fungus garden feeders. During and immediately after their mating flight, young queens are especially vulnerable to predators. In some areas man eats them (Conconi 1982). Once the nests become larger, however, the soldier caste in *Atta* aggressively defend them. This makes digging out an *Atta* nest a memorable experience and consequently, there are few predators of mature nests. Individual foraging workers are taken although they are spiny and heavily chitinized. Armadillos dig into mature nests, but usually destroy only a small percentage of the fungus garden (Schade 1973). Unless the queen is killed, the nest quickly regenerates. Some driver ants invade leaf-cutting ant nests and remove the brood but again this does not necessarily kill the colony (Fowler 1977).

Despite the activities of this wide range of predators, more than enough young nests survive to replace the old nests.

Considerable interest has been shown in the possibility of control using entomophagous fungi and Argollo Ferrão (1925) imported cultures of *Botrytis bassiana* and *Isaria densa* in an attempt to kill leaf-cutting ants. Natural epizootics of *Metarrhizium anisopliae* have been reported from *Atta sexdens* nests (Allen and Buren 1974) and of *Cordyceps unilateralis* from *Atta cephalotes* (Andrade 1980). Kermarrec and Mauleon (1975) studied the possible use of *Entomophthora coronata* against *Acromyrmex actospinosus*, finding it effective against worker ants isolated in dishes in the laboratory. The conidia were sufficiently resistant to make field application feasible but, as the fungus is a possible rhinophycomycosis agent for vertebrates, they warn that its use could be dangerous in tropical areas. However, leaf-cutting ants maintain fairly pure cultures in their fungus gardens despite constant bombardment with the spores and mycelial fragments of other fungi. Thus the effectiveness of entomophagous fungi in complete colonies may not be the same as that on individual workers without their fungal symbiont and imprisoned in conditions of temperature and humidity that they cannot control. Disease manipulation is clearly an area that needs further research.

Killing Foraging Workers

Confronted with a trail of leaf-cutting ants defoliating crop plants, the first reaction is to destroy them, hoping to deter subsequent workers following the trail. Belt (1874) sprinkled corrosive sublimate (mercuric chloride) in their tracks and for a while this stopped further damage. Schade (1973) mentions a similar use of Paris green (copper aceto-arsenite) and lead arsenate, while Mariconi (1970) records the use of HCH and paradichlorobenzene. Barreda (1922) recommends the application of calcium arsenate and chalk in the form of a ridge around each entrance hole. Fenjves (1950) advocates spraying chlordane around nests, along trails, and onto the trunks of susceptible trees. Such treatments may divert foraging workers away from a particular crop for a short time and, if all workers leaving the nest are poisoned, then in time the colony will decline by attrition. However, as a large *Atta* colony may contain several million workers (Jonkman 1980), and as the fungus gardens provide a buffer of stored food, getting rid of nests in this way is extremely slow.

Physically Destroying Nests

A traditional way of destroying leaf-cutting ant nests it to "puddle" the nest (Harrison et al. 1916, Dinther 1958). Provided the soil is not too sandy, this involves flooding the nest surface with water and then churning the surface up mechanically by treading or, more enjoyably, by dancing. In theory, the fungus gardens, brood, and ants are destroyed in the mud. This is only useful for species such as *Atta cephalotes*, which nest superficially. *Atta vollenweideri* nests, in areas of the Paraguayan Chaco liable to flooding in the rainy season, have their entrance holes opening onto the tops of raised turrets. As the nests penetrate the impermeable surface soils to considerable depths, local farmers are reputed to break open the surface mounds during floods to allow the water to run down into the nest and kill it. Nelson (1951) records that nests are sometimes dug out, and Bates (1884) reported that gun powder has been used to blow them up. All these methods are time-consuming and uncertain in their effects.

Chemically Destroying Nests

Almost all chemicals with any pretentions to insecticidal activity have been tested as gasses, smokes, dusts or liquids against leaf-cutting ant nests. It has been suggested that fungicides might be employed to kill the fungus gardens and so starve the ants, but in practice insects are easier to kill than are mats of fungal hyphae. Consequently, all the toxicants in use are insecticides, although carbon disulfide also acts as a fungicide (Walton et al. 1938). The ten chemicals most commonly used together with the formulations employed and key references containing details of dose rates and efficiency appear in Table 6.3. Since *Atta* nests are large and irregularly shaped it is difficult to get the toxicant to permeate the whole nest. If the queen is in an unaffected portion, the colony will recover, although the disturbance may cause it to change its nest site. This problem led Jacoby (1935, 1955) to carry out extensive research on nest structure, although the results have had little impact on control practice. Four methods of formulating the toxicants have been employed.

TABLE 6.3
References to Some of the More Commonly Used Chemicals for Destroying Leaf-Cutting Ant Nests

Chemical	Formulation			
	Dust	*Gas*	*Liquid*	*Smoke*
Aldrin	Amante 1967	—	Gonçalves 1957	Kennard 1965
B.H.C. (H.C.H.)	Vanetti 1958	—		Kennard 1965
Carbon disulfide	—		Vanetti 1958	—
Chlordane	Amante 1967		Fenjves 1950	Kennard 1965
Dieldrin	Vanetti 1958	—	Proctor 1957	Kennard 1965
Gasoline/kerosene	—	—	Valles and Meer 1974	—
Heptachlor	Amante 1967	—	—	—
Hydrogen cyanide	—	Amante 1967	—	—
Methyl bromide	—		Vanetti 1958	—
Sulfur and arsenic	—	Autuori 1942	—	—

Dusts

Dusts containing the toxicants in an inert carrier can be blown down the principal entrance holes of the nest using some form of pump (Mariconi 1970). The dust will then spread through the interior chambers and tunnels and, provided the toxicant is a stable contact insecticide, ants which walk over the contaminated areas die. The quantities required are based on the area of the excavated soil heaps over the nest; Mariconi and Amante (1966) recommend 30g per m^2. With species such as *Atta sexdens*, where the ants may emerge from holes more than 100 m from the site of the nest mound using long subterranean tunnels, care must be taken to ensure that the dust is pumped down into the central nest region. When the treatment is carried out carefully, toxic dusts can be cheap and effective.

Gases

Gases can diffuse readily into the structure of nests, killing by fumigation, although once the gas has dispersed there is little residual effect. One of the earliest methods of killing nests was to burn a mixture of sulfur and white arsenic on glowing charcoal and blow the resulting fumes into the nest holes. Mariconi (1970) reports on some of the apparatuses used. In a prize competition held in Brazil in 1920 for the most effective chemicals and machinery for killing *Atta sexdens* nests, there were 17 competitors (Anonymous 1921). Autuori (1942) found in the laboratory that burning sulfur alone to give sulfur dioxide gas gives equally good results. Methyl bromide and carbon disulfide have been the two most widely used fumigants. They are introduced as volatile liquids at rates of 4 and 75 cc/m^2, respectively (Mariconi and Amante 1966). The liquid is poured into natural entrance holes, or down artificial holes made with probing rods (Vanetti 1958). All holes are then sealed with soil to keep in the gas. When properly applied, methyl bromide gives excellent control. With carbon disulfide, the temptation to ignite the gases and watch the nest explode is often too great for some to resist (Pollard 1981), but leaving the gas to fumigate the nest is probably more effective.

Liquids

Gasoline and kerosene have been poured into nests where they undoubtedly exerted some fumigant action. Valles and Meer (1974) cite their use against the grassland pest *Acromyrmex landolti*. Normally, however, water is used as a carrier for persistant contact poisons. Early studies with DDT showed that it is not effective (Hambleton 1945). Although good results have been obtained with aldrin and chlordane emulsions, it is probable that dusts pumped down under pressure produce more effective coverage of the underground nest.

Smokes

Kennard (1965) carried out a series of trials in which nests were fumigated by pumping in toxic smokes generated by swing fog machines. He concludes that 2–8% of aldrin or dieldrin dissolved in gas oil gives the best results, and in a subsequent campaign between 1959 and 1963, 5,000 nests were destroyed in this way. The smoke was pumped down the principal nest entrances, until it could be seen issuing from all the subsidiary holes which were then blocked up. Brussel and Vreden (1967) found that smokes give the best kill of all the methods they compared. This method is time consuming and the application machinery is expensive, complex, and requires servicing.

Direct destruction of leaf-cutting ant nests by physical or chemical means involves finding the nest, and in the case of *Atta*, this can be located more than 200 m from where the damage is being done (Lewis and Norton 1973). When agricultural land borders thick forest, or when nests are situated on other people's property, location and subsequent control become difficult. In addition, small *Atta* nests are likely to be overlooked, and even mature *Acromyrmex* nests can be inconspicuous. Accordingly, an alternative method of control that does not involve nest finding is clearly preferable.

Destroying Colonies with Poison Baits

A poison bait that can be laid down where damage is occurring is sought out and carried back to the nest, and that can then be passed around by the ants so as to poison the workers and the queen would meet the requirements for an improved control technique. There have been reports of naturally occurring poison baits, that is, plants that are cut by the ants and that subsequently poison the colony. Santos (1926) claimed that *Sesamum indicum* kills *Atta* colonies that forage on it, but Gonçalves (1944) demonstrated that this is not true for laboratory cultures of *Atta sexdens*. Myers (1935) reported that *Melia azedarach* is toxic to *Atta*, and Weber (1972) reported a similar effect by an unspecified local Mexican herb. More recently, Mullenax (1979) claimed that the jackbean *Canavalia ensiformis* kills *Atta* if 5–15 kg of leaves are placed each night around the nests for three consecutive nights. He speculates that this is due to the action of fungicides such as dimethylhomopterocarpin contained in the leaves. To date, these observations have not been confirmed by detailed experiments.

Rodriguez (1928) suggested that wild plants known to be cut by ants should be sprayed with lead arsenate and placed on *Atta* nests, but there is no evidence that this was ever used in practice. Freire (1971) reported proposals in 1926 to use gelatin capsules containing potassium cyanide as a bait, but the practical use of baits seems to have begun around 1957 when wheat flour baits containing 2% aldrin were tested (Gonçalves 1960). This led to the introduction of the bait Tatuzinho, which achieved considerable success. Since then, at least 43 baits have been marketed for leaf-cutting ant control (Table 6.4).

Peregrine (1973), in his review of the use of toxic baits for controlling pests, outlines the advantages and disadvantages of baiting, and analyses their basic structure. The composition of leaf-cutting ant baits is examined below.

Carrier

Carriers form the physical structure of the bait and, since leaf-cutting ants normally carry dried stored products back to their nests, these have often been used. They commonly contain chemicals that cause foragers to retrieve them. Wheat flour and wheat middlings made into pellets (Gonçalves 1960); cereal grains and dried citrus pulp (Cherrett 1969); cassava flour and bagasse (Blanche 1965); soybean meal, and dried grasses for grass-cutting ants (Robinson et al. 1980) have all been used. The problems with using naturally occurring food stuffs as a matrix are: (1) the quality of the material varies from batch to batch, making it difficult to guarantee its effectiveness; (2) unless the bait is pelleted from finely ground materials, particles vary in shape and size, making it difficult to broadcast the bait from mechanical spreaders; (3) they deteriorate with age, especially when wet; and (4) the material is bulky, so that marketing involves shipping large quantities. These problems have led to a series of attempts to use synthetic matrices as bait carriers (Cherrett et al. 1973). Etheridge and Phillips (1976) tested the mineral vermiculite, foamed gelatin, and the urea-formaldehyde resin "Ufoam,"™ and conclude that vermiculite is the most satisfactory. It has proved to be a successful inert carrier in field trials (Phillips et al. 1979). Jutsum and Cherrett (1981) reduced the transportable bulk of bait matrices by adding the toxicants and attractants to one of the two liquid precursors of polyurethane foam. Once these two dense liquids are mixed and sprayed from a special nozzle, the droplets foam, enlarge greatly in size, and set within seconds. The foam particles thus formed are water resistant, can act as a slow release matrix for pheromones, are picked up by the ants, and give good kill in field trials.

TABLE 6.4.
Toxic Baits Used Against Leaf-Cutting Ants

Toxicant	Percentage AI	Brand names	Number	(%)
Aldrin	0.4–4.5	Blemco,® Blitz Super,® Camani,® Colonial,® Dynatox,® Formicida,® Pika-pau, Formicida Shell,® Formicidol Isca, Guyana,[a] Hormifin,® Hormitoks,® Isca Shell, Landrin Super,® Lantox,® Nitrosin Extra,® Paraguay,[a] Piragy,® Tatucito,® Tatuzhino,® Trinidad[a]	21	(48)
Aldrin + dieldrin	4.0	Parasol®	1	(2)
Dieldrin	0.25	Piragy 02®	1	(2)
Heptachlor	0.5–2.0	AG005, Agroeste,® Bas-Formid Isca, Dinagro,® Esso,® Formicida Hokko,® Formicida Merisca,® Formicida Rhodia,® Formicida Tamandua,® Isca atrativa Benzenex, Isca Formicida Agroceres, Tanajurex®	12	(29)
Mirex	0.45	AC Mirex,® Mirex 450,® Paramex, Tatuzhino Extra,® Paraguay[a]	5	(12)
Nonachlor	0.45	AG 450,® Arbinex Isca Atrativa, Formicidol®		(7)
			Σ43	

[a]Baits manufactured and distributed by the Departments of Agriculture of the countries concerned.

Attractants

It is assumed that leaf-cutting ants locate bait particles, pick them up, carry them back to the nest, process them and add them to their fungus garden primarily in response to chemicals. The sequence of behaviors involved is poorly understood. Here, the general term "attractants" has been used to cover all chemicals that make the bait acceptable (Peregrine 1973).

If the bait particle can be detected at some distance, location will be more efficient. Littledyke and Cherrett (1978) demonstrated that leaf-cutting ants respond positively in an olfactometer to certain vegetable oils and plant materials, and Cabello (1978) showed that such attractants can increase the number of times worker ants contact inert matrices. The effects of adding the trail pheromone, methyl 4-methylpyrrole-2-carboxylate, have been studied. When placed on filter paper disks at suitable concentrations, foraging ants are attracted, but the pheromone does not cause the disks to be picked up (Robinson and Cherrett 1978). Since adding the pheromone to baits such as dried citrus pulp, which already gives off volatile attractants, causes very little increase in the rate at which bait particles are discovered, the pheromone is not cost effective (Robinson et al. 1982).

The reasons why bait particles are picked up once they have been encountered have been investigated for citrus pulp by Cherrett and Seaforth (1970) and Mudd et al. (1978). They concluded that sugars are the most active fraction in promoting the pickup of filter paper discs. Acceptability declines during fractionation, although the fractions removed do not show activity when tested alone. Provided repellent components such as limonene (Cherrett 1972b) are removed, maximum rates of pickup seem to be stimulated by a complex mixture of chemicals.

Robinson and Cherrett (1974) tried unsuccessfully to extract pheromones from leaf-cutting ant brood to use on bait particles, to exploit the ants' brood-tending behavior. Their studies show that the responses are too complex and insufficiently understood to hope for any immediate application.

Toxicant

Cherrett and Lewis (1974) list five requirements for the ideal bait toxicant for use against social insects. (1) A delayed killing action so that the toxicant will be brought back to the nest and widely distributed before its effects begin to disrupt social activity. It is advantageous if the action occurs over a wide range of dosages. (2) Ready trophallactic transfer from one ant to another at dosages that will kill the recipient so as to maximize spread. (3) No repellency in the baits at effective dosages. (4) Specificity to the target organism and, in particular, a low toxicity without side effects if ingested by vertebrates. This is simplified if the toxicant is a stomach poison with no contact action. (5) A half-life short enough to minimize environmental accumulation and passage along food chains, but long enough to persist through storage, application, and the collection and trophallaxis by the ants.

Five toxicants have been used in leaf-cutting ant baits (Table 6.4), the most successful and widely used have been mirex and aldrin (Table 6.5).

Because of their environmental dangers attempts to find less persistent alternatives continue. Etheridge and Phillips (1976) screened 40 candidate compounds in the laboratory for a delayed killing action, but found that none is as good as mirex. The ten most promising compounds, with the more rapid-acting being micro encapsulated to slow their action, were field tested, seven giving good control of *A. sexdens*. None is as good as mirex against *A. cephalotes* (Phillips et al. 1976), and they have not yet been used commercially.

Bait is usually placed on the nest mound, scattered in active trails, or placed where damage is occurring (sometimes protective containers such as hollow bamboo stems are used). For large-scale treatment, random scattering of bait over large areas is possible. Cherrett and Merrett (1969), broadcasting a citrus pulp bait at 2 lb per acre in a citrus orchard, killed 100% of marked *Acromyrmex octospinosus* nests. They concluded that aerial application is a feasible control method. Lewis (1973) conducted trials using fixed wing aircraft, killing 91% of *A. octospinosus* col-

TABLE 6.5
The Most Frequently Cited Methods of Leaf-Cutting Ant Control in a 1972 Questionnaire

Control method	No. citations[a]
Mirex bait	15
Aldrin introduced into nest as a powder or emulsion	12
Chlordane introduced into nest as a powder or emulsion	10
Methyl bromide as a nest fumigant	7
Dieldrin introduced into nest as a powder or emulsion	6
Aldrin based baits	5
Heptachlor introduced into nest as a powder	4
Carbon disulfide as a nest fumigant	3
B.H.C. (H.C.H.) introduced into nest as a powder	2

[a]15 chemicals cited in all, only those mentioned by more than one country are recorded. 17 countries furnished data.

onies of an uncultivated island in the dry season using plain bait, and 85% in cultivated land on the Trinidad mainland in the wet season using waterproofed bait. Lewis and Norton (1973) calculated that this technique is probably more effective in killing nests than normal on-farm control methods, is about a quarter of the cost and uses only 10% to 50% as much organochlorine toxicant as is currently used. More recently, Phillips et al. (1979) have been able to obtain 100% kill of *Atta sexdens* nests with a random scattering of bait, and Robinson et al. (1980) obtain up to 100% kills of *Acromyrmex landolti* within 4 m of the bait swath when it is applied from a fertilizer spreader in bands in a pasture.

There are many ways of controlling leaf-cutting ant damage, but it is clear from Table 6.5 that the use of poison baits is the most favored. Baits are successful because they can be applied quickly, without expensive machinery or training, and give a high rate of kill the first time without retreatment. Bait is also safe to handle, and does not require protective clothing. Certainly, the bait-making industry is a substantial one and Cherrett (1982) records that in 1974, 3,000 tons of one bait, selling for $4.5 million, was being manufactured per annum in São Paulo, at a time when at least 11 other baits were available in the area.

THE ORGANIZATION OF CONTROL

As the damage caused by leaf-cutting ants has long been recognized, countries without these pests have been anxious to avoid their introduction. Leonard (1932) cites a Spanish decree of August 10, 1815 instituting a quarantine to prevent *Atta insularis* from Cuba being introduced into Puerto Rico. Given the number of leaf-cutting ant species and the range of habitats and climates they occupy in the New World, existing quarantine measures to prevent their introduction into the Old World, where the niche occupied by leaf-cutting ants appears to be empty, are amply justified.

Since ants from nests on one farm will forage into adjoining farms, and since all mature nests produce sexuals that disperse and found new colonies over a wide area, an obligation to control the leaf-cutting ants on one's land has been recognized by several governments. In Trinidad and Tobago, the Plant Protection Ordinance of 1911 permitted an order to be served on anyone with a nest on their land, compelling them to destroy it (Anonymous 1914a). Similar legislation was enacted in Guyana in 1914 (Anonymous 1914b). Mariconi (1970) quotes a series of Brazilian laws dating back to 1785 requiring ant nests to be exterminated. There were fines and 30 days imprisonment for noncompliance! Although drafting such legislation is relatively easy, enforcing it is not, although 17 notices were served in Trinidad in 1915 (Freeman 1916). Governments genuinely concerned about losses have usually tried to instigate control campaigns.

Freeman (1925) reports that 1100 nests were treated with 137 gallons of carbon disulfide during 1924 in Trinidad. Oliveira (1934) reports that 141,100 nests were killed during 1930 by control teams in São Paulo, while Carvalho (1945) records the treatment of 16,600 nests in the same area, using about 12,700 kg of carbon disulfide. Kennard (1965) killed 5,000 nests in Guyana over the period 1959–1963, and claims that the best results are achieved when large areas are cleared and treated in their entirety. Blanche (1965) launched a scheme to eradicate the recently introduced infestation of Acromyrmex octospinosus in Guadeloupe, but to date it has not been successful in its objective. Between 1961 and 1965, the United States Aid Scheme launched a large campaign in Paraguay to eradicate Atta from selected areas of the country. The cost of materials alone was $130,000 and 12 years later no evidence of the scheme could be detected. Robinson (1979) concludes that complex, expensive control schemes in developing countries should be avoided in favor of producing simple, locally made baits. These can be distributed to farmers indefinitely, free or at a very low price as has been done in Guyana (Pollard 1981).

At present, eradication of leaf-cutting ants from any sizeable area is not practical even if it were desirable, and so any control schemes should aim at controlling damage within acceptable limits. As Kennard (1965) points out, there are benefits if large areas are subjected to ant control at one time, as foraging from neighboring property and the annual influx of newly fertilized queens is reduced. It is also sensible to kill ant nests whenever a new area is brought under cultivation. Otherwise, the nests often survive the process of clearing the previous vegetation, and are left waiting for the new crops (Cherrett 1981). Effective leaf-cutting ant control requires adequate supplies of cheap effective bait, and the knowledge and motivation to use it. In this sense, area campaigns can serve an important educational role, bringing the problem to the attention of farmers, reminding them of the most effective current techniques, and coordinating individual effort.

THE OUTLOOK

In the forseeable future, the extension and intensification of agriculture will continue to change the species composition of the leaf-cutting ant fauna. In some cases, as in the upsurge of Acromyrmex landolti in grasslands (Cherrett 1981), there is no economic solution at present. In all cases potential damage from leaf-cutting ants will rise, but whether or not actual damage does will depend upon maintaining the present high standards of control, at least in the developed agricultural sectors. If mirex bait is withdrawn on environmental grounds there will, at the present time, be difficulties in replacing it with a product as cheap and effective.

Unless some unexpected breakthrough occurs in biological control, the use of toxic baits as the control technique of choice is likely to continue. However, many improvements can be envisaged. More specific attractants, possibly pheromones, may be found and better methods of slow release discovered. Improvements in bait formulation are long overdue; these might involve the adoption of plastic matrices (Jutsum and Cherrett 1981), or might borrow techniques from the food-processing industry such as those used to produce crunchy-flavored snacks from starch. In either case the objective will be a more closely controlled bait with defined weathering properties and containing known attractants. In order to counteract changes in ant preferences (Cherrett 1972b) and retain bait acceptability against the constantly changing background of newly flushing vegetation that competes with it, mixtures of attractants and flavors may be required (the "licorice all sorts" formulation). The search for new, safer, and better toxicants for fire ant control is intensifying, and it is likely that the leaf-cutting ant control program will continue to lean heavily on this research. The use of hormones, growth inhibitors, and modifiers of such behavior as caste regulation, brood recognition, and egg laying are all potential sources of toxicants. Although such toxicants may work in vitro, they may not work in a nest complete with fungus gardens, as juvenile hormone analog testing shows (Little et al. 1977). If Cherrett is correct in thinking that an important reason for the evolution of the fungus-culturing habit is detoxification of potential ant poisons (Cherrett 1980), then this potential detoxification applies to poison baits just as much as to the poisons in natural vegetation. Perhaps only very stable, inert compounds such as aldrin and mirex, with their long persistence in the environment, can survive the metabolic activities of the fungus garden to poison the ants.

REFERENCES

Allen, G. E. and W. F. Buren. 1974. Microsporidian and fungal diseases of Solenopsis invicta Buren in Brazil. J. N. Y. Entomol. Soc. 82: 125–130.

Amante, E. 1967. Competição entre heptachloro pó e outros formicidas chlorados em pó, no combate à formiga saúva Atta sexdens rubropilosa Forel, 1908, e A. laevigata F. Smith (1858). Biológico 33: 80–84.

Amante, E. 1972. Influência de alguns fatores microclimaticos sôbre a formiga saúva Atta laevigata (F. Smith 1858), Atta sexdens rubropilosa Forel, 1908,

Atta bisphaerica Forel, 1908 e *Atta capiguara* Gonçalves, 1944, em formigueiros localizados no estado de São Paulo. Doctoral Thesis, School of Agriculture, University of São Paulo (Piracicaba).

Andrade, C. F. S. de. 1980. Natural epizootic caused by *Cordyceps unilateralis* (Hypocreales. Euascomycetes) in adults of *Camponotus* sp. in the region of Manaus, Amazonas, Brazil. *Acta Amaz.* 10: 671–678.

Anonymous 1914a. Parasol ants. *Bull. Dept. Agric. Trin. Tobago* 13: 280.

Anonymous 1914b. An ordinance to prevent and suppress diseases of plants and plant pests in the Colony of British Guiana. Georgetown No. 12. 18th July 1914, British Guiana.

Anonymous 1921. Servicio de inspecção e defesa agricola Resumo do relatoria. *Bolm. de Agricultura* 22: 319–329.

Argollo Ferrão, V. A. 1925. A formiga saúva. *Correio Agric.* 3: 325.

Autuori, M. 1942. O enxôfre no combate as saúvas e outras formigas cortadeiras. (Baixo rendimento em anidrido sulfurosa nas combustões de enxôfre e carvão). *Biológico* 8: 249–251.

Autuori, M. 1947. Contribuição para o conhecimento da saúva (*Atta* spp.) IV O sauveiro depois da lª revoada (*Atta sexdens rubropilosa* Forel, 1908). *Arq. Inst. Biol. São Paulo, Brasil* 18: 39–70.

Barreda, L. de la. 1922. La hormiga arriera. *Boln. Agric.* 1: 14pp.

Bates, H. W. 1884. "The Naturalist of the River Amazon," 5th ed., John Murray, London.

Belt, T. 1874. "The Naturalist in Nicaragua." Bumpus, London.

Blanche, D. 1961. La fourmi-manioc. Ministère de l'Agriculture, Service de la Protection des Végétaux, France. Edit. SEP, Paris (1er), 23 pp.

Blanche, D. 1965. Appats empoisonnés contre la fourmi-manioc a la Guadeloupe. pp. 449–454. In: Congrès de la protection des cultures tropicales. *C. R. Trav. Fac. Sci. Marseille*, France.

Brussell, E. W. van and G. van Vreden. 1967. Nieuwe methoden ter bestrijding van draagmieren (*Atta* spp.) in Suriname. *Surin. Landb.* 2: 74–81.

Cabello, L. A. 1978. Control de hormigas cortadoras: Evaluacion en el laboratorio de sustancias atractivas como posibles agregados en matrices no atractivas de cebos hormigacidas. Informes Cientificos (Instituto de Ciencias Basicas, Universidad Nacional de Asuncion Paraguay). pp. 1, 7.

Carvalho, J. C. 1945. O combate às formigas. *Biológico* 11: 227–231.

Cherrett, J. M. 1968. Some aspects of the distribution of pest species of leaf-cutting ants in the Caribbean. *Proc. Amer. Soc. Hort. Sci.* (*Trop. Reg.*) 12: 295–310.

Cherrett, J. M. 1969. Baits for the control of leaf-cutting ants. I.-Formulation. *Trop. Agric., Trin.* 46: 81–90.

Cherrett, J. M. 1972a. Some factors involved in the selection of vegetable substrate by *Atta cephalotes* (L.) in tropical rain forest. *J. Anim. Ecol.* 41: 647–660.

Cherrett, J. M. 1972b. Chemical aspects of plant attack by leaf-cutting ants. *In* "Phytochemical Ecology," pp. 13–24 (J. B. Harbourne, ed.), Academic Press, New York.

Cherrett, J. M. 1980. Possible reasons for the mutualism between leaf-cutting ants and their fungus. *Biol. Ecol. Méditer.* 7: 113–122.

Cherrett, J. M. 1981. The interaction of wild vegetation and crops in leaf-cutting ant attack. In "Pests, Pathogens and Vegetation," pp. 315–325 (J. M. Thresh, ed.), Pitman Press, London.

Cherrett, J. M. 1982. The economic importance of leaf-cutting ants. In "The Biology of Social Insects," pp. 114–118 (M. D. Breed, C. D. Michener, and H. E. Evans, eds.), Proc. 9th Congr. IUSSI, Westview Press, Boulder, Colo.

Cherrett, J. M. 1983. Resource conservation by the leaf-cutting ant Atta cephalotes in tropical rain forest. In "The Tropical Rain Forest," pp. 253–263 (S. L. Sutton, T. C. Whitmore, and A. C. Chadwick eds.), Blackwell Scientific, Oxford.

Cherrett, J. M. and A. R. Jutsum. 1983. The effect of some ant species, especially Atta cephalotes (L.), Acromyrmex octospinosus (Reich) and Azteca sp. on citrus growing in Trinidad. In "Social Insects in the Tropics" (P. Jaisson, ed.), Proc. 1st Int. Symp. Int. Union Study Soc. Insects and Soc. Mex. Ent. Vol. 2

Cherrett, J. M. and T. Lewis. 1974. Control of insects by exploiting their behavior. In "Biology in Pest and Disease Control," pp. 130–146 (D. Price Jones and M. E. Solomon, eds.), Blackwell Scientific, Oxford.

Cherrett, J. M. and M. R. Merrett. 1969. Baits for the control of leaf-cutting ants. 3. Waterproofing for general broadcasting. Trop. Agri. Trin. 46: 221–231.

Cherrett, J. M. and D. J. Peregrine. 1976. A review of the status of leaf-cutting ants and their control. Ann. Appl. Biol. 84: 124–128.

Cherrett, J. M., D. J. Peregrine, P. Etheridge, A. Mudd and F. T. Phillips. 1973. Some aspects of the development of toxic baits for the control of leaf-cutting ants. Proc. 7th Congr. IUSSI London, pp. 69–73.

Cherrett, J. M., G. V. Pollard, and J. A. Turner. 1974. Preliminary observations on Acromyrmex landolti (For.) and Atta laevigata (Fr. Smith) as pasture pests in Guyana. Trop. Agri. Trin. 51: 69–74.

Cherrett, J. M. and C. E. Seaforth. 1970. Phytochemical arrestants for the leaf-cutting ants, Atta cephalotes (L.) and Acromyrmex octospinosus (Reich) with some notes on the ants' response. Bull. Entomol. Res. 59: 615–625.

Cherrett, J. M. and B. G. Sims. 1968. Some costs for leaf-cutting ant damage in Trinidad. J. Agri. Soc. Trin. 68: 313–322.

Conconi, J. Ramos Elorduy de. 1982. Los insectos como fuente de proteinas en el futuro. Ed. Limusa, Mexico City.

Cramer, H. H. 1967. Plant protection and world crop production. "Bayer" Pflanzenschutz, Leverkusen.

Dinther, J. B. M. van. 1958. Draagmieren in Suriname en hun bestrijding. Surin. Landb. 6: 18–27.

Eberhard, G. and O. Kafury. 1974. La ecologia de la hormiga Azteca trigona, una possible defensa contra las arrieras, Memorias II. Congreso de la Sociedad Colombiana de Entomologia, Cali, Columbia, pp. 33–37.

Etheridge, P. and F. T. Phillips. 1976. Laboratory evaluation of new insecticides and bait matrices for the control of leaf-cutting ants. Bull. Entomol. Res. 66: 569–578.

FAO. 1979. Production Yearbook, Rome.

Fenjves, P. 1950. Einige Probleme der angewandten Entomologie in Venezuela. Mitt. Schweiz. Entomol. Ges. 23: 135–154

Fowler, H. G. 1977. Field response of Acromyrmex crassispinus (Forel) to aggression by *Atta sexdens* (Linn.) and predation by *Labidus praedator* (Fr. Smith). *Aggr. Behav.* 3: 385–391.

Fowler, H. G. and S. W. Robinson. 1975. Estimaciones acerca de la accion de *Acromyrmex landolti* (Forel) sobre el pastoreo y la ganaderia en el Paraguay. *Revta. Soc. Cient. Paraguay* 15: 64–69.

Freeman, W. G. 1916. Plant protection ordinance. *Rept. Dept. Agric.* Trinidad and Tobago for 1915, Port of Spain.

Freeman, W. G. 1925. Plant pathology. *Rept. Dept. Agric.* Trinidad and Tobago for 1924, Port of Spain.

Freire, J. 1971. Emprego de iscas granuladas no contrôle da saúva. Imprensa Universitaria, Universidade Federal de Viçosa, Minais Gerais, Brazil.

Gonçalves, A. J. L. 1957. Algunas experiencias com aldrin no combate a saúva vermelha (*Atta sexdens rubropilosa* Forel). *Rev. Soc. Bras. Agron.* 12: 1–4.

Gonçalves, A. J. L. 1960. O emprêgo das iscas no combate as formigas cortadeiras. *Bol. Campo.* 16: 3–10.

Gonçalves, C. R. 1944. O gergelim no combate à saúva. *Bolm Fitossanit.* 1: 19–27.

Hambleton, E. J. 1945. Experiments with DDT on leaf-cutting ants in Ecuador. *J. Econ. Entomol.* 38: 282.

Harrison, J. B., C. K. Bancroft, and G. E. Bodkin. 1916. The cultivation of limes. III. *J. Bd. Agri. Brit. Guiana* 9: 122–129.

Jacoby, M. 1935. Erforschung der Struktur des *Atta*—Nestes mit Hülfe des Cementausguss-Verfahrens. *Revta. Entomol.* (*Rio de Janeiro*) 5: 420–424.

Jacoby, M. 1955. Die Erforschung des Nestes der Blattschneider-Ameise *Atta sexdens rubropilosa* Forel. *Z. Ang. Entomol.* 37: 129–152.

Jonkman, J. C. M. 1979. Distribution and densities of nests of the leaf-cutting ant *Atta vollenweideri* Forel, 1893 in Paraguay. *Z. Ang. Entomol.* 88: 27–43.

Jonkman, J. C. M. 1980. Average vegetative requirement, colony size and estimated impact of *Atta vollenweideri* on cattle raising in Paraguay. *Z. Ang. Entomol.* 89: 135–143.

Jutsum, A. R. and J. M. Cherrett. 1981. A new matrix for toxic baits for control of the leaf-cutting ant *Acromyrmex octospinosus* (Reich). *Bull. Entomol. Res.* 71: 607–616.

Jutsum, A. R., J. M. Cherrett, and M. Fisher. 1981. Interactions between the fauna of citrus trees in Trinidad and the ants *Atta cephalotes* and *Azteca* sp. *J. Appl. Ecol.* 18: 187–195.

Kennard, C. P. 1965. Control of leaf-cutting ants (*Atta* spp.) by fogging. *Expl. Agri.* 1: 237–240.

Kermarrec, A. and H. Mauleon. 1975. Quelques aspects de la pathogénie d'Entomophthora coronata Cost. Kervork. pour la fourmi-manioc de la Guadeloupe: *Acromyrmex octospinosus* (Attini). *Ann. Parasitol.* 50: 351–360.

Leonard, M. D. 1932. An early quarantine in Puerto Rico. *J. Econ. Entomol.* 25: 930–931.

Leston, D. 1978. A neotropical ant mosaic. *Ann. Entomol. Soc. Amer.* 71: 649–653.

Lewis, T. 1973. Aerial baiting to control leaf-cutting ants in Trinidad. 2. Field application, nest mortality and the effects on other animals. *Bull. Entomol. Res.* 63: 275–287.

Lewis, T. and G. A. Norton. 1973. Aerial baiting to control leaf-cutting ants in Trinidad. 3. Economic implications. *Bull. Entomol. Res.* 63: 289–303.

Little, C. H., A. R. Jutsum, and J. M. Cherrett. 1977. Leaf-cutting ant control. The possible use of growth regulating chemicals. Proc. 8th Congr. IUSSI, Wageningen, Netherlands, pp. 89–90.

Littledyke, M. and J. M. Cherrett. 1976. Direct ingestion of plant sap from cut leaves by the leaf-cutting ants *Atta cephalotes* (L.) and *Acromyrmex octospinosus* (Reich). *Bull. Entomol. Res.* 66: 205–217.

Littledyke, M. and J. M. Cherrett. 1978. Olfactory responses of the leaf-cutting ants *Atta cephalotes* (L.) and *Acromyrmex octospinosus* (Reich) in the laboratory. *Bull. Entomol. Res.* 68: 273–282.

Lugo, A. E., E. G. Farnworth, D. Pool, P. Jerez, and G. Kaufman. 1973. The impact of the leaf-cutter ant *Atta colombica* on the energy flow of a tropical wet forest. *Ecology* 54: 1292–1301.

Mariconi, F. A. M. 1970. As saúvas. São Paulo: Editôra Agronômica "Ceres."

Mariconi, F. A. M. and E. Amante. 1966. Recomendações atuais de combate às saúvas. *Rev. Agri.* 41: 41–45.

Mendes Filho, J. M. 1979. Tecnicas de combate as formigas. Piracicaba, IPEF, Circular Technica 75: 14 pp.

Mudd, A., D. J. Peregrine and J. M. Cherrett. 1978. The chemical basis for the use of citrus pulp as a fungus garden substrate by the leaf-cutting ants *Atta cephalotes* (L.) and *Acromyrmex octospinosus* (Reich). *Bull. Entomol. Res.* 68: 673–685.

Mullenax, C. H. 1979. The use of jackbean (*Canavalia ensiformis*) as a biological control for leaf-cutting ants (*Atta* spp.). *Biotropica* 11: 313–314.

Myers, J. G. 1935. Second report on an investigation into the biological control of West Indian insect pests. *Bull. Entomol. Res.* 26: 181–252.

Nelson, H. S. 1951. Chlordane in the control of leafcutter ants *Atta* spp. The San Tome Experimental Station, Venezuela. Bull. No. 1., Mene Grande Oil Co.

Oliveira, F. M. L. 1934. Combate à saúva. Bolm. Agric. (São Paulo) 35: 541–610.

Pena, O. N. S. 1974. Variedas de pastos resistentes al ataque de la hormiga "indanera" *Acromyrmex landolti* Forel 1884. Inf. Dpto. Proteccion de Cultivos CRIANO, Tarapoto, Peru, 12 pp.

Peregrine, D. J. 1973. Toxic baits for the control of pest animals. *Pest. Art. News Summary* 19: 523–533.

Phillips, F. T., P. Etheridge, and A. P. Martin. 1979. Further laboratory and field evaluations of experimental baits to control leaf-cutting ants in Brazil. *Bull. Entomol. Res.* 69: 309–316.

Phillips, F. T. P. Etheridge, and G. C. Scott. 1976. Formulation and field evaluation of experimental baits for the control of leaf-cutting ants in Brazil. *Bull. Entomol. Res.* 66: 579–585.

Pollard, G. V. 1981. A review of the distribution, economic importance and control of leaf-cutting ants in the Caribbean region with an analysis of current control programmes. Proc. First Meeting Soc. Plant Prot. in the Caribbean. IICA Misc. Publ. 378, Trinidad and Tobago, pp. 43–61.

Proctor, J. H. 1957. Coushi ant control with chlordane and aldrin. Brit. Guiana Dept. Agric. Farmers Leaflet 5.

Ribeiro, G. T. and R. A. Woessner. 1979. Teste de eficiência com seis sauvicidas no controle de saúvas (*Atta* spp.) na Jari, Para, Brasil. *An. Soc.. Entomol. Brasil* 8: 77–84.

Robinson, S. W. 1979. Leaf-cutting ant control schemes in Paraguay, 1961–1977. Some failures and some lessons. *Pest. Art. News Summary* 25: 386–390.

Robinson, S. W., A. Aranda, L. Cabello and H. G. Fowler. 1980. Locally produced toxic baits for leaf-cutting ants for Latin America; Paraguay, a case study. *Turrialba* 30: 71–76.

Robinson, S. W. and J. M. Cherrett. 1974. Laboratory investigations to evaluate the possible use of brood pheromones of the leaf-cutting ant *Atta cephalotes* (L.) as a component in an attractive bait. *Bull. Entomol. Res.* 63: 519–529.

Robinson, S. W. and J. M. Cherrett. 1978. The possible use of methyl 4-methyl-pyrrole-2-carboxylate, an ant trail pheromone as a component of an improved bait for leaf-cutting ant control. *Bull. Entomol. Res.* 68: 159–170.

Robinson, S. W. and H. G. Fowler, 1982. Foraging and pest potential of Paraguay-an grass-cutting ants (*Atta* and *Acromyrmex*) to the cattle industry. *Z. Ang. Entomol.* 93: 42–54.

Robinson, S. W., A. R. Jutsum, J. M. Cherrett, and R. J. Quinlan. 1982. Field evaluation of methyl 4-methylpyrrole-2-carboxylate, an ant trail pheromone, as a component of baits for leaf-cutting ant control. *Bull. Entomol. Res.* 72: 345–356.

Rockwood, L. L. 1975. The effects of seasonality on foraging in two species of leaf-cutting ants (*Atta*) in Guanacaste Province, Costa Rica. *Biotropica* 7: 176–193.

Rodriguez, A. J. 1928. Un raro procedimiento para combatir la hormiga arriera. *Bol. Mens. Defensa Agric. Mexico* 2: 122–123.

Sampaio de Azevedo, A. G. 1894. Saúva ou Manhúaára. Monographia como subsidio a historia da fauna paulista. Tip. Diário Oficial, São Paulo, 74 pp.

Santos, L. F. dos. 1926. O gergelim extinctor da formiga saúva. *Correio Agric. Bahia* 4: 9–10.

Schade, F. H. 1973. The ecology and control of the leaf-cutting ants of Paraguay. *In* "Paraguay Ecological Essays," pp. 77–95 (J. R. Gorham, ed.), Academy of the Arts and Sciences of the Americas, Miami, Florida.

Stradling, D. J. 1978. The influence of size on foraging in the ant *Atta cephalotes* and the effect of some plant defence mechanisms. *J. Anim. Ecol.* 47: 173–188.

Townsend, C. H. T. 1923. Um inseto de um bilhão de dollares e sua eliminação. A formiga saúva. *Alm. Agric. Brasil.* (São Paulo) 12: 253–254.

Troppmair, H. 1973. Considerações sôbre as condições naturais e alguns aspectos da geografia agraria do municipio de descalvado (S.P.). Doctoral thesis, Faculty of Philosophy, Science and Letters, Rio Claro, São Paulo, Brazil.

Valles, C. R. P. and F. T. van der Meer. 1974. El control quimico de la hormiga "indanera" (*Acromyrmex landolti* Forel, 1884), en pasto "Castilla," *Panicum maximum* Jacq. II CONIAP—Lima, Peru. 75: 9 pp.

Vanetti, F. 1958. Resultados dos tratamentos de sauveiros no periodo de 1949 a 1958. *Revta Ceres Viçosa* 10: 149–163 and 252–268.

Vaughan, R. B. 1946. Silvicultura e formigas. *Chac Quint.* (São Paulo) 74: 350–353.

Walton, E. V., L. Seaton, and A. A. Mathewson. 1938. The Texas leaf-cutting ant and its control. Circ. U.S. Dept. Agric. No. 494, 18 pp.

Weber, N. A. 1972. Gardening ants the Attines. Mem. Amer. Phil. Soc. 92: 1–146.

Weyrauch, W. 1942. Las hormigas cortadoras de hojas del Valle de Chanchamayo. Bol. Dir. Agric. Ganad. (Lima) 15: 204–259.

Wille, J. 1929. Die Blattschneiderameisen Südbrasiliens und versuche zu ihrer Bekämpfung. Tropenpflanzer 32: 404–426.

Wille, J. 1943. Entomologia agricola del Peru. Min. of Agri., Lima, Peru.

Chapter 7

THE BIOLOGY, PHYSIOLOGY, AND ECOLOGY OF IMPORTED FIRE ANTS

S. Bradleigh Vinson and Les Greenberg

INTRODUCTION

Ants represent one of the most successful groups of social insects, being found almost everywhere but reaching their greatest diversity in the tropics. Most species are very important components of the ecosystem (see Chapter 11). However, there are a few species that are important because they also conflict with man's interests. These fall into three major groups: (1) those that destroy agricultural crops, such as the leaf-cutters; (2) those that destroy structures of importance to man, such as the carpenter ants; and (3) those that sting and cause discomfort, such as the fire ant. The problems caused by stinging ants are hard to document. While some stinging ants can damage crops and structures such as telephone cables, and occasionally even cause some deaths, they are generally only nuisance pests. Of these, the *Solenopsis* group, generally referred to as fire ants, have become a particular problem in the southern United States. The *Solenopsis* group probably evolved in South America, and several species migrated into the Northern hemisphere after the formation of the land bridge between North and South America. The major fire ant problem occurs in the southern United States where one species, *Solenopsis invicta*, known as the red imported fire ant, was accidentally introduced and has rapidly reached high densities across the southeastern United States (see Chapter 8).

The control of the imported fire ant depends on a knowledge of its biology, particularly since entomologists and pest control technologists have realized that the indiscriminate use of pesticides is often not a suitable solution. This review describes the biology of the imported fire ant and tries to identify aspects of its biology that may facilitate its control. Some of the control approaches and problems may be applicable to other ant species.

THE NATIVE Solenopsis COMPLEX IN THE UNITED STATES

Prior to the introduction of the imported fire ant, several related species of fire ants were common in the southern United States. Of those, the most common and economically important was the tropical fire ant, Solenopsis geminata (F), (Clark, 1931). As noted by E. O. Wilson and Brown (1958) and Brown (1961) this species was probably introduced into North America during pre-Columbian times from its original range in central and northern South America. Solenopsis geminata is found from Texas to South Carolina and primarily near the more coastal regions of its range (Creighton, 1950; Hung and Vinson, 1978; Moody et al., 1981). However, as noted by Hung et al. (1977) and Moody et al. (1981), the tropical fire ant has been largely replaced throughout its original range in the United States by the imported fire ant.

Very little is known about the biology of S. geminata, although it probably is similar to the imported fire ant. The mounds of S. geminata are generally located in open grasslands and may be as large as those of the imported fire ant, although more rounded than dome-shaped. Colonies are thought to contain a single queen, although multiple queen colonies have been noted (Banks et al. 1973, Hung et al. 1977, Adams et al. 1976). Travis (1941) reported that captive queens produced over a thousand eggs per day. These eggs became adults in 44 days at summer temperatures. Sexuals have been noted in colonies in North Florida and South Georgia from spring through December (Travis 1941) and in Texas from May through July with nuptial flights occurring in late afternoon (Hung et al. 1977).

The second native species, the southern fire ant, Solenopsis xyloni McCook, ranges from California to North Carolina and tends to live further from the coast than S. geminata (Hung and Vinson 1978, Moody et al. 1981). This species has been largely replaced by the imported fire ant where the latter occurs. The mounds of S. xyloni are flat, with many irregular piles of dirt around the entrances (Hess 1958). Although single-queen colonies appear to be the rule (Smith 1958), multiple-queen colonies have been reported (Summerlin et al. 1976). Sexual forms have been noted from June through September with nuptial flights occurring from 3:30 PM to 5:30 PM (Roe 1973).

INTRODUCTION AND TAXONOMIC STATUS
OF THE IMPORTED FIRE ANT

Creighton (1930) first described the imported fire ant in the United States as *Solenopsis saevissima richteri* Forel (Creighton 1950). He suggested that the imported fire ant was introduced into Mobile, Alabama, about 1918. Wilson (1951) recognized the existence of two forms of the imported fire ant: an original dark form and a light form which he suggested was either a mutation or a second introduction. He later provided information to support a second introduction into Mobile in the 1930s, suggesting that the light form was a hybrid between two South American species, *Solenopsis saevissima* and *Solenopsis saevissima richteri* (Wilson 1952, 1953). Wilson and Brown (1958) suggested that the two color forms interbred completely in the United States and produced a graded series of intermediate forms which they referred to as *Solenopsis saevissima*, a name continued by Snelling (1963) and Ettershank (1964).

Buren (1972) and Buren et al. (1974) accepted the double importation hypothesis and demonstrated that there are two morphologically distinct species of imported fire ants in the United States. The original dark brown form was recognized as *Solenopsis richteri* Forel, the black imported fire ant (hereafter called the "black IFA"). The light form was described as *Solenopsis invicta* Buren (Buren 1972), the red imported fire ant (hereafter called the "red IFA").

Distribution and Spread of the Fire Ants

Wilson (1952) postulated that the borders of Argentina, Uruguay, Paraguay, Bolivia and Brazil encompassed the range of *S. richteri*. However, it now appears that the distribution of *S. richteri* is more restricted to southernmost Brazil and Uruguay and northeastern Argentina (Buren et al. 1974). *Solenopsis invicta* is located in the Pantanal in the state of Mato Grosso, Brazil (Lennartz 1973, R. N. Williams et al. 1975, and Whitcomb 1980).

The black IFA was probably first imported into Mobile, Alabama prior to 1918 (Creighton 1930). The black IFA remained in and around the Mobile area through the 1920s and 1930s. In the 1940s imported fire ants began to spread out from the Mobile area. This movement was due to a second introduction, the red IFA, *S. invicta*. Lennartz (1973) speculated that *S. invicta* was imported into Mobile between 1933 and 1945. While Lennartz examines old cargo records of shipping into the Mobile area, she was unable to associate the red IFA introduction with any specific cargo.

The specific cargo involved or the specific dates of importation of the two species into Mobile will probably never be known. It is of interest to note that shipping activity from South America to other locations along the Gulf Coast, particularly New Orleans, was at least as great as to Mobile, yet both introductions occurred in Mobile. Lofgren et al. (1975) suggest that S. richteri may have been able to outcompete the Argentine ant, Iridomyrmex humilis, in Mobile. Buren et al. (1974) believe that S. invicta was able to invade because S. richteri preconditioned the Mobile area for invasion.

The spread of S. invicta since the 1950s has been spectacular. Wilson and Brown (1958) noted that in 1953 the IFA had invaded a number of localities from Louisiana to the Carolinas and was usually associated with nurseries. By 1974 Buren et al. (1974) reported that S. invicta had infested more than 52 million acres of land from Texas to North Carolina. During this period S. richteri had moved from Mobile, Alabama to occupy an area in Northeast Mississippi and Northwest Alabama, having been replaced by S. invicta in Mobile. Why S. richteri has persisted in those areas is not known.

New infestations of S. invicta diminished after the fire ant quarantine of 1956. But by 1975 the red IFA occupied most of the land between East Texas (Hung and Vinson 1978) to the southernmost portion of North Carolina (Bass and Hays 1976) and south through North Florida and across the Gulf Coast. When the red IFA was first reported in Texas in 1953 (Culpepper 1953), it was quickly eradicated. The red IFA again moved into six East Texas counties in 1957 (Hung and Vinson 1978, Summerlin and Green 1977) and moved westward across the state at about 30 miles per year from 1957 to 1978. The red IFA is still moving west and south through Texas, although at a reduced rate (Francke et al. 1983, Moody et al. 1981).

There is much speculation on the ultimate range of the red IFA. Pimm and Bartell (1980) examined the range of red IFA infestation with respect to temperature and humidity. They found that in Texas the ant had recently spread primarily into areas that were warm and either dry or wet, although the spread into the latter was minimal since most of those areas were already occupied. The ants did not spread into cold and dry areas, indicating that the northward limit may have been reached. Morrill (1977a) and Morrill et al. (1978) reported 94% mortality of colonies during the cold winter of 1976–1977 in the Piedmont area of northern Georgia, while only 20% mortality was recorded in southern Georgia. Mortality of the black IFA in Mississippi due to freezing conditions was discussed by Green (1959), who noted that wet and cool weather preceding a subzero period may increase ant mortality due to exposure and even drowning. Pimm and Bartell (1980) suggest that the red IFA will not proceed north of the 0°F isotherm, at least in Texas. However, the ant can infest buildings (Green 1952, Bruce et al. 1978) and could possibly survive further north under these conditions.

How far west the red IFA may move is not clear. We observed 70% mortality of colonies in Brazos County, Texas during the summer drought of 1979 (S. B. Vinson, unpublished). The red IFA appears sensitive to hot, dry conditions and may be unable to invade the drier southwest except for irrigated land or watered urban areas. As noted by Hung and Vinson (1977) and Tschinkel (1982), the ultimate spread will not only depend on the extreme abiotic factors, but also on the frequency of rains, dry spells, and warm periods. Such favorable periods must be long enough for a founding colony to build up strength. Biotic factors may also be important in limiting the range of the IFA, but little is known concerning the impact of predators, parasites, pathogens, or competitors. Some of these will be discussed later.

The spread of the red IFA occurred in four ways. The early spread from Mobile was largely due to the transport of sod and nursery plants from infested to uninfested areas. This resulted in the distribution of small colonies throughout the Southeast. The second mode of spread is through nuptial flights. Mating takes place 150 to 300 ft in the air (Markin et al. 1971) and wind can scatter them. How far a mated female can fly is unknown. Banks et al. (1973) noted queens dispersing up to 12 miles and Wojcik (1983) indicated that queens can "fly" in excess of 20 miles.

A third mode of spread is through the ability of colonies to float down rivers during floods (K. L. Hays 1959 and Morrill 1974a). If the red IFA establishes in the upper watershed, and its nests are flooded, it can form living rafts of ants surrounding the queen. The fourth mode of spread is accidental transport by man due to the behavior of newly mated females. After mating the females descend to within one or two meters above ground, where they fly along apparently searching for suitable landing sites. It appears that these females are attracted to reflective surfaces (D. E. Ball and S. B. Vinson, unpublished). These include trains and trucks, where hundreds of mated queens have been collected (Vinson, personal observations). In Texas, red IFA infestations are often found along the railroad and highway right of ways many kilometers ahead of the general infestation. As of 1980 the ants infested over 230 million acres (Carter 1981).

REPRODUCTION AND COLONY FOUNDING

Production of Sexuals

Lofgren et al. (1975) reported that 8.7% of the biomass of colonies examined in May was composed of reproductive brood, while it was only 2.5% in September. Markin and Dillier (1971) showed that the biomass of females was greatest in September; for males it was greatest in June.

Markin et al. (1973) also reported that males could be found seven months after colony founding, while Lofgren et al. (1975) found both sexual forms in colonies within five months. Rhoades and Davis (1967) also reported the presence of reproductives within five months, although they did not indicate the sex.

Although Markin et al. (1973) reported that alate production decreased in colonies 1 year to 6 months old, Lofgren et al. (1975) reported that 76% of the 1-year-old colonies contained sexuals, a point also noted by E. O. Wilson (1962). In summary, the available data suggest that colonies can produce sexuals within six months of founding, depending on the season and environmental conditions. Sexual production appears to continue for several years or more.

Morrill (1974b) studied the occurrence and size of mating flights and found that while they occur throughout the year, they begin to increase in May, peak in mid-June and decrease through August. The average number of alates leaving a mound at the June flight peak was 690. The average number of alates produced per hectare per year was nearly 500,000 in Florida. Roe (1973) found similar results in Arkansas. Over 3,000 alates were produced from one mound in the Arkansas study and one mound released over 7,600 alates in a 152-day study period (Roe 1973). Bass and Hays (1979) reported that mounds tend to release one sex at a time, but the overall ratio is 2.6 males to one female. In South Carolina, nuptial flights occur all year, but peak in June and July. While mounds usually produced predominantly one sex or the other (90% of the mounds produced over 74% of one sex), the sex ratio of a particular mound changed during the year (Morrill 1974b). None of the colonies produced only one sex throughout the year and mound density had no effect on total alate production per acre, so the total alate production was consistent. Morrill (1974b) also reported that a sex ratio of nearly 1 : 1 was maintained for his study area in contrast to the previously mentioned sex ratio of 2.6 : 1.

Mating Behavior

Mating flights may be very local or cover large areas depending on the conditions that trigger mating flight activity (Markin et al. 1971). Glancey et al. (1976a) reported that sexual maturity of the adult was reached in seven to ten days postemergence. The males, which are dark black and have relatively small heads, have four large paired testes that atrophy as the sperm mature. If alates of the proper age are present, mating or nuptial flights will occur 1 or 2 days after a rain, if the ambient air temperature is between 20°C and 32°C (Rhoades and Davis 1967, Markin et al. 1971), and if the wind gusts are below 15 mph, preferably less than 5 mph

(Markin et al. 1971). Rhoades and Davis (1967) found that flights only occurred if the relative humidity was over 80% and the soil temperature was above 18°C at a depth of 4 in. They did not find that cloudiness had any effect. The mating flight usually occurs between 10:00 AM and 2:00 PM.

Where mating occurs, and whether males attract females or the reverse, is unknown. Markin et al. (1971) collected flying males from their mating aggregations using nets fixed to an aircraft and reported that males were located at a height of between 150 to 300 m. Females start to land within an hour and continue to land for three to four hours. Males land later in the afternoon, between 3:00 PM and 6:00 PM. The males usually die within a day.

Colony Founding

Markin et al. (1971) suggested that females do not land at random. Mated females can be found in large numbers in swimming pools, the backs of pickup trucks, and recently cultivated fields or roads. They are less obvious in heavily vegetated areas. We found that swimming pools measuring 8 ft across and covered with tin foil or water collected more queens after a mating flight than similar pools covered in soil or dark cloth. The results suggest that mated queens prefer to land on shiny surfaces (Ball et al. 1984b). After landing the queens shed their wings and begin searching for a suitable place to begin a tunnel. The process of dealation has attracted much attention. More than one condition may initiate dealation. Following mating, dealation is relatively fast, usually occurring in 10 to 15 minutes. (D. E. Ball and S. B. Vinson, unpublished). Dealation also occurs with virgin queens in the parent colony if the queen is removed or if the virgin queens are removed from the queen's influence of the parent colony (Fletcher and Blum 1981a). However, dealation under these conditions often requires several days (Fletcher and Blum 1981b). Dealation also follows CO_2 exposure or the application of juvenile hormones; however, the time for dealation is usually somewhere between the two extremes described above (Kearney et al. 1977, Barker 1979). The sequence of physiological events that leads to dealation remains unknown.

There are also dramatic changes in the biochemical makeup of the queen during colony initiation. Toom et al. (1976a) reported that amino acids increased in the hemolymph after mating, presumably due to the breakdown of the flight muscles (Jones et al. 1978). R. G. Jones and S. B. Vinson (unpublished) found that ^{14}C-labeled amino acids are distributed to the ovaries, eggs, and maxillary and labial glands of the queen. They also show up in the larvae. These results suggest that the flight muscle breakdown provides protein to the developing first larvae via secretions of the labial and/or maxillary glands and possibly by the eggs that are consumed by the larvae.

Glycogen changes were also described in newly mated queens by Toom et al. (1976b). They found a 60% decrease in glycogen stores during flight and a second decrease after oviposition (Toom et al. 1976c). They also found that lipids decreased during the first nine days after oviposition. Vinson et al. (1980) reported that prior to the mating flight queens gorge themselves on lipids, which fill the crop and postpharyngeal glands. As the queen establishes a new colony the lipids move forward to the esophagus to form the esophageal crop in the space provided by the degenerating wing muscles. Glancey et al. (1981b) suggested that the esophageal crop, which formed as the muscles degenerated, contained lipids that were provided by the degenerating wing muscles.

After shedding her wings a founding queen begins to select a site to initiate a colony. The queen typically selects sites under stones or small objects, but she may select a site in the open. Usually this is accomplished by a single queen, as is typical of other nonparasitic myrmicine ants. However, in situations where many queens occur a number of queens may be found co-founding (Markin et al. 1972). Co-founding was studied by Tschinkel and Howard (1983), who showed that co-founding improves the likelihood of colony survival. More workers were more rapidly produced when five queens were present than one or two. Queens and colonies are most vulnerable during colony founding, although there have been no specific studies to determine the factors involved. Nuptial flights are most common in the spring and fall. As noted by Lofgren et al. (1975), successful founding may depend on soil physical factors, climate, vegetation, food availability, inter- and intraspecific competition for food and space, and the presence of predators, parasites, and pathogens. Successful overwintering may depend on whether the nest has enough workers to dig deep enough to escape frosts (Lofgren and Weidhaas 1972, Markin et al. 1973) and droughts.

In newly infested land that has not recently been disturbed, the number of red IFA colonies increases slowly (S. B. Vinson, personal observations; Summerlin and Green 1977). However, other workers (Green 1967, Markin et al. 1973) have reported over 2,500 colonies per acre in newly infested land. These authors studied land where the red IFA had once been eliminated. Heavy resurgence of ants is common in such cases (Summerlin et al. 1977).

While several thousand colonies may begin in an area, most do not survive and the number of colonies stabilizes at between 20 and 30 mounds/ha (Green 1967). The reason for the decline in colony number is not known, but has been attributed to competition, food availability, and the merging of colonies. Several authors (Markin et al. 1973, Rhoades and Davis 1967) have suggested that small colonies may fuse with one another after abandoning their queen. This point has not been proven and is related to the occurrence of polygynous colonies.

Reproduction

The male testes develop during the larval stage, reach their peak in the pupal stage and begin to degenerate in the adult. The degeneration is complete in 7 to 10 days (Glancey et al. 1976a). Tice (1967) described the development and histology of the male reproductive system of S. richteri, and Ball and Vinson (1985) described it for S. invicta. In the latter, the sperm mature during the pupal period and after adult eclosion they migrate into the seminal vesicles where they are held until mating. Hung and Vinson (1975) clarified some of the terminology used to describe the various male organs. The accessory gland was studied by Ball and Vinson (unpublished) and several lipoproteins were found to be secreted, filling the lumen of the accessory gland prior to mating. They found the lipoprotein was transferred to the female during mating where its functions remain unknown. They did show that the accessory gland lipoprotein disappeared in the female in several days and did not accompany the sperm into the spermatheca. Ball et al. (1984) also examined the aedeagal bladder and found that the lumen contains monoglycerol esters of dodecanoate and decanoate in a 10 : 1 ratio five days after emergence. However, the material in the aedeagal bladder remains after mating and its function is also unknown.

Thompson and Blum (1967) described the sperm of the red IFA. Hung et al. (1974) found sterile males in multiple queen colonies. These males were diploid and their testes did not develop. The presence of sterile diploid males was confirmed by Glancey et al. (1976a) who reported that the chromosome numbers of the red and black IFA are 16.

The female reproductive system was described by Hermann and Blum (1965). Fletcher and Blum (1981a) found that in orphaned colonies some of the virgin queens dealate, begin oocyte development, and assume the egg-laying function of the missing queen. They suggested that the queen or substitute queen produces an inhibitory pheromone that prevents dealation and oocyte development (Fletcher and Blum 1983). Barker (1978) reported that the brain and corpus allatum were necessary for oocyte maturation. He found that oocyte development could be restored in allatectomized females by the application of juvenile hormones. However, in normal queens the application of juvenile hormone analogs suppresses egg laying (Troisi and Riddiford 1974, Vinson and Robeau 1974, Robeau and Vinson 1976). There is no obvious explanation for this discrepancy. In studies concerning egg production, Fletcher et al. (1980) reported that physogastric queens produced about 200 eggs per day. However, the queens in multiple queen colonies often produce only 20 or 30 per day (Fletcher et al. 1980). This would suggest that there is only partial inhibition in multiple queen colonies. Inhibiting oocyte maturation may eventually have fire ant control applications.

BIOLOGY AND COLONY DEVELOPMENT

Eggs usually hatch in about 8 to 10 days (Fincher and Lund 1967, Petralia and Vinson 1978). O'Neal and Markin (1975b) described four larval instars based on head capsule size and mandibular morphology. Their four instars correspond to the third and fourth instars of Petralia and Vinson (1979a), who used additional characters to also describe four instars. The first three instars are primarily fed liquid food regurgitated by workers while the fourth instar is capable of feeding on solid food brought in by the workers (Petralia and Vinson 1978). The solid food is placed on the ventral area of the fourth instar larvae where it appears to be extraorally digested. These authors also described the anatomy of the region where the food is placed (Petralia and Vinson 1979b).

Fincher and Lund (1967) found that only minims were produced from founding queens, and that they required 27–37 days to become adults at 28°C–30°C. Isolated winged queens produced only males (an average of 60 per female) requiring 63–68 days to become adults. Markin and Dillier (1971) and Markin et al. (1973) reported that newly established single queen colonies produced 10–20 minimum workers in about 30 days. This number increased to 85 in 60 days and over 200 in 70 days. By five months over 1,000 workers were counted, most of which were minor workers, although a few major workers were observed. The minim workers could no longer be located. After 1 year colonies contained over 11,000 workers, and over 230,000 were seen in mature colonies 2½ to 3 years old. As noted by Lofgren et al. (1975) colony size is probably underestimated since 10% of the worker force may be absent from the colony at any given time. Markin and Dillier (1971) and Markin et al. (1973) only counted workers in the mound.

There are different worker castes in a fire ant mound. The first workers (minims) produced by a founding queen have head capsules measuring 0.5 mm (L. Greenberg, personal observation). As the colony matures larger workers are found, head widths ranging up to 1.5 mm (Wood and Tschinkel 1981). There are no distinct worker castes. They reported that the proportion of large workers increases with colony age. However, the observed distribution depends on whether the colonies are multiple- or single-queen colonies. Greenberg et al. (1985) showed that the workers of multiple-queen colonies are generally smaller and that the larger size classes are absent. Wilson (1978) noted that the larger workers could handle larger food and discussed division of labor based on size. Mirenda and Vinson (1981) found that division of labor was also related to age. The youngest ants are nurses, middle-aged ants work in and defend the nest or are recruited to food, and the oldest ants forage. However, their behavior is also influenced by size. The smaller ants go through the behavioral changes slower than larger ants and spend more time as nurses, while the larger ants spend more time as foragers (Mirenda and Vinson 1981). Further, there is some behavioral flexibility. Young workers can assume the work of older ants more easily than vice versa (Sorensen et al., 1983d).

In the laboratory at 29.5 to 32° larval development requires six to ten days, while the pupal stage takes seven to eight days. The time required from oviposition to adult stage takes seven to eight days. The time required from oviposition to adult is 20–21 days (O'Neal and Markin 1975a). Major workers developing in the field from May through July require 22–23 days, while 24–30 days are needed for the development of queens (Markin et al. 1972).

The life span of a red IFA colony is unknown. For a single-queen colony it will be the same as the queen's age, which is also unknown. On the other hand, Tschinkel and Howard (1978, 1980) showed that orphaned colonies could replace their queen. They observed that these replacement queens provided workers, but they considered it a case of thelytoky (production of females parthenogenetically). There is some doubt as to the validity of this conclusion. Fincher and Lund (1967) and Voss (1981) reported that only males are produced from virgin females.

Mound Development

Markin et al. (1973) reported that the founding queen's burrow extends a few inches into the ground and is often sealed. The burrow is opened and deepened by the newly emerged workers. Distinct chambers are formed by 60 days and by 90 days a small mound (3–7 cm by 5–7 cm high) is formed. This small mound consists of numerous chambers and five to 15 side tunnels branching at various depths from the original burrow. In one to two years a mound may provide 40 liters of space with tunnels extending 2 or 3 meters from the mound just beneath the soil surface. The subsurface tunnels have been referred to as foraging tunnels; along their length, which may branch, there are small openings to the outside. Tunnels may also extend down into the soil a meter or more to the water table. The mound structure will depend on the soil type. Mounds are flat in sandy soil but more dome-shaped in clays. Large mature mounds are often 0.4 m high by 0.5 m wide. Some very large mounds have been measured, particularly in *S. richteri* (Green 1967).

Many ant species appear to be capable of regulating the temperatures of their mounds (Petal 1978). There is no evidence, at present, that the red IFA can regulate either the temperature or humidity of their mound. However, they do move the brood and queen in response to these factors. On a hot summer day the queen, brood and workers are usually found deep within the mound, and are reluctant to come to the surface, even when disturbed. In the spring and fall during the cool mornings the brood and queen are brought to the surface, usually on the sunny side where they can be easily collected. In the cooler months the queen and brood are often 16 to 40 cm beneath the surface and may be relatively inactive.

PHYSIOLOGY AND BEHAVIOR

Feeding, Food Distribution and Nutrition

The red IFA has been described as omnivorous (Hays and Hays 1959), but its primary diet consists of insects and other small invertebrates (Green 1962, 1967, Wilson and Eads 1949, N. L. Wilson and Oliver 1969, Hays and Hays 1959). However, among these the ant does exhibit some preferences (Ricks and Vinson 1970). The ant also feeds on small animals (Travis 1938, Lyle and Fortune 1948), seeds, plant fats and oils (Hays and Arant 1960, Lofgren et al. 1961, 1964), and even living plants (N. L. Wilson and Oliver 1969, Smittle et al. 1983; also see Chapter 8).

The red IFA has been called an "oil loving" species due to its fondness for plant oils (Hays and Arant 1960, Lofgren et al. 1961, 1964), a fact used in the development of the current baits used for control (Banks et al. 1978). Vinson et al. (1967) determined that linolenic and linoleic acids and their tri- or diglycerides act as phagostimulants, as do several phospholipids containing these fatty acids. These fatty acids and their esters are responsible for the major "attractiveness" of the oils (Vinson et al. 1967). In an effort to determine the factors that stimulate the red IFA to feed on insects, Ricks and Vinson (1970) found that the ants prefer water extracts of insects. Most amino acids are no more attractive than water, with the exception of leucine and valine. Several sugars are attractive, including melezitose, sucrose, and glucose. Xylose, ribose, mannose, arabinose, and galactose are rejected, as are all the B vitamins examined (Ricks and Vinson 1970). Smittle et al. (1983) found the red IFA feeding on okra pods; this activity appears to be due to sugars (R. K. Vander Meer, personal communication). The red IFA also tends aphids (Hays and Hays

1959, Nielsson et al. 1971, Scarborough and Vinson, in prep.) attracted by sugars present in the honeydew. The planthopper *Oliarus vicarius*, a subterranean root feeder, has also been found in mounds of the red IFA (Sheppard et al. 1979) and may be tolerated if they produce honeydew. The red IFA sometimes collects nectar from plants and their abundance in cotton fields has been correlated to nectar flow (Agnew and Dean 1982). Although Lyle and Fortune (1948) reported that the IFA had no interest in sugars, the preference for sugar and other foods may change with the season (Ricks and Vinson 1972a). D. Bogar (unpublished data), using corncob grits impregnated with soybean oil, sucrose solutions, or egg yolk protein, found that field colonies prefer proteins in the spring, and oils throughout the summer and fall. Sucrose, although less preferred over the other two, was preferred in the summer. In Bogar's study colonies varied markedly in their responses, as was the case with Glunn et al. (1981). The latter reported that recruitment levels in the field were, in decreasing order, soybean oil, serum (protein) and 0.1 M sucrose. In the laboratory they reported a decreasing preference of sucrose, soybean oil, and protein.

The nutritional needs of the red IFA are not well defined. Bhatkar and Whitcomb (1970) developed an artificial diet for ants. However, this diet was not adequate to the red IFA unless 10% soybean oil was included (Vinson 1980). Improved diets for the red IFA have since been published (D. F. Williams et al. 1980) consisting of insect parts, 50% honey water provided separately, and a source of water.

Food distribution and exchange have received considerable attention in the red IFA. E. O. Wilson (1971) observed that the complex system of food exchange not only provided nourishment to the colony but also played an important role in its social organization. Early studies of food distribution either involved a dye or insecticide incorporated into various foods (Lofgren et al. 1975; Vinson 1968, O'Neal and Markin 1973, Petralia and Vinson 1978, Glancey et al. 1973b). These efforts were reviewed by Vinson (1980, 1983). In recent years, ^{125}I-labeled protein (Sorensen and Vinson 1981) or ^{125}I mixed with food or bound to soybean oil (Howard and Tschinkel 1981a, b) have been used to elucidate some of the complexities of food flow. Sorensen and Vinson (1981) found that powdered egg yolk was fed to larvae more quickly than liquid protein, honey, or soybean oil. (After 48 hours 85%, 45–50%, 40%, and 30%, respectively, of these foods, had reached the larvae.) Howard and Tschinkel (1981a) found that sugar is utilized primarily by the workers, amino acids are directed preferentially to the queen and larvae, and soybean oil is shared equally among workers and larvae.

The kind of food also influences its distribution in the nest. Petralia and Vinson (1978) reported that the first 3 larval instars were fed liquid food while the fourth was also fed solid food. Studies on the size of food particles consumed by workers and larvae (Glancey et al. 1981c) revealed that fourth instar larvae could swallow particles as large as 45 μm while workers could filter out particles as small as 1 μm. The authors also reported that workers contain a system of ridges and hairs in the buccal tube that act as a filter to screen out particles from the liquid food. The solids from buccal pellets that average 130 to 140 μm are often fed to the fourth instar larvae (Petralia and Vinson 1978). Solid foods (proteins) brought in by foragers are ultimately fed to the larvae, reaching them in three hours (Sorensen et al. 1981). Liquid foods are passed among nest mates several times and reach larvae in 48 hr or longer (Sorensen et al. 1981).

The different behavioral subcastes (Mirenda and Vinson 1981) play distinct roles in distributing food. Sorensen et al. (1983b) found that food passes sequentially from foragers to nurses, and then to the larvae and queen. Solid foods are relayed more rapidly than liquid foods (Sorensen et al. 1981). Horton et al. (1975) noted that foraging workers return to their nest and give the food to a nest mate. They then go out to forage again, rather than enter the nest. The castes have different turnover rates of food, the reserves being slowest and acting as a temporary storage caste (Sorensen et al. 1981). Wilson (1978) found that smaller workers retain sugar water longer than larger workers. Glancey et al. (1973b) concluded that the larger workers store oil. On the other hand, contradictory evidence was given by Howard and Tschinkel (1981b) who concluded that smaller workers were least suitable for retaining sugar water. In a study of role flexibility Sorensen et al. (1983d) found that foragers gather more food than either reserves or nurses. Isolated foragers and reserves assume new tasks of tending brood or foraging. Isolated nurses start to forage only after a delay of several days.

Howard and Tschinkel (1980) found that colony size and starvation both influence food flow in a colony. Sugars are consumed in greater amounts and shared with more nest mates when the colony is starved. Small colonies also distribute sugars differently when compared to large colonies. In a later study Howard and Tschinkel (1981a) found that starvation stimulates food dispersal, but some foods are more effective at this than others. Amino acids are preferentially given to larvae. Sorensen et al. (1983a) found that starvation of foragers and nurses influences the amount of sugar and protein, respectively, brought into the colony.

Colony starvation also effects brood cannibalism (Sorensen et al. 1983a). Not only does starvation increase cannibalism, but the type of food available to the colony influences the rate of cannibalism.

Temperature also affects food flow. Howard and Tschinkel (1981a) showed that food intake is lower at 25°C than at 35°C. Markin et al. (1974) found that foraging activity is greatest when the air temperature is between 21°C and 32°C and when the soil temperature is not greater than 43°C. Foraging often increases sharply after a period of rain (Rhoades and Davis 1967). D. Bogar (unpublished) found that in Texas fire ants forage mostly at night during the summer, when daytime soil temperatures are high. In early spring and late fall they forage more during the day. Morrill (1977b) found that red IFAs feeding on white flies, *Trialeurodes vaporariorum*, in a greenhouse forage 24 hours a day.

In addition to temperature, starvation, type of food, time of year and presence of brood, competition also affects foraging activity of the red IFA. Urbani and Kannowski (1974) reported that during the day over 95% of baits distributed over a red-IFA-infested pasture were first discovered by *S. invicta*, and that they retrieved 100% of the food. These results demonstrate the efficiency of *S. invicta* in locating and dominating a food resource. However, Urbani and Kannowski (1974) also found that when baits were placed on a red IFA mound, only 70% were exploited by *S. invicta*. The remaining 30% were exploited by *Monomorium minimum* (Buckley), which emits a strong repellent that is effective against *S. invicta* but not *S. xyloni* (Urbani and Kannowski 1974). The competitive strategies used by *Monomorium minimum* have been discussed by Adams and Traniello (1981), and the repellent has been identified by Blum et al. (1980). Feener (1981) found that *Pheidole dentata* would outcompete *Solenopsis texana* for food by recruiting many major workers quickly. Buren (1983) discussed the possibility of using certain ant species that are competitors or predators of *S. invicta*. However, Feener (1981) also found that the phorid fly *Apocephalus* spp., which is parasitic on major *P. dentata* workers, caused a shift in the behavior of the major workers. The fly's presence reduces the recruitment of major *P. dentata*, allowing *S. texana* to take over the food resource.

According to Wilson et al. (1971) the fire ant has distinct foraging territories. On the other hand, Summerlin et al. (1975) showed that food was transferred between colonies in both Florida and Texas. These two sets of results are not easily reconciled. One possibility is that the first author studied monogynous colonies while the latter examined polygynous colonies.

Digestion, Absorption, and Excretion

In addition to the postpharyngeal gland, the red IFA has maxillary, mandibular, and labial glands associated with feeding. All of these glands, with the exception of the postpharyngeal, are, relatively speaking, best developed in workers (Phillips and Vinson 1980). However, our knowledge of the function of these glands is limited. Ricks and Vinson (1972b) examined their enzyme activity and found lipase and invertase associated with the mandibular gland of workers, while lipase and amylase were found in the labial gland. Sorensen et al. (1983c) examined the distribution of proteinase in the colony and found that it was correlated with the presence of larvae. Petralia and Vinson (1980) described the internal anatomy of larvae and reported the presence of a well-developed labial gland. Additional studies (Petralia et al. 1980) revealed the presence of a proteinase and amylase in the larval labial gland, and the authors suggested that the protein found in colonies may come from the larvae. Their speculation was supported by Sorensen et al. (1983c), who reported that in the absence of larvae, protein levels in workers dropped regardless of the type of food provided, and increased again when larvae were added.

In an effort to determine the fate of lipids in the digestive tract, Vinson et al. (1980) injected ^{14}C-labeled triolein and oleic acid into the crops of recently mated queens. They found that the label moved forward to the esophageal crop and postpharyngeal gland, and that some was regurgitated to the developing larvae. If a queen with a ligated esophagus was fed ^{14}C-labeled lipids, the label appeared in the postpharyngeal gland and hemolymph, suggesting that the postpharyngeal glands are involved in lipid absorption. Postpharyngeal glandectomized queens were treated behaviorally as normal queens for two months, although they lost weight and died even though their crop and midgut still contained food (Vinson et al. 1980).

Vander Meer et al. (1981a) analyzed the hydrocarbons in the crop and postpharyngeal gland and found that the hydrocarbons are the main components in the postpharyngeal gland, and in higher concentrations than in the crop. They also found that the postpharyngeal gland hydrocarbons consisted almost entirely of four uncommon methyl-branched compounds identified by Thompson et al. (1981) from the cuticle. These results led Vander Meer et al. (1981a) to suggest that the postpharyngeal gland secreted them.

Adult excretion has not been studied in the red IFA, but Petralia and Vinson (1980) described the anatomy of the larval excretory system. In a later study, Petralia et al. (1982) identified the larval excretory product as water, salts, and uric acid. Although O'Neal and Markin (1973) suggested that the uric acid was consumed by adults, Petralia et al. (1982) found that uric acid is carried out of the colony by workers. Under water stress, Petralia et al. (1982) did find that excreted water was consumed by workers.

BEHAVIOR AND SEMIOCHEMICALS

Vander Meer (1983), in his review of semiochemicals (i.e., chemicals used in communication) in the red IFA, noted that many behaviors involve chemical communication. Although a number of pheromone systems exist, only four have been examined in any detail (Blum 1980). These are the brood pheromones, trail pheromones, queen pheromones, and nest mate recognition pheromones.

Brood Pheromone

Glancey et al. (1970) showed that a brood pheromone could be extracted from larvae with hexane. Walsh and Tschinkel (1974) demonstrated pheromone activity with sexual pupae, as well as worker and sexual larvae. They also indicated that the pheromone probably elicited a behavior only on contact. Bigley and Vinson (1975) utilized the bioassay procedure of Glancey et al. (1970), but used sexual prepupae and identified the major active component as triolein. Vander Meer (1983) criticized their work, suggesting that other components may be present, a possibility not excluded by Bigley and Vinson (1975). Furthermore, sexual and worker larvae and pupae are often placed separately in a nest, suggesting that there is a complex of brood pheromones yet to be isolated and identified from the red IFA.

Trail Pheromone

The trail pheromone has probably been the most extensively studied and was among the first investigated. Wilson (1959) recognized the Dufour's gland as the source of a chemical deposited on the substrate by workers that elicits trail following by nest mates. Callahan et al. (1959) described the morphology of the Dufour's gland and accompanying poison gland. The behavior involved in depositing a trail was examined in some detail by Wilson (1962).

The specificity of the trail pheromone has also been examined (Wilson 1962, Barlin et al. 1976, Jouvenaz et al. 1978). In general their results are in agreement in that *S. richteri* and *S. invicta*, the two imported species, responded to each other's trails but not to the trails of the native species, while the two native species *S. geminata* and *S. xyloni*, respond to each other, but not to the imported species. Wilson started the chemical study of the trail pheromone (1959). He purified a steam distillate of the pheromone; it lost activity when it was chromatographed. Additional purification was reported by Walsh et al. (1965). Barlin et al. (1976)

studied trail pheromones of several *Solenopsis* species and suggested an empirical formula of $C_{13}H_{21}$ for *Solenopsis invicta*. Both Vander Meer et al. (1981b) and H. J. Williams et al. (1981a, b) described the trail pheromone of *S. invicta*. Both authors reported the presence of Z, E-α-farnesene, but H. J. Williams et al. (1981b) considered it a product of Z, Z, Z-allofarnesene. Vander Meer et al. (1981b) additionally reported the presence of E, E-α-farnesene, Z, E-homofarnesene, and Z, Z-homofarnesene, all of which had weak trail-following activity. Vander Meer (1983) has suggested that these may act as recruitment pheromones. Whether the Z, E-α-farnesene or Z, Z, Z-allofarnesene is the "true" trail pheromone is unresolved. However, the colonies used by H. J. Williams et al. (1981b) contained multiple queens, while those of Vander Meer et al. (1981b) were single-queen colonies. Whether this is of any significance is unknown.

Queen Pheromone

Jouvenaz et al. (1974) described a *S. invicta* queen attractant. Glancey (1980) and Vander Meer et al. (1980) developed bioassays for this attractant. Vander Meer also showed that the active chemicals were stored in the poison sac. D. F. Williams et al. (1981) observed that workers respond to dead queens for more than one week, and suggested that this activity was due to the pheromone's slow release. Glancey et al. (1981a) found that the queen pheromone is also produced by virgin dealated queens. Lofgren et al. (1983) found that queen extracts are attractive to workers at concentrations as low as 0.01 queen equivalents, although they reported that the queen's attractiveness decreases after 30 minutes of isolation from a colony. Tumlinson (1980) reported that concentrated hexane extract distillates remained active for several days.

Fletcher and Blum (1981a,b) described another *S. invicta* pheromone that inhibits the dealation and egg laying of virgin females. If the original queen is removed the inhibition is removed and some virgin females alates then dealate. Within a few weeks all but one of the dealate queens are executed (Fletcher and Blum 1983). Fletcher and Blum (1981b) developed a bioassay for this primer pheromone.

Nest Mate Recognition Pheromones

There is considerable evidence that ants recognize odors of nest mates (Wilson 1971, Mintzer 1982). Wilson et al. (1971) showed that colonies of the red IFA often have specific territories on the ground where they forage, and there is evidence that food in one colony does not end up in neighboring colonies. However, there have been exceptions where food from one colony does end up in many surrounding colonies (Summerlin et al. 1976, Bhatkar 1979). Some of these discrepancies may be due to differences between single- and multiple-queen colonies. The latter have been found in Texas and Mississippi (Mirenda and Vinson 1982, Fletcher et al. 1980, Glancey et al. 1973a, 1975). Mirenda and Vinson (1982) showed that workers from multiple-queen colonies are less aggressive toward non-nest-mate workers than those from single-queen colonies and that workers may move between colonies.

Hubbard (1974) showed that red IFA workers prefer to dig in soil from their own colony. This led Hubbard (1974) to suggest that the ants must transfer specific chemicals to the soil. The cuticle may be the source of these colony-specific odors. Lok et al. (1975), in a study of the cuticular lipids of the red IFA, reported that 65–75% of them are hydrocarbons; triglycerides, free fatty acids, wax esters, and sterols comprise the remainder of the lipids. Of these, the triglycerides are of interest because triolein and oleic acid are present. Wilson et al. (1958) reported that some chemicals release necrophoric behavior, that is, causing workers to pick up dead ants and carry them out of the colony. These authors speculated that bacterial action degrades cuticular triglycerides, releasing fatty acids. One of these, oleic acid, elicits necrophoric behavior. Blum (1970) confirmed that oleic and other fatty acids release the necrophoric behavior of *Solenopsis invicta*. Howard and Tschinkel (1976) found that dead ants are removed in random directions from the mound unless a slope exists, in which case the dead ants are carried downhill. They also noted that the workers drop the dead ants on reaching a refuse pile, suggesting that a contact chemical signal causes the behavior.

In addition to fatty acids the cuticle contains hydrocarbons that may be involved in nest-mate recognition. The major hydrocarbons have been identified by Nelson et al. (1980) and are composed of saturated monomethyl and dimethyl branched hydrocarbons, the pattern being distinctly different for S. richteri and S. invicta. The same hydrocarbons were found by Thompson et al. (1981) in the crop and postpharyngeal gland, indicating that they could be readily transferred during trophallaxis. Vander Meer et al. (1981a) examined the hydrocarbons of the postpharyngeal glands and crop and found that the postpharyngeal glands contain a greater concentration of them. They hypothesized that the hydrocarbons are secreted by the postpharyngeal glands. The role of hydrocarbons in nest-mate recognition is speculative, but Vander Meer and Wojcik (1982) demonstrated that the ant hydrocarbons play a role in the acceptance of the myrmecophilus beetle, *Myrmecaphodius excavaticollis* (Blanchard) in ant colonies. The cuticular hydrocarbons of the beetle reared away from the ants are distinctive. When the beetle was reared with *Solenopsis invicta* the hydrocarbon pattern of the beetle showed its own pattern plus that of S. *invicta*. When the beetle was reared with S. *richteri* it had its own hydrocarbon pattern plus that of S. *richteri* (Vander Meer and Wojcik 1982). Thus, the hydrocarbons may play an important role in nest-mate recognition.

FIRE ANT VENOM

Space limitations do not permit detailed discussion of this topic. Several deaths have been traced to red IFA stings (F. K. James, personal communication). For detailed discussion of the venom, its effects and components, see Adams and Lofgren (1981), Rhoades et al. (1978), Parrino and Kandawalla (1981), James (1976), Lind (1982), T. N. Jones and Blum (1982), Matsumura et al. (1982), and Buffkin and Russell (1972).

ECOLOGY

The ecology of the red IFA has received little study, even though understanding the conditions that limit the insect's spread is imperative if we hope to control it. There have been a few studies of S. *invicta* in South America. Buren et al. (1974) and Wojcik (1983) state that the red IFA constructs smaller mounds in South America than in the United States. In the United States the red IFA apparently nests in all soil types (USDA 1958, Wangberg et al. 1980). The type of soil may affect mound size. Mounds in sandy soil dry areas tend to be small (Buren 1982, Wojcik 1983), while mounds in wet areas tend to be large, probably to escape the high water tables (Wojcik 1983, USDA 1958).

Eisenberg (1972) studied the arrangement of IFA mounds using a nearest neighbor analysis. He found that the mound distribution was uniform rather than random. The mean distance to a nearest neighbor was 5.9 m, as compared to 4.47 m for a random expectation. The reasons given for the nonrandom spacing were competition for food and space. Wilson et al. (1971) showed that colonies have distinct defended territories. Such territories are not static since colony movement has been frequently observed (Eisenberg 1972, Hays et al. 1982). These observations undoubtedly apply only to single-queen colonies.

The question of how to control the red IFA is still controversial. Obviously any control program will initially reduce the ant population. But long-term effects of insecticide use are less certain. Summerlin et al. (1977) studied the effects of an eradication effort in Brazos County, Texas. Initially there were many ant species and a low population of the red IFA. After treatment, several species, including the red IFA, decreased in number. However, within one year the red IFA and *Conomyrma insana* became the dominant species at the expense of the others. Buren (1983) believes that insecticide usage eventually causes a resurgence of the red IFA. Hence, insecticides do not solve the problem. In 1981 Texas treated Madison County with Amdro® ; within six months the red IFA returned with more mounds than before treatment (Vinson, personal observation).

The red IFA may prefer disturbed habitats. Summerlin et al. (1976) surveyed for the presence of the red IFA in various habitats and found more colonies on pastures than cropland, with both having many more than forested areas. Tschinkel (1982) discusses the fire ant as a weed-like insect. Like a weed the red IFA appears to be a good invader of disturbed ecosystems, but once in such an ecosystem the red IFA has a significant impact. Risch and Carroll (1982) have shown that S. geminata is a keystone predaceous ant that has a significant impact on reducing the arthropod fauna. Like S. geminata, the IFA simplifies the ecosystem, particularly with regard to other ant species. Although Rhoades (1963) was unable to observe any effect of the IFA on other arthropods, he was primarily concerned with the impact of control practices. Glancey et al. (1976b), in contrast, found fewer ant species in an area where the IFA is well established and numerous than in comparable areas with incipient fire ant infestations.

Abiotic Factors Affecting Distribution

The 0°F isotherm has been proposed as the limiting factor to the northern spread of the red IFA in the United States (Anon. 1972, Pimm and Bartell 1980). The availability of moisture, as discussed earlier in the sections concerning importation and spread, may also be important. Vinson (Texas Department of Agriculture 1980 report), in a study of the temperature and humidity tolerance of the red IFA, noted that the ants are susceptible to hot and dry conditions, but less so than either S. geminata or S. xyloni.

Support for the view that the red IFA will not invade the drier desert regions is provided by the native Solenopsis fauna. Both S. xyloni and S. geminata, to which the red IFA is similar, are replaced by S. aurea in the desert regions of the United States (Moody et al. 1981).

Biotic Factors Affecting Distribution

The role that predators, parasites, diseases and competitors play in limiting the spread of the IFA has not been adequately studied in either the United States or South America. Newly mated queens appear to be most vulnerable to biotic factors. Queens on their nuptial flights are consumed by dragonflies, birds, and robber flies (Whitcomb et al. 1973, Glancey

1981), while tiger beetles, mites, spiders, and antlions are important predators after landing (Edwards et al. 1974, Lucas and Brockman 1981). Other red IFA colonies may also be a major source of mortality. However, these opportunistic predators have a doubtful impact on the population.

Leston (1980), discussing the role of biotic factors in limiting the spread of the red IFA, considered the predators, parasites, and pathogens to be of little importance. On the other hand, he believes that the competing ants are important. The feasibility of importing competing ants for control has been discussed by Buren (1980, 1983). With regard to other ant species, both *Conomyrma insana* (Hung 1974) and *Lasius niger* (Bhatkar et al. 1972) do attack and kill workers. However, they probably affect only the youngest colonies. O'Neal (1974) reported that the army ant, *Neivamyrmex opacithorax* (Emery), as well as *Paratrechina melanderi arenivaga* (Wheeler) and *S. molesta* (Say), consume eggs of the IFA, but their impact is unknown. Nickerson et al. (1975) observed *C. insana* predating on founding queens.

Subterranean ants may play a significant role in reducing the red IFA (Thompson 1980). *Diplorhoptrum* species such as *S. molesta* have been found in areas where the red IFA is not firmly established (J. M. Lammers, personal observation). The role that these and other competing ant species play in retarding the spread of the red IFA is not clear. The effects may be indirect through competition for the limited food supply. Where native ants are abundant, they may reduce the ant-carrying capacity to levels where the red IFA is unable to obtain enough food to develop strong colonies. However, if the red IFA becomes well established it can likely outcompete most other species.

Ant parasites may be limited in their ability to control the red IFA directly, but may be very important in weakening their competitive advantage. For reviews of this topic see R. N. Williams (1980), Williams and Whitcomb (1974), Feener (1981), Wojcik et al. (1978), Neece and Bartell (1981), Silviera-Guido et al. (1973), Summerlin (1978), Jouvenaz (1983), Knell et al. (1977), Allen and Buren (1974).

REFERENCES

Adams, C. T. and C. S. Lofgren. 1981. Red imported fire ants: frequency of sting attacks on residents of Sumter County, Georgia. *J. Med. Entomol.* 18: 378–382.

Adams, C. T., W. A. Banks, and J. K. Plumley. 1976. Polygyny in the tropical fire ant, *Solenopsis geminata*, with notes on the imported fire ant, *Solenopsis invicta*. *Fla. Entomol.* 59: 411–415.

Adams, E. S., and J. F. A. Traniello. 1981. Chemical interference competition by *Monomorium minimum*. *Oecologia* 51: 265–270.

Agnew, C. W., and D. A. Dean. 1982. Influence of cotton nectar on red imported fire ants and other predators. *Environ. Entomol.* 2: 629–634.

Allen, G. E. and W. F. Buren. 1974. Microsporidian and fungal diseases of *Solenopsis invicta* Buren in Brazil. *J. N.Y. Entomol. Soc.* 82: 125–130.

Anonymous. 1972. Ecological ranges for the imported fire ant based on plant hardiness. *Coop. Econ. Insect Rep.* 22(7).

Ball, D. E. and S. B. Vinson. 1985. Anatomy and histology of the male reproductive system of the imported fire ant, *Solenopsis invicta. Int. J. Morph. Embryol.* 13: 283–294.

Ball, D. E., S. B. Vinson, and H. J. Williams. 1984. Chemical analysis of the male aedeagal bladder of the fire ant, *Solenopsis invicta* Buren. New York Entomol. Soc. 92: 365–370.

Banks, W. A., B. M. Glancy, C. E. Stringer, D. P. Jouvenaz, C. S. Lofgren, and D. E. Weidhaas. 1973. Imported fire ants: Eradication trials with mirex bait. *J. Econ. Entomol.* 66: 785–789.

Banks, W. A., C. S. Lofgren, and D. P. Wojcik. 1978. A bibliography of imported fire ants and the chemicals and methods used for their control. U.S. Dept. Agric., Agric. Res. Serv., ARS-S-180.35 pp.

Barker, J. F. 1978. Neuroendocrine regulation of oocyte maturation in the imported fire ant, *Solenopsis invicta. Gen. Comp. Endocrinol.* 35: 234–237.

Barker, J. F. 1979. Endocrine basis of wing casting and flight muscle histolysis in the fire ant *Solenopsis invicta. Experientia* 35: 552–554.

Barlin, M. R., M. S. Blum, and J. M. Brand. 1976. Fire ant trail pheromones: Analysis of species specificity after gas chromatographic fractionation. *J. Insect Physiol.* 22: 839–844.

Bass, J. A. and S. B. Hays. 1976. Geographic location and identification of fire ant species in South Carolina. *J. Ga. Entomol. Soc.* 11: 34–36.

Bass, J. A. and S. B. Hays. 1979. Nuptial flights of the imported fire ant in South Carolina. *J. Ga. Entomol. Soc.* 14: 158–161.

Bhatkar, A. P. 1979. Evidence of intercolonial exchange in fire ants and other Myrmicinae, using radioactive phosphorus. *Experientia* 35: 1172–1173.

Bhatkar, A. and W. H. Whitcomb. 1970. Artificial diet for rearing various species of ants. *Fla. Entomol.* 53: 229–232.

Bhatkar, A., W. H. Whitcomb, W. F. Buren, P. Callahan, and T. Carlysle. 1972. Confrontation behavior between *Lasius neoniger* and the imported fire ant. *Environ. Entomol.* 1: 274–279.

Bigley, W. S. and S. B. Vinson. 1975. Characterization of a brood pheromone isolated from sexual brood of the imported fire ant. *Ann. Entomol. Soc. Amer.* 68: 301–304.

Blum, M. S. 1970. The chemical basis of insect sociality. *In* "Chemicals controlling insect behavior," pp. 61–94 (Beroza, M., ed.), Academic Press. New York.

Blum, M. S. 1980. Pheromones of adult fire ants (*Solenopsis* spp.). *Proc. Tall Timbers Conf. Ecol. Anim. Control Habitat Manag.* 7: 55–60.

Blum, M. S., T. H. Jones, B. Holldobler, H. M. Fales, and T. Jaouni. 1980. Alkaloidal venom mace: Offensive use by a thief ant. *Naturwissenschaften* 67: 144–145.

Brown, W. L., Jr. 1961. Mass insect control programs: Four case histories. *Psyche* 68: 75–109.

Bruce, W. A., L. D. Cline, and G. L. LeCato. 1978. Imported fire ant infestation of buildings. *Fla. Entomol.* 61: 230.

Buffkin, D. C. and F. E. Russell. 1972. Some chemical and pharmacological properties of the venom of the red imported fire ant, *Solenopsis saevissima richteri. Toxicon* 10: 526.

Buren, W. F. 1972. Revisionary studies on the taxonomy of the imported fire ants. *J. Ga. Entomol. Soc.* 7: 1–26.

Buren, W. F. 1980. The importance of fire ant taxonomy. *Proc. Tall Timbers Conf. Ecol. Anim. Control Habitat Manag.* 7: 61–66.

Buren, W. F. 1982. Red imported fire ant now in Puerto Rico. *Fla. Entomol.* 65: 188–189.

Buren, W. F. 1983. Artificial fauna replacement for imported fire ant control. *Fla. Entomol.* 66: 93–100.

Buren, W. F., G. E. Allen, W. H. Whitcomb, F. E. Lennartz, and R. N. Williams. 1974. Zoogeography of the imported fire ants. *J. N.Y. Entomol. Soc.* 82: 113–124.

Callahan, P. S., M. S. Blum, and J. R. Walker. 1959. Morphology and histology of the poison glands of the imported fire ant (*Solenopsis saevissima* v. *richteri.* Forel). *Ann. Entomol. Soc. Amer.* 52: 573–590.

Carter, L. W. 1981. Final programmatic environmental impact statement. U.S.D.A., APHIS-ADM-81-01-F. 240 pp.

Clark, S. W. 1931. The control of fire ants in the lower Rio Grande Valley. *Tex. Agr. Exp. Sta.* B-435. 12 p.

Creighton, W. S. 1930. The New World species of the genus *Solenopsis. Proc. Amer. Acad. Arts Sci.* 66: 39–151, plates 1–8.

Creighton, W. S. 1950. "The ants of North America." *Bull. Mus. Comp. Zool. Harvard Univ.* 104: 1–585.

Culpepper, G. H. 1953. Status of the imported fire ant in the southern states in July 1953. *U.S. Dep. Agric. Bur. Entomol. Plant Q.* [Rep.] E-867. 8 pp.

Edwards, G. B., J. F. Carroll, and W. H. Whitcomb. 1974. *Stoidis aurata,* a spider predator of ants. *Fla. Entomol.* 57: 337–346.

Eisenberg, R. M. 1972. Partition of space among colonies of the fire ant, *Solenopsis saevissima.* I. Spatial Arrangement. *Tex. J. Sci.* 24: 39–43.

Ettershank, G. 1964. A generic revision of the world Myrmicinae related to *Solenopsis* and *Pheidologeton.* Ph.D. dissertation, Cornell University, Ithaca. [Published in 1966 in *Aust. J. Zool.* 14: 73–171.]

Feener, D. H., Jr. 1981. Competition between ant species: Outcome controlled by parasitic flies. *Science* 214: 815–817.

Fincher, G. T. and H. O. Lund. 1967. Notes on the biology of the imported fire ant. *Solenopsis saevissima richteri* Forel in Georgia. *J. Ga. Entomol. Soc.* 2: 91–94.

Fletcher, D. J. C. and M. S. Blum. 1981a. Pheromonal control of dealation and oogenesis in virgin queen fire ants. *Science* 212: 73–75.

Fletcher, D. J. C. and M. S. Blum. 1981b. A bioassay technique for an inhibitory primer pheromone of the fire ant, *Solenopsis invicta* Buren *J. Ga. Entomol. Soc.* 16: 352–356.

Fletcher, D. J. C., and M. S. Blum. 1983. Regulation of queen number by workers in colonies of social insects. *Science* 29: 312–314.

Fletcher, D. J. C., M. S. Blum, T. V. Whitt, and N. Temple. 1980. Monogamy and polygamy in the fire ant. *Ann. Entomol. Soc. Amer.* 73: 658–661.

Francke, O. F., J. C. Cokendolpher, A. H. Horton, S. A. Phillips, L. R. Potts. 1983. Distribution of fire ants in Texas. *Southw. Entomol.* 8: 32–41.

Glancey, B. M. 1980. Biological studies of the queen pheromone of the red imported fire ant. *Proc. Tall Timbers Conf. Ecol. Anim. Control Habitat Manag.* 7: 149–154.

Glancey, B. M. 1981. Two additional dragonfly predators of queens of the red imported fire ant, *Solenopsis invicta* Buren. *Fla. Entomol.* 64: 194–195.

Glancey, B. M., C. E. Stringer, C. H. Craig, P. M. Bishop, and B. B. Martin. 1970. Pheromone may induce brood tending in the fire ant, *Solenopsis saevissima*. *Nature (London)* 226: 863–864.

Glancey, B. M., C. H. Craig, C. E. Stringer, and P. M. Bishop. 1973a. Multiple fertile queens in colonies of the imported fire ant, *Solenopsis invicta*. *J. Ga. Entomol. Soc.* 8: 237–238.

Glancey, B. M., C. E. Stringer, C. H. Craig, P. M. Bishop, and B. B. Martin. 1973b. Evidence of a replete caste in the fire ant, *Solenopsis invicta*. *Ann. Entomol. Soc. Am.* 66: 233–234.

Glancey, B. M., C. E. Stringer, C. H. Craig, and P. M. Bishop. 1975. An extraordinary case of polygyny in the red imported fire ant. *Ann. Entomol. Soc. Amer.* 68: 922.

Glancey, B. M., M. K. Vandenburgh, and M. K. St. Romain. 1976a. Testes degeneration in the red imported fire ant, *Solenopsis invicta*. *J. Ga. Entomol. Soc.* 11: 83–88.

Glancey, B. M., D. P. Wojcik, C. H. Craig, and J. A. Mitchell. 1976b. Ants of Mobile County, Ala., as monitored by bait transects. *J. Ga. Entomol. Soc.* 11: 191–197.

Glancey, B. M., A. Glover, and C. S. Lofgren. 1981a. Pheromone production by virgin queens of *Solenopsis invicta* Buren. *Sociobiolog* 6: 119–127.

Glancey, B. M., A. Glover, and C. S. Lofgren. 1981b. Thoracic crop formation following dealation by virgin females of two species of *Solenopsis*. *Fla. Entomol.* 64: 454.

Glancey, B. M., R. K. Vander Meer, A. Glover, C. S. Lofgren, and S. B. Vinson. 1981c. Filtration of microparticles from liquids ingested by the red imported fire ant, *Solenopsis invicta* Buren. *Insects Soc.* 28: 395–401.

Glunn, F. J., D. F. Howard, and W. R. Tschinkel. 1981. Food preference in colonies of the fire ant *Solenopsis invicta*. *Insectes Soc.*, 28: 217–222.

Green, H. B. 1952. Biology and control of the imported fire ant in Mississippi. *J. Econ. Entomol.* 45: 593–597.

Green, H. B. 1959. Imported fire ant mortality due to cold. *J. Econ. Entomol.* 52: 347.

Green, N. B. 1962. On the biology of the imported fire ant. *J. Econ. Entomol.* 55: 1003–1004.

Green, H. B. 1967. The imported fire ant in the southeastern United States. *Agric. Exp. Stat., Miss. St. Univ.* Bulletin 737.

Greenberg, L., D. J. C. Fletcher, and S. B. Vinson. 1985. Monogyny and polygyny in the imported fire ant: correlation with worker size and mound location.

J. Kans Entomol. Soc. 58: 9–18.

Hays, K. L. 1959. Ecological observations on the imported fire ant, *Solenopsis saevissima richteri* Forel, in Alabama. *J. Ala. Acad. Sci.* 30: 14–18.

Hays, S. B., and K. L. Hays. 1959. Food habits of *Solenopsis saevissima richteri* Forel. *J. Econ. Entomol.* 52: 455–457.

Hays, S. B., and F. S. Arant. 1960. Insecticidal baits for control of the imported fire ant, *Solenopsis saevissima richteri. J. Econ. Entomol.* 53: 188–191.

Hays, S. B., P. M. Horton, J. A. Bass, and D. Stanley. 1982. Colony movement of imported fire ants. *J. Ga. Entomol. Soc.* 17: 266–274.

Hermann, H. R., and M. S. Blum. 1965. Morphology and histology of the reproductive system of the imported fire ant queen, *Solenopsis saevissima richteri. Ann. Entomol. Soc. Amer.* 58: 81–89.

Hess, C. C. 1958. The ants of Dallas County, Texas, and their nesting sites; with particular reference to soil texture as an ecological factor. *Field Lab.* 26: 1–72.

Horton, P. M., S. B. Hays, and J. R. Holman. 1975. Food carrying ability and recruitment time of the red imported fire ant. *J. Ga. Entomol. Soc.* 10: 207–213.

Howard, D. F. and W. R. Tschinkel. 1976. Aspects of necrophoric behavior in red imported fire ant, *Solenopsis invicta. Behaviour* 56: 157–180.

Howard, D. F. and W. R. Tschinkel. 1980. The effect of colony size and starvation on food flow in the fire ant, *Solenopsis invicta. Behav. Ecol. Sociobiol.* 7: 293–300.

Howard, D. F., and W. R. Tschinkel. 1981a. The flow of food in colonies of the fire ant *Solenopsis invicta*: A multifactorial study. *Physiol. Entomol.* 6: 29–306.

Howard, D. F. and W. R. Tschinkel. 1981b. Internal distribution of liquid foods in isolated workers of the fire ant, *Solenopsis invicta. J. Insect Physiol.* 27: 67–74.

Hubbard, M. D. 1974. Influence of nest material and colony odor on digging in the ant, *Solenopsis invicta. J. Ga. Entomol. Soc.* 9: 127–132.

Hung, A. C. F. 1974. Ants recovered from refuse pile of the pyramid ant, *Conomyrma insana* (Buckley). *Ann. Entomol. Soc. Amer.* 67: 522–523.

Hung, A. C. F. and S. B. Vinson. 1975. Notes on the male reproductive system in ants. *J. N.Y. Entomol. Soc.* 83: 192–197.

Hung, A. C. F. and S. B. Vinson. 1977. Interspecific hybridization and caste specificity of protein in fire ant. *Science* 196: 1458–1460.

Hung, A. C. F. and S. B. Vinson. 1978. Factors affecting the distribution of fire ants in Texas, *Southwestern Naturalist* 23: 205–214.

Hung, A. C. F., S. B. Vinson, and J. W. Summerlin. 1974. Male sterility in the red imported fire ant, *Solenopsis invicta* Buren. *Ann. Entomol. Soc. Amer.* 67: 909–912.

Hung, A. C. F., M. R. Barlin, and S. B. Vinson. 1977. The fire ants of Texas. Bull. Texas Agricultural Experiment Station (B-1185).

James, F. K., Jr. 1976. Fire ant sensitivity. *J. Asthma Res.* 13: 179–183.

Jones, R. G., W. L. Davis, A. C. F. Hung, and S. B. Vinson. 1978. Insemination induced histolysis of the flight musculature in fire ants (*Solenopsis* spp.). *Amer. J. Anat.* 151: 603–610.

Jones, T. N. and M. S. Blum. 1982. Ant venom alkaloids from *Solenopsis* and *Monomorium* species. *Tetrahedron* 38: 1949–1958.

Jouvenaz, D. P. 1983. Natural enemies of fire ants. *Fla. Entomol.* 66: 111–121.

Jouvanez, D. P., W. A. Banks, and C. S. Lofgren. 1974. Fire ants: Attraction of workers to queen secetions. *Ann. Entomol. Soc. Amer.* 67: 442–444.

Jouvenaz, D. P., C. S. Lofgren, D. A. Carlson, and W. A. Banks. 1978. Specificity of the trail pheromones of four species of fire ants, *Solenopsis* spp. *Fla. Entomol.* 61: 244.

Kearney, G. P., P. M. Toom, and G. L. Blomquist. 1977. Induction of dealation in virgin female *Solenopsis invicta* with juvenile hormones. *Ann. Entomol. Soc. Amer.* 70: 699–701.

Knell, J. D., G. E. Allen, and E. I. Hazard. 1977. Light and electron microscope study of *Thelohania solenopsae* n. sp. in the red imported fire ant, *Solenopsis invicta*. *J. Invert. Pathol.* 29: 192–200.

Lennartz, F. E. 1973. Modes of dispersal of *Solenopsis invicta* from Brazil into the continental United States—A study in spatial diffusion. M. S. thesis. Univ. Florida, Gainesville, 242 pp.

Leston, D. 1980. Applied myrmecology theory and the fire ant. *Proc. Tall Timbers Conf. Ecol. Anim. Control Habitat Manag.* 7: 163–165.

Lind, N. K. 1982. Mechanism of actions of fire ant *Solenopsis* venoms. I. Lytic release a histamine from mast cells. *Toxicon* 20: 831–840.

Lofgren, C. S., and D. E. Weidhaas. 1972. On the eradication of imported fire ants: A theoretical appraisal. *Bull. Entomol. Soc. Amer.* 18: 17–20.

Lofgren, C. S., F. J. Bartlett, and C. E. Stringer. 1961. Imported fire ant toxic bait studies. The evaluation of various food materials. *J. Econ. Entomol.* 54: 1096–1100.

Lofgren, C. S., F. J. Bartlett, and C. E. Stringer. 1964. The acceptability of some fats and oils as food to imported fire ants. *J. Econ. Entomol.* 57: 601–602.

Lofgren, C. S., W. A. Banks, and B. M. Glancey. 1975. Biology and control of imported fire ants. *Annu. Rev. Entomol.* 20: 1–30.

Lofgren, C. S., B. M. Glancey, A. Glover, J. Rocca and J. Tumlinson. 1983. Behavior of workers of *Solenopsis invicta* to the queen-recognition pheromone: Laboratory studies with an olfactometer and surrogate queens. *Ann. Entomol. Soc. Amer.* 76: 44–50.

Lok, J. B., E. W. Cupp, and B. J. Blomquist. 1975. Cuticular lipids of the imported fire ants, *Solenopsis invicta* and *Solenopsis richteri*. *Insect Biochem.* 5: 821–829.

Lucas, J. R. and H. J. Brockman. 1981. Predatory interactions between ants and antlions. *J. Kans. Entomol. Soc.* 54: 228–232.

Lyle, C., and J. Fortune. 1948. Notes on an imported fire ant. *J. Econ. Entomol.* 41: 833–834.

Markin, G. P. and J. H. Dillier. 1971. The seasonal life cycle of the imported fire ant, *Solenopsis saevissima richteri*, on the gulf coast of Mississippi. *Ann. Entomol. Soc. Amer.* 64: 562–565.

Markin, G. P., J. F. Dillier, S. O. Hill, M. S. Blum, and H. R. Hermann. 1971. Nuptial flight and flight ranges of the red imported fire ant, *Solenopsis saevissima richteri*. *J. Ga. Entomol. Soc.* 6: 145–156.

Markin, G. P., H. L. Collins, and J. H. Dillier. 1972. Colony founding by queens of the red imported fire ant, *Solenopsis invicta*. *Ann. Entomol. Soc. Amer.* 65: 1053–1058.

Markin, G. P., J. H. Dillier, and H. L. Collins. 1973. Growth and Development of colonies of the red imported fire ant, Solenopsis invicta. Ann. Entomol. Soc. Amer. 66: 803–809.

Markin, G. P., J. O'Neal, J. H. Dillier, and H. L. Collins. 1974. Regional variation in the seasonal activity of the imported fire ant, Solenopsis saevissima richteri. Environ. Entomol. 3: 446–452.

Matsumura, Y., K. Maruoka, and H. Yamamota. 1982. Stereoselective synthesis of solenopsin A and B. Tetrahedron Lett. 23: 1929–1932.

Mintzer, A. 1982. Nestmate recognition and incompatibility between colonies of the acacia-ant Pseudomyrmex Ferruginea. Behav. Ecol. Sociobiol. 10: 165–168.

Mirenda, J. T. and S. B. Vinson. 1981. Division of labour and specification of castes in the red imported fire ant, Solenopsis invicta Buren. Anim. Behav. 29: 410–420.

Mirenda, J. T., and S. B. Vinson. 1982. Single and multiple queen colonies of imported fire ants in Texas. Southw. Entomol. 7: 135–141.

Moody, J. V., O. F. Francke, and F. W. Merichel. 1981. The distribution of fire ants Solenopsis in Western Texas. J. Kans. Entomol. Soc. 54: 469–480.

Morrill, W. L. 1974a. Dispersal of red imported fire ants by water. Fla. Entomol. 57: 39–42.

Morrill, W. L. 1974b. Production and flight of alate red imported fire ants. Environ. Entomol. 3: 265–271.

Morrill, W. L. 1977a. Overwinter survival of red imported fire ant in Central Georgia. Environ. Entomol. 6: 50–52.

Morrill, W. L. 1977b. Red imported fire ant foraging in a greenhouse. Environ. Entomol. 6: 416–418.

Morrill, W. L., P. B. Martin, and D. C. Shepard. 1978. Overwinter survival of the red imported fire ant: Effect of various habitats and food supply. Environ. Entomol. 7: 262–264.

Neece, K. C. and D. P. Bartell. 1981. Insects associated with Solenopsis spp. in Southeastern Texas. Southw. Entomol. 6: 307–311.

Nelson, D. R., C. L. Fatland, R. W. Howard, C. A. McDaniel, and G. J. Blomquist. 1980. Re-analysis of the cuticular methyl alkanes of Solenopsis invicta and Solenopsis richteri. Insect Biochem. 10: 409–418.

Nickerson, J. C. E., W. H. Whitcomb, A. P. Bhatkar, and M. A. Naves. 1975. Predation of founding queens of Solenopsis invicta by workers of Conomyrma insana. Fla. Entomol. 58: 75–82.

Nielsson, R. J., A. P. Bhatkar, and H. A. Denmark. 1971. A preliminary list of ants associated with aphids in Florida. Fla. Entomol. 54: 245–248.

O'Neal, J. 1974. Predatory behavior exhibited by three species of ants on the imported fire ants: Solenopsis invicta Buren and Solenopsis richteri Forel. Ann. Entomol. Soc. Amer. 67: 140.

O'Neal, J. and G. P. Markin. 1973. Brood nutrition and parental relationships of the imported fire ant Solenopsis invicta. J. Ga. Entomol. Soc. 8: 294–303.

O'Neal, J., and G. P. Markin. 1975a. Brood development of the various castes of the imported fire ant, Solenopsis invicta J. Kans. Entomol. Soc. 48: 152–159.

O'Neal, J. and G. P. Markin. 1975b. The larval instars of the imported fire ant,

Solenopsis invicta. *J. Kans. Entomol. Soc.* 48: 141–151.

Parrino, J., and N. M. Kandawalla. 1981. Treatment of local skin response to imported fire ant sting. *South. Med. J.* 74: 1346.

Petal, J. 1978. The role of ants in ecosystems. *In* "Productive ecology of ants and termites," pp. 293–325 (M. V. Brian, ed.), Cambridge University Press, London.

Petralia, R. S. and S. B. Vinson. 1978. Feeding in the larvae of the imported fire ant, *Solenopsis invicta*: Behavior and morphological adaptations. *Ann. Entomol. Soc. Amer.* 71: 643–648.

Petralia, R. S. and S. B. Vinson. 1979a. Developmental morphology of larvae and eggs of the imported fire ant, *Solenopsis invicta. Ann. Entomol. Soc. Amer.* 72: 472–484.

Petralia, R. S. and S. B. Vinson. 1979b. Comparative anatomy of the ventral region of ant larvae, and its relation to feeding behavior. *Psyche* 86: 375–394.

Petralia, R. S. and S. B. Vinson. 1980. Internal anatomy of the fourth instar larva of the imported fire ant, *Solenopsis invicta Int. J. Insect Morphol. Embryol.* 9: 89–106.

Petralia, R. S., A. A. Sorensen, and S. B. Vinson, 1980. The labial gland system of larvae of the imported fire ant, *Solenopsis invicta*; Ultrastructure and enzyme analysis. *Cell Tissue Res.* 206: 145–156.

Petralia, R. S., H. J. Williams, and S. B. Vinson, 1982. The hind gut ultrastructure and excretory products of larvae of the imported fire ant, *Solenopsis invicta* Buren. *Insectes Soc.* 29: 332–345.

Phillips, S. A., Jr. and S. B. Vinson. 1980. Comparative morphology of glands associated with the head among castes of the red imported fire ant, *Solenopsis invicta* Buren. *J. Ga. Entomol. Soc.* 15: 215–226.

Pimm, S. C. and D. P. Bartell. 1980. Statistical model for predicting range expansion of the red imported fire ant, *Solenopsis invicta*, in Texas. *Environ. Entomol.* 9: 653–658.

Rhoades, R. B., D. Kalof, F. Bloom, and H. J. Wittig. 1978. Cross reacting antigens between imported fire ants and other Hymenoptera species. *Ann. Allergy* 40: 100–104.

Rhoades, W. C. 1963. A synecological study of the effects of the imported fire ant eradication program. II. Light trap, soil sample, litter sample, and sweep net methods of collecting. *Fla. Entomol.* 46: 301–310.

Rhoades, W. C. and D. R. Davis. 1967. Effects of meteorological factors on the biology and control of the imported fire ant. *J. Econ. Entomol.* 60: 554–558.

Ricks, B. L. and S. B. Vinson. 1970. Feeding acceptability of certain insects and various water-soluble compounds to two varieties of the imported fire ant. *J. Econ. Entomol.* 63: 145–148.

Ricks, B. L. and S. B. Vinson. 1972a. Changes in nutrient content during one year in workers of the imported fire ant. *Ann. Entomol. Soc. Amer.* 65: 135–138.

Ricks, B. L. and S. B. Vinson. 1972b. Digestive enzymes of the imported fire ant, *Solenopsis richteri. Entomol. Exp. Appl.* 15: 319–334.

Risch, S. J. and C. R. Carroll. 1982. Effect of a keystone predaceous ant, *Solenopsis geminata*, on arthropods in a tropical agroecosystem. *Ecology* 63: 1979–1983.

Robeau, R. M. and S. B. Vinson. 1976. Effects of juvenile hormone analogues on caste differentiation in the imported fire ant, Solenopsis invicta. J. Ga. Entomol. Soc. 11: 198–202.

Roe, R. A., II. 1973. A biological study of Solenopsis invicta Buren, the red imported fire ant, in Arkansas, with notes on related species. M.S. thesis, University of Arkansas, Fayetteville, 135 pp.

Sannasi, A. and M. S. Blum. 1969. Pathological effects of fire ant venom on the integument and blood of house fly larvae. J. Ga. Entomol. Soc. 4: 103–110.

Sheppard, C., P. B. Martin, and F. W. Mead. 1979. A planthopper associated with red imported fire ant mounds. J. Ga. Entomol. Soc. 14: 140–144.

Silviera-Guido, A., J. Carbonell, and C. Crisci, 1973. Animals associated with Solenopsis (fire ants) complex, with special reference to Labauchena daguerri. Proc. Tall Timbers Conf. Ecol. Anim. Control Habitat Manag. 4: 41–52.

Smith, M. R. 1958. Family Formicidae. In "Hymenoptera of America north of Mexico," pp. 108–162 (K. V. Krombein, ed.), Agric. Monogr. 2, 1st suppl., U.S. Govt. Printing Office, Washington, D.C.

Smittle, B. J., C. T. Adams, and C. S. Lofgren. 1983. Red imported fire ants: Detections of feeding on corn, okra and soybeans with radioisotopes. J. Ga. Entomol. Soc. 18: 78–82.

Snelling, R. R. 1963. The United States species of fire ants of the genus Solenopsis, subgenus Solenopsis Westwood, with synonymy of Solenopsis aurea (Wheeler). Occas. Pop. Bur. Entomol. Calif. Dept. Agric. 3, 15 pp.

Sorensen, A. A. and S. B. Vinson. 1981. Quantitative food distribution studies within laboratory colonies of the imported fire ant, Solenopsis invicta Buren. Insectes Soc. 28: 129–160.

Sorensen, A. A., R. Kamas, and S. B. Vinson. 1980. The biological half-life and distribution of iodine and radioiodinated protein in the imported fire ant, Solenopsis invicta. Entomol. Exp. Appl. 28: 247–258.

Sorensen, A. A., J. T. Mirenda, and S. B. Vinson. 1981. Food exchange and distribution by functional subcastes of the imported fire ant, Solenopsis invicta Buren. Insectes Soc. 28: 383–394.

Sorensen, A. A., T. M. Busch, and S. B. Vinson. 1983a. Factors affecting brood cannibalism in laboratory colonies of the imported fire ant, Solenopsis invicta Buren. J. Kansas Ent. Soc. 56: 140–150.

Sorensen, A. A., T. M. Busch, and S. B. Vinson. 1983b. Behaviour of worker subcastes in the fire ant Solenopsis invicta, in response to proteinaceous food. Physiological Entomol. 8: 83–92.

Sorensen. A. A., R. S. Kamas, and S. B. Vinson. 1983c. The influence of oral secretions from larvae on levels of proteinases in colony members of Solenopsis invicta Buren. J. Insect Physiol. 29: 163–168.

Sorensen, A. A., T. M. Busch, and S. B. Vinson. 1983d. Behaviour of worker subcastes in the fire ant, Solenopsis invicta, in response to food. Physiol. Entomol. 8: 83–92.

Sonnet, P. E. 1967. Fire ant venom: Synthesis of a reported component of solenamine. Science 156: 1759–1760.

Summerlin, J. W. and L. R. Green. 1977. Red imported fire ant, a review on invasion, distribution and control in Texas, U.S.A. Southw. Entomol. 2: 94–101.

Summerlin, J. W., W. A. Banks, and K. H. Schroeder. 1975. Food exchange between mounds of the red imported fire ant. *Ann. Entomol. Soc. Amer.* 68: 863–866.

Summerlin, J. W., J. K. Olson, and J. O. Fick. 1976. Red imported fire ant: Levels of infestation in different land management areas of the Texas coastal prairies and an appraisal of the control program in Fort Bend County, Texas. *J. Econ. Entomol.* 69: 73–78.

Summerlin, J. W., A. C. F. Hung, and S. B. Vinson. 1977. Residues in nontarget ants, species simplification, and recovery of populations following aerial applications of mirex. *Environ. Entomol.* 6: 193–197.

Teson, A., and A. M. M. de Remes Lenicov. 1979. Estrepsipteros, Parasitoides de hymenopteros. *Rev. Soc. Entomol. Arg. Tomo.* 38: 115–122.

Thompson, C. R. 1980. *Solenopsis (Diplorhoptrum)* of Florida. Ph.D. dissertation, Univ. of Florida, Gainesville. 115 pp. [Dissert. Abstr. Int. B, 41(9):3306]

Thompson, M. J., B. M. Glancey, W. E. Robbins, C. S. Lofgren, S. R. Dutky, J. Kochansky, R. K. Vander Meer, and A. R. Glover. 1981. Major hydrocarbons of the post-pharyngeal glands of mated queens of the red imported fire ant, *Solenopsis invicta. Lipids* 16: 485–495.

Thompson, T. E. and M. S. Blum. 1967. Structure and behavior of spermatozoa of the fire ant *Solenopsis saevissima Ann. Entomol. Soc. Amer.* 60: 632–642.

Tice, J. E. 1967. The anatomy and histology of some of the systems of the male of the red imported fire ant *Solenopsis saevissima richteri* Forel. 122 pp. Ph.D. dissertation, Fordham University, New York.

Toom, P. M., E. W. Cupp, and C. P. Johnson. 1976a. Amino acid changes in newly inseminated queens of *Solenopsis invicta. Insect Biochem.* 6: 327–331.

Toom, P. M., E. Cupp, C. P. Johnson, and I. Griffin. 1976b. Utilization of body reserves for minim brood development by queens of the imported fire ant, *Solenopsis invicta. J. Insect Physiol.* 22: 217–220.

Toom, P. M., C. P. Johnson, and E. W. Cupp, 1976c. Utilization of body reserves during preoviposition activity by *Solenopsis invicta. Ann. Entomol. Soc. Amer.* 69: 145–148.

Travis, B. V. 1938. The fire ant (*Solenopsis* spp.) as a pest of quail. *J. Econ. Entomol.* 31: 649–52.

Travis, B. V. 1941. Notes on the biology of the fire ant, *Solenopsis geminata* (F.) in Florida and Georgia. *Fla. Entomol.* 24: 15–22.

Troisi, S. J. and L. M. Riddiford. 1974. Juvenile hormone effects of metamorphosis and reproduction of the fire ant, *Solenopsis invicta. Environ. Entomol.* 3: 112–116.

Tschinkel, W. R. 1982. History and biology of fire ants. Proceedings of the symposium on the imported fire ant. Pp. 16–35. Environmental Protection Agency; USDA; Animal, Plant Health Inspection Service, Atlanta, Georgia.

Tschinkel, W. R. and D. F. Howard. 1978. Queen replacement in orphaned colonies of the fire ant, *Solenopsis invicta. Behav. Ecol. Sociobiol.* 3: 297–310.

Tschinkel, W. R., and D. F. Howard. 1980. Replacement of queens in orphaned colonies of the fire ant, *Solenopsis invicta. Proc. Tall Timbers Conf. Ecol. Anim. Control Habitat Manag.* 7: 135–147.

Tschinkel, W. R., and D. F. Howard 1983. Colony founding by pleometrosis in the fire ant, *Solenopsis invicta. Behav. Ecol. Sociobiol.* 12: 103–113.

Tumlinson, J. H. 1980. Chemistry of the queen pheromone of the red imported fire ant. *Proc. Tall Timbers Conf. Ecol. Animal Control Habitat Manag.* 7: 155–161.

Urbani, C. B., and P. B. Kannowski. 1974. Patterns in the red imported fire ant settlement of a Louisiana pasture: Some demographic parameters, interspecific competition and food sharing. *Environ. Entomol.* 3: 755–760.

U.S. Department of Agriculture. 1958. Observations on the biology of the imported fire ant. Rep. ARS-33-49. 21 p.

Vander Meer, R. K. 1983. Semiochemicals and the red imported fire ant (*Solenopsis invicta* Buren). *Fla. Entomol.* 66: 139–161.

Vander Meer, R. K., and D. P. Wojcik. 1982. Chemical mimicry in the myrmecophilous beetle *Myrmecaphodius excavaticollis*. *Science* 218: 806–808.

Vander Meer, R. K., B. M. Glancey, C. S. Lofgren, A. Glover, J. H. Tumlinson, and J. Rocca. 1980. The poison sac of red imported fire ant queens: source of a pheromone attractant. *Ann. Entomol. Soc. Amer.* 73: 609–612.

Vander Meer, R. K., B. M. Glancey, and C. S. Lofgren. 1981a. Biochemical changes in the crop, oesophagus and postpharyngeal gland of colony-founding red imported fire ant queens (*Solenopsis invicta*). *Insect Biochem.* 12: 123–127.

Vander Meer, R. K., D. F. Williams, and C. S. Lofgren. 1981b. Hydrocarbon components of the trail pheromone of the red imported fire ant, *Solenopsis invicta. Tetrahedron Lett.* 22: 1651–1654.

Vinson, S. B. 1968. The distribution of an oil, carbohydrate and protein food source to members of the imported fire ant colony. *J. Econ. Entomol.* 61: 712–714.

Vinson, S. B. 1972. Imported fire ant feeding on *Paspalum* seed. *Ann. Entomol. Soc. Amer.* 65: 988.

Vinson, S. B. 1980. The physiology of the imported fire ant. *Proc. Tall Timbers Conf. Ecol. Anim. Control Habitat Manag.* 7: 67–85.

Vinson, S. B. 1983. The Physiology of the imported fire ant revisited. *Fla. Entomol.* 66: 126–139.

Vinson, S. B. and R. Robeau. 1974. Insect growth regulator: Effects on colonies of the imported fire ant. *J. Econ. Entomol.* 67: 584–587.

Vinson, S. B., S. A. Phillips, and H. J. Williams. 1980. The function of the postpharyngeal glands of the red imported fire ant, *Solenopsis invicta* Buren. *J. Insect. Physiol.* 26: 645–650.

Vinson, S. B., J. L. Thompson, and H. B. Green. 1967. Phagostimulants for the imported fire ant, *Solenopsis saevissima* var. *richteri. J. Insect Physiol.* 13: 1729–1736.

Voss, S. H. 1981. Trophic egg production in virgin fire ant queens. *J. Ga. Entomol. Soc.* 16: 437–440.

Walsh, J. P. and W. R. Tschinkel. 1974. Brood recognition by contact pheromone in the red imported fire ant, *Solenopsis invicta. Anim. Behav.* 22: 695–704.

Walsh, C. T., J. H. Law, and E. O. Wilson. 1965. Purification of the fire ant (*Solenopsis saevissima*) trail substance. *Nature (London)* 207: 320–321.

Wangberg, J. K., J. D. Ewig, Jr., and C. K. Pinson. 1980. The relationship of *Solenopsis invicta* to soils of Eastern Texas, U.S.A. *Southw. Entomol.* 5: 16–18.

Whitcomb, W. H. 1980. Expedition into the Pantanal. *Proc. Tall Timbers Conf. Ecol. Anim. Control Habitat Manag.* 7: 113–122.

Whitcomb, W. H., A. Bhatkar, and J. C. Nickerson. 1973. Predators of *Solenopsis invicta* queens prior to successful colony establishment. *Environ. Entomol.* 2: 1101–1103.

Williams, D. F., C. S. Lofgren, and A. Lemire. 1980. A simple diet for rearing laboratory colonies of the red imported fire ant. *J. Econ. Entomol.* 73: 176–177.

Williams, D. F., C. S. Lofgren, and R. N. Vander Meer. 1981. Tending of dead queens by workers of *Solenopsis invicta*. *Fla. Entomol.* 64: 545–547.

Williams, H. J., M. R. Strand, and S. B. Vinson. 1981a. Synthesis and purification of the allofarnesenes. *Tetrahedron* 37: 2763–2767.

Williams, H. J., M. R. Strand, and S. B. Vinson. 1981b. Trail pheromone of the red imported fire ant, *Solenopsis invicta* Buren. *Experientia* 37: 1159–1160.

Williams, R. N. 1980. Insect natural enemies of fire ants in South America with several new records. *Proc. Tall Timbers Conf. Ecol. Anim. Control Habitat Manag.* 7: 123–134.

Williams, R. N. and W. H. Whitcomb. 1974. Parasites of fire ants in South America. *Proc. Tall Timbers Conf. Ecol. Anim. Control Habitat Manag.* 5: 49–59.

Williams, R. N., M. de Menezes, G. E. Allen, W. F. Buren, and W. H. Whitcomb. 1975. Observacoes ecologicas sobre a formiga lava-pe, *Solenopsis invicta* Buren. 1972. *Rev. Agr. Piracicaba* 50: 7–22.

Wilson, E. O., Jr. 1971. "The insect societies." Belknap Press, Cambridge, Mass. 548 pp.

Wilson, E. O., Jr. 1952. The *Solenopsis saevissima* complex in South America. *Mem. Inst. Oswaldo Cruz.* 50: 60–68.

Wilson, E. O., Jr. 1953. Origin of the variation in the imported fire ant. *Evolution* 7: 262–263.

Wilson, E. O., Jr. 1959. Source and possible nature of the odor trail of fire ants. *Science* 129: 643–644.

Wilson, E. O., Jr. 1962. Chemical communication among workers of the fire ant *Solenopsis saevissima* (Fr. Smith) 1. The organization of mass-foraging. 2. An information analysis of the odour trail. 3. The experimental induction of social responses. *Anim. Behav.* 10: 134–164.

Wilson, E. O., Jr. 1971. "The insect societies." Belknap Press, Cambridge, Mass. 548 pp.

Wilson, E. O. 1978. Division of labor in fire ants based on physical castes (*Solenopsis*). *J. Kans. Entomol. Soc.* 51: 615–636.

Wilson, E. O. and W. L. Brown. 1958. Recent changes in the introduced populations of the fire ant (*Solenopsis saevissima*) (Fr. Smith). *Evolution* 12: 211–218.

Wilson, E. O., N. I. Durlach, and L. M. Roth. 1958. Chemical releasers of necrophoric behavior in ants. *Psyche* 65: 108–114.

Wilson, E. O. and J. H. Eads. 1949. A report on the imported fire ant, *Solenopsis saevissima* var. *richteri* Forel in Alabama. *Ala. Dep. Conserv. Spec. Rep.*, 53 pp., 13 plates [Mimeographed.]

Wilson, N. L. and A. D. Oliver. 1969. Food habits of the imported fire ant in pasture and pine forest areas in southern Louisiana. *J. Econ. Entomol.* 62: 1268–1271.

Wilson, N. L., J. H. Dillier, and G. P. Markin, 1971. Foraging territories of imported fire ants. *Ann. Entomol. Soc. Amer.* 64: 660–665.

Wojcik, D. P. 1983. Comparison of the ecology of red imported fire ants in North and South America. *Fla. Entomol.* 66: 101–111.

Wojcik, D. P., W. A. Banks, and D. H. Habeck. 1978. Fire ant myrmecophiles: Flight periods of *Myrmecaphodius excavaticollis* and *Euparia castanea. Coleopt. Bull.* 31: 335–338.

Wood, Lois A. and W. R. Tschinkel. 1981. Quantification and modification of worker size variation in the fire ant *Solenopsis invicta. Insectes Soc.* 28: 117–128.

Chapter 8

THE ECONOMIC IMPORTANCE AND CONTROL OF IMPORTED FIRE ANTS IN THE UNITED STATES
Clifford S. Lofgren

INTRODUCTION

The red imported fire ant, *Solenopsis invicta* Buren and, to a lesser extent, the black imported fire ant, *Solenopsis richteri* Forel have infested, since their introduction in the early 1900s, all, or part of, nine southern states and Puerto Rico. Their spread was enhanced, prior to initiation of a federal-state quarantine in the early 1950s, by concealment of queens or small colonies in nursery stock that was shipped from the original infested area around Mobile, Alabama (USDA 1958). Wherever the imported fire ants (IFA) have become established they have been the source of innumerable complaints from the local residents because of their mound-building, stinging, and feeding habits. These complaints resulted in the promulgation by the United States Congress in 1957 of a federal-state cooperative program for the quarantine, control, and eradication of IFA. Since that time there has been a continuing debate as to the effectiveness and the advisability of such a program. In fact, no consensus has been reached to date on the issue of the economic importance of the IFA, as illustrated by the indeterminant statements on the subject expressed by the various panels of the Symposium on the Imported Fire Ant sponsored by the Environmental Protection Agency and the United States Department of Agriculture (Tschirley 1982). Despite the apparent confusion there is a body of research to draw on and some general conclusions about the economic status of the IFA can be made.

Control procedures for the IFA that utilized various pesticides have also been controversial because of their environmental impact; however, advances in research have revealed several new chemicals that may be used effectively and safely for reducing local IFA population levels, where necessary. In addition new research on pheromones and biocontrol suggests that techniques that are more target specific may become available. These techniques may eventually lead to comprehensive pest management strategies.

ECONOMIC IMPORTANCE

Economics is defined as the management of income and expenditures. Since over $172,000,000 U.S. dollars (U.S. dollars are used throughout this chapter) were spent by federal, state, and other governmental agencies (Canter 1981) for control of the IFA from 1957 to 1981, there can be little question that they are economic pests. The problem comes in defining the actual types of economic losses caused by the IFA. This is further complicated since, in some instances, the ants are actually beneficial. Lofgren et al. (1975) concluded that IFAs affected a broad spectrum of "things," but were not major pests of any one crop or commodity. In the light of some recent findings, this conclusion may be only partially correct. It is apparent that a thorough understanding of the biology and ecology of the IFA is necessary before their economic importance can be judged.

Tschinkel (1982) builds an interesting case for considering IFA as a weed species, but he views them strictly on a theoretical ecological basis that does not consider their place in the practical everyday world. The definition of a weed species is, according to Webster's New World Dictionary, "any undesired, uncultivated plant (or animal), especially one growing in profusion and crowding out a desired crop, spoiling a lawn, etc." When IFA are considered from this perspective, the weed comparison is valid because the IFA are uninvited guests that sting people and their animals, deface lawns, parks, and interfere with farming operations. This view is also too simplistic since imported fire ants are opportunistic, omnivorous feeders who prey on various animal and plant species and thus, depending on the species preyed on, their feeding activity can be beneficial as well as harmful. In fact, if one follows the logic of this argument, we can understand that IFAs can be beneficial or harmful to the same plant or animal species depending on the time of year and/or developmental stage of the species, environmental conditions, or the status of the ant colony itself. It is this point that appears to have been overlooked by many scientists and is the cause of much of the confusion over the pest or nonpest status of the IFA. For example, the literature is replete with examples of beneficial or harmful behaviors attributable to IFAs, but the importance of any one of these cannot be determined without viewing the impact of the ants on the entire life cycle of the particular crop or animal involved. Thus, the fact that IFAs prey on velvet bean caterpillars on soybeans may be of little interest to a farmer if, at an earlier date, the ants had destroyed a fourth of the plants by predation on the germinating seeds. In other words, any evaluation of the economic importance of the ants must consider the total behavior of the ants with the bottom line being the balance after all debits and credits have been totalled.

Another aspect of imported fire ant biology that relates to their economic importance is the long-term impact of their predation on various fauna. An example of this is a recent report by Mount (1981) that predation by S. invicta over a period of 10 to 20 years may be a serious threat to certain ground-dwelling reptiles and birds. The impact of IFA on other ant species has been noted by some researchers (Whitcomb et al. 1972, Roe 1973). Changes of this type can be very subtle, and thus difficult to detect and verify experimentally.

The preceding discussion illustrates the difficulties in making an accurate assessment of economic damages or benefits associated with the IFA. However, with this background, the current knowledge of the economics of IFA behavior can be reviewed with the understanding that the full impact for many of these behaviors has not been determined and other economic behaviors remain to be discovered. The areas of impact can be subdivided into four categories for ease of discussion: (1) public health, (2) agriculture, (3) wildlife and environment, and (4) miscellaneous.

Public Health

Imported fire ant workers are highly aggressive and will attack and sting any animal that disturbs their nest. Persons may also be stung by foraging worker ants away from their nests if they are inadvertently disturbed. The venom consists primarily of alkaloids (2,6-disubstituted piperidines). The identification and synthesis of these compounds has been described by MacConnell et al. (1970). Baer et al. (1977) have shown that the venom also contains a small proteinaceous component with three highly active allergenic antigens.

The typical response to the venom of the IFA follows a predictable pattern (Lockey 1974, James 1976). Initially the venom causes a burning and itching sensation, which accounts for the common name attached to these insects. This is followed immediately by the formation of a wheal and flare that may become as large as 10 mm in diameter. Several hours later, a small vesicle containing a clear fluid develops at the sting site. The vesicle eventually becomes cloudy and a white pustule forms, usually after 24 hours. The pustule is sterile and is a result of the necrotizing effects of the venom alkaloids. If not broken, the pustule remains for a few days to as much as a week. The latter is a special problem for laborers, farmhands, and children since secondary infections may occur.

By far, the most serious potential consequence of the stings is the development of a hypersensitivity reaction to the proteinaceous allergens in the venom. The extreme potency of these allergens is illustrated by Rhoades et al. (1977), who stated that less than 1% of the average volume of venom from the sting (0.07–0.10 μl) is composed of protein. This translates to a maximum protein weight of about 0.001 μg. In comparison, the protein content of the average bee sting is 50 μg.

Numerous cases of severe allergic reactions have been reported (Brown 1972, Triplett 1973, Rhoades et al. 1977, Lockey 1974) and Lockey (1979) cited five fatalities attributable to IFA stings within a period of two years. Rhoades et al. (1977) determined that the incidence rate of new cases for the city of Jacksonville was 3.8 per 100,000 persons per year. However, they considered this figure to be low since it was based only on reports from two allergists who served the city; patients reporting to emergency rooms, general practitioners, etc., were not included in the survey. On the basis of these data, Lofgren and Adams (1982) estimated that at least 1,460 new cases of allergy to imported fire ant venom occur each year. They estimated that the cost of desensitization therapy for these persons at $205,860 per year. Because of the limitations of the survey by Rhoades et al. (1977) it is possible that the number of new cases and cost could easily be twice this estimate. If hyperallergic persons are not desensitized it is imperative that they have ready access to allergy treatment kits.

While systemic allergic reactions are the most dramatic reaction to IFA venom, lesser primary or secondary complications may require medical treatment. For example, Dr. R. F. Triplett conducted a mail survey for the years 1969 to 1971 in Mississippi. In his unpublished survey there were reports of secondary infections, debridements, skin grafts, and amputations.

Clemmer and Serfling (1975) surveyed a total of 240 households (777 persons) by telephone in Metarie, Louisiana. Sting attacks were reported for 29% of the persons during the months of June–August, 1973 and 1.3% required medical consultation. Similar surveys were conducted in Lowndes and Sumter Counties in Georgia. In Lowndes County, Yeager (1978) sampled 156 families, including equal numbers of urban and rural residents. He found that one out of every five residents could expect to be stung each month and less than 5% of those stung required medical care. The Sumter County study (Adams and Lofgren 1981) was conducted in a predominately rural area. A total of 213 sting attacks were reported during one year (1976) on 95 of 272 survey participants (35%). Two individuals (1%) classified their sting reaction as severe, 26 (12%) as moderate, and 183 (87%) as mild.

Adams and Lofgren (1982) reviewed data on patients reporting for medical treatment for arthropod sting or bite attacks at Ft. Stewart, Georgia. IFA were responsible for 161 (49%) of a total of 329 patients treated from April 1 to September 30, 1979. This represents 0.7% of the post population including military personnel (12,000) and dependents (11,000). Only 7% of the sting attacks were caused by bees and wasps. Eight persons (5%) exhibited symptoms of shock due to IFA stings and 11 (7%) developed secondary infections. Five patients were hospitalized for one day each. Direct medical cost for outpatient visits was $24.10 while the cost of hospitalization was $176.45 per day. Total estimated cost attributable to IFA was $5,070.

For purposes of calculating potential medical costs, Lofgren and Adams (1982) used the preceding data to project expenses for the 38.4 million people living in the IFA-infested area. They averaged the sting rates for the studies by Clemmer and Serfling (1975) and Adams and Lofgren (1981) to obtain an average sting attack rate of 32%. Also, they averaged the percentage of persons requesting medical treatment in the Clemmer and Serfling (1975) study (1.3%) and the percentage of the Ft. Stewart population reporting to the dispensary for treatment of IFA stings (0.7%) to obtain a medical treatment rate of 1.0%. Based on these data, the annual cost for medical treatment (one office visit per patient at $23.10) was approximately $2.84 million ($38.4 \times 10^6 \times 0.32 \times 0.01 \times \$23.10$).

Obviously, other research and surveys need to be conducted, but the preceding data indicate that the actual and potential costs to residents of the infested states are significant. Of equal importance is the worry and concern experienced by parents for young children who may suffer severe injury through encounters with IFA in their yards or playgrounds and the constant fear of death endured by those persons unfortunate enough to develop severe allergic reactions to the venom. These fears undoubtedly provide much of the driving force for the continued interest in federal-state programs for wide-scale control of IFA.

Agriculture

Summarized in Tables 8.1 and 8.2 are literature reports of harmful and beneficial environmental interactions of the IFA that affect agriculture. The listings are not intended to be complete tabulations, but major research efforts to document the economic impact of the ants are cited. To realize that these tabulations are incomplete, one needs only to review the papers of Hays and Hays (1959) and Wilson and Oliver (1969), which contain extensive lists of the various food materials (primarily inverte-

TABLE 8.1
Reports of Predation by Imported Fire Ants on Beneficial Insects and Wildlife

Animal	Impact of Predation	Reference
Invertebrates		
Ground beetle, *Vacusus vicinus*	Significant population decrease	Howard and Oliver (1978)
Carabids, *Galeritula sp. Pterostichus chalcites*	Significant population decrease	Brown and Goyer (1982)
Rove beetles, *Oxytelis* spp.	Significant population decrease	Howard and Oliver (1978)
Ants, *S. xyloni, S. Geminata*	Almost completely displaced	Roe (1973)
		Whitcomb et al. (1972)
Various species	Displaced from soybean fields	Wilson and Eads (1949)
Honeybees, *Apis mellifera*	No data	Cockerman and Oertel (1954)
		Vinson (1982)
Ground nesting bees (not identified)	No data	Morrill (1977), Wilson and Oliver (1969)
Collembola (not identified)	No data	
Spiders, *Trachelas deceptus*	Significant population decrease	Howard and Oliver (1978)
Lycosa riparia-helluo complex	Significant population decrease	Howard and Oliver (1978)
Unidentified species	No data	Morrill (1977), Reagan (1981)
Isopoda (not identified)	No data	Morrill (1977), Wilson and Oliver (1969)
Annelida (not identified)	No data	Morrill (1977), Wilson and Oliver (1969), Reagan (1981)

232

Vertebrates

Six-lined racerunner *Cnemidophorous sexlineatus*	Egg destruction in simulated field tests	Mount et al. (1981)
Chicken turtle, *Deirochelys reticularia*	Ants fed on baby turtles	Mount (1981)
Gopher tortoise, *Gopherus polyphemus*	Hatchlings killed	Landers et al. (1980)
Box turtle, *Terrapene carolina*	One hatchling killed	Landers et al. (1980)
Map turtle, *Graptemys barnouri*	Second year hatchling killed	Mount (1981)
Chameleon, *Anolis carolinensis*	Egg perforated	Mount (1981)
Five-lived skink, *Eumeces fasciatus*	Population decline in Alabama coastal plain associated with IFA	Mount (1981)
Snakes, 7 species	Population decline in Alabama coastal plain associated with IFA	Mount (1981)
Nighthawk, *Chordeiles minor*	Population decline in Alabama coastal plain associated with IFA	Mount (1981)
Ground dove, *Columbigallina passerina*	Population decline in Alabama coastal plain associated with IFA	Mount (1981)
Eastern meadowlark, *Sturnella magna*	Population decline in Alabama coastal plain associated with IFA	Mount (1981)
Bobwhite quail, *Colinus virginianus*	Chicks killed under certain conditions, no evidence of population decline	Johnson (1962)
Cottontail rabbit, *Sylvilagus floridanus*	Penned nestlings killed, no evidence of population decline	Hill (1969)
Wood duck, *Aix sponsa*	Ducklings and pipped eggs destroyed	Ridlehuber (1982)

TABLE 8.2
Reports of Predation by Imported Fire Ants on Pest Arthropods

Arthropod	Impact of Predation	Reference
Lepidoptera		
Sugarcane borer, *Diatraea saccharalis*	Significant reduction in damaged internodes	Negm and Hensley (1969), Reagan et al. (1972)
Bollworm, *Heliothis zea*	No data	Sterling et al. (1979)
Tobacco budworm, *Heliothis virescens*	Significant egg predator	McDaniel and Sterling (1979), McDaniel and Sterling (1982)
Nantucket pine tip moth, *Rhyacionia frustrana*	Not significant	Wilson and Oliver (1970)
Velvetbean caterpillar, *Anticarsia gemmatalis*	No data	Whitcomb et al. (1972), Buschman et al. (1977)
Greater wax moth, *Galleria mellonella*	No data	Williams (1976)
Soybean looper, *Pseudoplusia includens*	No data	Whitcomb et al. (1972)
Green cloverworm, *Plathypena scabra*	No data	Whitcomb et al. (1972), Snodgrass (1976)
Coleoptera		
Bollweevil, *Anthonomus grandis*	Significant reduction in damaged squares	Sterling (1978), Jones and Sterling (1979), McDaniel and Sterling (1979)
Small southern pine engraver, *Ips avulsus*	No data	Sterling et al. (1979)
Banded cucumber beetle, *Diabrotica balteata*	No data	Sterling et al. (1979)

Pest	Result	Reference
Pecan weevil, *Curculio caryae*	Significant predation on larvae in simulated field test	Dutcher and Sheppard (1981)
Sugarcane rootstalk borer weevil, *Diaprepes abbreviatus*	Minor predation in simulated field test	Whitcomb et al. (1982)
Alfalfa weevil, *Hypera postica*	Predation in greenhouse tests	Morrill (1978)
Cowpea curculio, *Chalcodermus aeneus*	Significant predation in small field plots	Russell (1981)
Diptera		
Stable fly, *Stomoxys calictrans*	Significant predation on immatures in simulated field test	Summerlin and Kunz (1978)
Face fly, *Musca autumnalis*	Significant predation on larvae in simulated field test	Combs (1982)
Horn fly, *Haematobia irritans*	Significant predation in laboratory and field tests	Summerlin et al. (1977), Howard and Oliver (1978)
Homoptera		
Greenhouse whitefly, *Trialeurodes vaporariorum*	Significant in one greenhouse	Morrill (1977)
Pea aphid, *Acyrthosiphon pisum*	Predation in greenhouse tests	Morrill (1978)
Hemiptera		
Rice stink bug, *Oebalus pugnax*	No data	Sterling et al. (1979)
Green stink bug, *Nezara viridula*	Predation in field cage tests	Kryspin and Todd (1982)
Other		
Striped earwig, *Labidura riparia*	Significant reduction in lawns	Gross and Spink (1969)
Termites, *Reticulitermes* sp.	Large number attacked but not of economic significance	Green (1967), Wilson and Oliver (1969)
Lone star tick, *Amblyomma americanum*	Simulated field studies and surveys	Harris and Burns (1972), Burns and Malancon (1977)

brates) that are harvested, dead or alive, as food. However, these reports on the food habits of IFA do not cite any significant amount of feeding by IFA on plants. This seems strange since IFA require large amounts of carbohydrates in their diet (Williams et al. 1980b). Only one life stage of the IFA (fourth instar larvae) ingests large food particles (Glancey et al. 1981b). Consequently, one would expect that the liquid portion of many foods would be extracted and carried back to the nest in the workers' crop or esophagus. Evidence that this does occur has been obtained recently in studies of IFA feeding on ^{32}P-labeled crops (Smittle et al. 1983, Harlan and Banks 1983, Adams et al. 1983). This research emphasizes the point made earlier in this chapter that the economic impact on any crop must be determined on the basis of the cumulative effects of the IFA throughout growth and harvest. For this review, however, the beneficial and harmful effects are discussed independently.

Harmful Effects

The first effort to document the harmful effects of IFA on agricultural crops was made by Wilson and Eads (1949). They conducted a survey in Mobile, Baldwin, and Washington Counties in southern Alabama that consisted of a systematic poll of 174 farmers and on-site observations of feeding by IFA on various crops. Their direct observations revealed that ants fed on the seeds or seedlings of corn, peanuts, beans, Irish potatoes, and cabbage. According to the farmers polled, the crops most seriously affected were corn, Irish potatoes, soybeans, sweet potatoes, and cabbage. Similar observations of feeding by IFA on various field crops were reported by researchers at a USDA research laboratory located at Mobile, Alabama (USDA 1958). In subsequent years there were few reports of damage to row or cultivated crops; however, there were numerous complaints about IFA infestations of pastures and hayfields.

At least three scientific committees reviewed the IFA problem during the 1960s and 1970s (National Academy of Science, Mills 1967; Council for Agricultural Science and Technology, Anonymous 1976; and the Georgia Academy of Science, Bellinger et al. 1965). All concluded that IFA were not a significant problem in row or cultivated crops. Lofgren and Adams (1982) theorized that lack of complaints of IFA damage to cultivated crops was associated with the extensive use of chlorinated hydrocarbon insecticides in cultivated fields for the control of insects, such as cutworms, wireworms, and white-fringe beetles. They stated that IFA could become a major crop pest following the withdrawal of registrations of chlorinated hydrocarbon insecticides for crop pests by the Environmental Protection Agency in the early 1970s. They concluded that once these compounds were no longer available, residues in the soil would dissipate within a few years and IFA could readily infest the fields.

Evidence for the preceding conclusion occurred in the mid-1970s when farmers in Georgia began complaining about interference of the mounds of the IFA with soybean harvesting operations. Because IFA increase the diameter and height of their mounds in the fall of the year in response to wetter and/or cooler soil conditions, the farmers were faced with either (1) pushing the mounds over with their combines, thus losing some of the beans and risking equipment damage of (2) raising the header bar to avoid impacting the mound tumulus and thus not harvesting all of the beans. Their studies (Adams et al. 1976, Adams et al. 1977) revealed minor losses (< 1 bu per acre; 67.3 kg per ha), although at the time of the study, the loss amounted to $6 to $12 per ha.

Subsequent to these early reports, Lofgren and Adams (1981) noted a greater loss of soybeans than could be attributed solely to interference of the mounds with harvesting. They conducted a series of tests involving eight paired fields of soybeans (one field of each pair was treated with mirex to eliminate the fire ant). At harvest the average reduction in yield of soybeans amounted to 5.86 bu per acre (394 kg per ha) in fields with 20 to 72 mounds per acre (49 to 176 per ha).

Later studies conducted by Adams et al. (1983) and Apperson and Powell (1983) produced results that were consistent with this report. In addition, the study by Adams et al. (1983) presented data that indicated that the majority of the reduction in yield in soybeans could be attributed to destruction of seeds and seedlings by the IFA. However, they also presented evidence that the IFA fed on the growing plants as indicated by the use of ^{32}P.

If these research findings are typical throughout the infested states, the impact on soybean production would be very great. For example, Lofgren and Adams (1982) estimated for 1981 that if (1) 25% of the soybean acreage in the infested states was heavily infested with IFA, (2) the average reduction in yield was 400 kg per ha, and (3) the sale price of soybeans was $0.22 per kg, the total loss of soybeans in 1981 could have been 560,734 metric tons at a value of almost $125 million.

While the effect of IFA on soybeans has received the most attention, research has also been or is being conducted, on other crops. The impact of IFA on citrus in Florida has received considerable attention over the last two years as the potential of IFA damage to young citrus trees has begun to be realized. There had been little prior concern about IFA in citrus groves because until the 1970s they could easily be controlled either with residual chlorinated hydrocarbon insecticides or with mirex bait. In the last few years the population of IFA in citrus groves has literally exploded in some areas of Florida, particularly along the east coast. Brown (1982) reported a loss of more than 50% of the young citrus trees (oranges and grapefruit) in some groves because of IFA. Early reports stated that the trees were killed because the IFA chewed the bark, girdling the trees at or just below the soil surface. Feeding on branches, new terminal growth, flowers, and young fruit has been noted also.

Recent observations indicate that the ants feed on sap oozing from the citrus trees either as a result of mechanical or freeze damage or from direct feeding by the ants. This bleeding phenomenon is most prevalent during periods of warm weather in the winter months. The ants also carry dirt and pack it into these areas of bleeding. This could be a source of infectious plant diseases.

Brown (1982) noted in mature groves that the IFA construct their nests along the drip line of the trees. This causes a problem for fruit pickers since their ladders are usually placed on or adjacent to the ant nests. When the ants are disturbed they crawl onto the ladder and sting the fruit picker. The pickers can also be stung by ants foraging in the tree.

A few instances of damage to corn by IFA have been reported (see Table 8.3). For example, the report by Glancey et al. (1979) involved a situation in which a field had been flooded prior to corn planting. Apparently, IFA colonies floated into the field on flood water. Later, when the farmer planted, the IFA fed heavily on the young corn plants causing a severe reduction in stand and a monetary loss of $4,000 to $10,000. Harlan et al. (1981) demonstrated in greenhouse and field tests that IFA feed on corn plants and, in the case of their greenhouse test, they destroyed large numbers of the seedlings. However, a companion field trial showed only a slight, but significant, reduction in the number of plants and no reduction in corn yield.

Consistent economic damage is caused by IFA to mature okra plants. Feeding by IFA on the flowers of this common southern garden plant has been observed since the early report by Wilson and Eads (1949). Attraction to the okra flowers is so consistent that it can be used as a method for detecting IFA in the field. The IFA feed around the calyx, but some feeding may occur on the pod. In some cases the flower or pod is destroyed but most often the feeding results in a misshapen, scarred, and curved pod with no market value.

The incidence of damage to other cultivated crops has been sporadic, although on the basis of previous findings with soybeans, it would not be surprising if reduction in plant density and yield of other cultivated crops will be found in the future, particularly on some of the minor vegetable crops. For example, Adams (1983) reported on a 20- to 30-acre field of eggplants near Ocala, Florida in which 50% of the plants were destroyed by IFA resulting in a potential estimated crop loss of $56,000 to $90,000. In another instance C. T. Adams and W. A. Banks (personal communication) reported loss of potatoes near Hastings, Florida that was attributable to direct feeding by IFA on the new potatoes.

IFA and their mounds in hayfields and meadows have been the

source of consistent complaints since the 1940s and farmers complain of death of newborn calves and pigs in pastures as a result of IFA attacks. This latter behavior is very difficult to document. However, the author has seen piglets being killed by IFA in one instance. Some researchers have theorized that IFA are not a problem in pastures and have attempted to demonstrate that the IFA are beneficial because the mound construction brings nutrients to the surface and creates better soil conditions for the growth of grass (Herzog et al. 1976, Blust et al. 1982).

The problem with the IFA in the hayfields is understandable since, as in the case of soybeans, the mounds interfere with harvest operations. Anyone who has seen a field infested with 100 to 200 IFA mounds per hectare can visualize why the mounds would increase the cost of, and the time required for hay harvest. This is especially true in the heavier, clay soils. In addition, hay bales left on the ground are quickly invaded by IFA, which causes a problem for the field workers. In recent years there is evidence that problems with hay have diminished because of the use of rotary mowers and round balers or hay stackers.

The tending of aphids or scale insects by IFA has been reported for a number of plants including weeds as well as crops (USDA 1958, Wilson and Eads 1949, Wilkinson and Chellman 1979). The latter is the only report that cites actual detrimental effects of this behavior, that is, a 40% reduction in growth of 3-year-old slash pine trees in a grove in north Florida. The wide-spread occurrence of IFA-homopteran associations suggests that other harmful effects may occur.

Beneficial Effects

Since the IFA are omnivorous, it is not surprising that their feeding behavior should have beneficial as well as detrimental aspects. Thus, it is not unexpected that Long et al. (1958) noted that there was an increase in damage by the sugarcane borer (*Diatraea saccharalis*) in sugarcane fields in south Louisiana subsequent to applications of insecticide (heptachlor) for IFA control. Their observations were substantiated by a number of other researchers over the following 15 years and Reagan (1981) reported on a pest management program for a sugarcane borer in Louisiana that depends partially on the predatory activity of IFA. He stated that the sugarcane borer is responsible for more than 90% of all sugarcane crop losses ascribed to insect damage and that losses in sugar yields average 13 to 15% where infestations are not held below the economic threshold. The pest management program ascribes 25% of the control of sugarcane borers to beneficial predators of which IFA are the most important.

In studies in Florida sugarcane fields, Adams et al. (1981) found that

TABLE 8.3
Reports of Economic Damage Caused by Imported Fire Ants

Crop or item affected	Extent of damage	Reference
Soybeans (germinating seeds and yield)	Extensive (60 to 670 kg/ha yield loss)	Lofgren and Adams (1981), Apperson and Powell (1983), Adams et al. (1983)
Soybeans (interference with harvest)	Moderate (<67 kg/ha yield loss)	Adams et al. (1976), Adams et al. (1977)
Citrus (girdling young trees; annoyance of fruit pickers)	Severe in many Florida groves; many young trees killed; occasional increased cost of picking fruit	Lyle and Fortune (1948), USDA (1958), Brown (1982)
Corn (germinating seeds and seedlings)	Severe in isolated fields	Lyle and Fortune (1948), Wilson and Eads (1949), Glancey et al. (1979), Wilson and Eads (1949), USDA (1958)
Hay (mounds interfere with harvest, damage machinery; worker ants invade hay bales)	Moderate	Adams (1983)
Eggplant (maturing plant destroyed; dirt carried onto plant)	Severe in one Florida field (potential loss $56,000 to $90,000)	Wilson and Eads (1949), USDA (1958)
Okra (feeding on flowers and young plants; malformed pods)	Moderate	Wilson and Eads (1949)
Peanuts (germinating seeds)	Minor	Wilson and Eads (1949), USDA (1958), Adams and Banks (1984 pers. comm.)
Potatoes (young plants, tubers)	Minor	

Item	Estimate	Reference
Sweet potatoes (damage not described)	Minor	Wilson and Eads (1949)
Cotto (damage not described)	Minor	Wilson and Eads (1949)
Strawberries (mounding on plants; tending aphids)	Minor	Wilson and Eads (1949), USDA (1958)
Watermelon (seedlings)	Minor	Wilson and Eads (1949)
Cabbage (seedlings)	Minor	Wilson and Eads (1949), USDA (1958)
Fish (ingestion of ant workers or newly-mated queens)	Moderate in isolated ponds	Crance (1965), Green (1967)
Slash pine (tending scale insects)	Growth reduced 40% in one plantation in Florida	Wilson and Eads (1949), Wilkinson and Chellman (1979)
Longleaf pine (seedlings)	Thirty-three percent of germinating seeds destroyed in test plots	Campbell (1974)
Homes and other buildings (nuisance; consume food)	No estimate	Bruce, et al. (1978), Wilson and Eads (1949)
Air conditioners (electrical shorts and dirt in unit)	Estimate of $1,000 to $15,000 from one firm in Marshall, Texas	Poulan (pers. comm.)
Telephone pedestals (annoyance and damage to equipment)	No estimate	Carlson (pers. comm.)
Airport runway lights (electrical short)	Minor	Lofgren et al. (1975)
Asphalt roads (collapse because of nest excavation under asphalt)	Minor, one area in North Carolina	Grothaus (pers. comm.)
Newborn calves, pigs, and chickens	No estimate	Wilson and Eads (1949), Hunt (1976)

241

the elimination of IFA with mirex caused a slight but significant increase in damage from sugarcane borer. In contrast to the studies in Louisiana, where the IFA were the only significant ant predator, other ant species were found. They concluded that a multiple predator ant complex was more effective than one dominated by IFA. This research indicates the problem associated with IFA domination of an environment, since they are very competitive and may eliminate or reduce populations of other predatory ants (Whitcomb et al. 1972).

The possible beneficial aspect of IFA for control of the boll weevil in cotton fields was reported by Sterling (1978). He exposed weevil-infested cotton bolls in cotton fields infested with IFA and found that in one set of experiments the ants consumed up to 85% of the weevils, with a weekly average of 60%. Subsequently, Jones and Sterling (1979) reported on another series of tests conducted in east Texas in which they used the chemical exclusion technique to eliminate most of the fire ants from one plot with mirex. Predation by IFA reduced the number of emerging F_1 beetle adults with the result that weevil-damaged squares never exceeded 16.6% through June and early July in the infested plots. At the same time the incidence of weevil-damage squares in the ant-free plot reached a maximum of 38.8%. They concluded from their data that "in some instances, the presence of IFA will result in lower weevil abundance than would have occurred in their absence."

McDaniel and Sterling (1979, 1982) conducted simulated field tests with the tobacco budworm, Heliothis virescens, that showed IFA were effective natural predators of ^{32}P-labeled eggs placed on cotton plants. However, they stated that the importance of IFA might be overestimated (1) because the ^{32}P was probably distributed to their nest mates via trophallaxis and (2) the eggs were clumped on the cotton plants and this could have encouraged recruitment of other IFA workers. Despite these problems, it was evident that the IFA were a valuable component of the predaceous arthropod fauna.

Predation by IFA can also affect other components of the predator complex of H. virescens as indicated by Lopez (1982) who found that IFA severely affected the survival of the braconid parasite, Cardiochiles nigriceps. These data are in contrast to an earlier paper by Sterling et al. (1979), who reported that IFA were very specific predators and failed to reduce entomophagus insect and spider abundance in an agroecosystem. The validity of the data in the latter paper are suspect for a number of reasons, the principle of which are: (1) they used a sampling system that they recognized as having an inherent bias and inaccuracy as to types of insects collected, (2) the numbers of insects collected were so small that it was almost impossible to determine significant differences in the populations between the IFA-infested and free plots, (3) pitfall traps were used only in the IFA-free plot, and (4) the number of ground predators collected, such as rove beetles, ground beetles and earwigs, was insignificant, suggesting that populations of these insects may have already been severely reduced. Brown and Goyer (1982) found evidence for this possibility when they determined that certain species of carabids were reduced significantly in soybean fields by IFA. Since IFA are widespread in distribution and highly competitive for food and living space, they may, over a period of several years, alter the numbers and diversity of other species occupying the same ecological niche. Thus, a census of the fauna of a specific area at some point in time could give the appearance that the IFA are not detrimental when, in fact, they may have already caused changes in the diversity and density of some other animal species.

IFA readily invade cow dung in pastures; thus, it is not surprising that they are predacious on a number of insects that utilize dung for their growth and development. Predation by IFA has been reported on the stable fly, *Stomoxys calcitrans* (Summerlin and Kunz 1978), the face fly, *Musca autummalis* (Combs 1982), and the hornfly (*Haematobia irritans*) (Summerlin et al. 1977, Howard and Oliver 1978).

Harris and Burns (1972) and Burns and Melancon (1977) have reported that IFA are predators of the lone star tick, *Amblyomma americanum*. The latter authors reported significant reductions in populations of the lone star tick in woodland pastures in northwestern Louisiana. However, the overall significance of IFA on tick populations need to be studied in more detail and in a greater variety of habitats. For example, lone star ticks are an important pest on a number of wild animals and are particularly abundant along deer trails. It would be interesting to know if tick populations in these habitats are reduced as significantly as reported for pasture land.

Wildlife and Environmental Effects

Reports of the effect of fire ants on wildlife date back to the late 1930s when there were reports of *Solenopsis geminata* being predacious on the nestlings of the bobwhite quail. Reports of this type were also noted for the IFA in the late 1940s by Wilson and Eads (1949) and USDA (1958). During the 1960s and 1970s most wildlife experts viewed IFA as of little importance on bobwhite quail or other types of wildlife. During the past 20 years the author has heard numerous reports of IFA attacks, primarily on ground-nesting birds. However, the relative importance of isolated instances of predation by IFA cannot be judged by reports of this type.

Recently, Mount et al. (1981) provided data on predation of IFA on eggs of a lizard, *Cnemidophorous sexlineatus*, and Mount (1981) made general observations on population levels of vertebrates in the Alabama coastal plains. He suggested that long-term predation by IFA may be reducing natural populations of a number of vertebrates and that a lapse of 10 to 20 years is necessary between the time an area becomes heavily infested and the time the impact of the IFA becomes obvious to the field naturalists. The most striking declines in populations have occurred among some species of snakes and lizards, particularly five-lined skinks.

Aside from declines in populations of lizards and snakes, Mount (1981) also reports observations of another biologist who concluded that populations of the nighthawk, the ground dove, and the eastern meadowlark have declined in the Alabama coastal plain, but not in northern Alabama.

Predation by fire ants on ducklings and pipped eggs of the wood duck was reported by Ridlehuber (1982) in 3 of 20 nest boxes located in east-central Texas. Observations of natural cavities in trees over land showed that 14 of 20 sites (70 percent) were visited regularly by ants. None were used as nesting sites by wood ducks.

Miscellaneous Effects

This section is included to provide illustration of the versatility of the IFA in causing problems for man. IFA are known to sporadically invade homes and other buildings on occasion, mostly in search of food or water (Bruce et al. 1978). An example of the potential problems associated with IFA in buildings is the fact that during the writing of this chapter the air-conditioning system in the author's building was inactivated twice because IFA invaded the electrical control box.

As minor as this problem may seem, it did involve the expenditure of time and money since a maintenance man was required to correct the problem at a labor cost of about $15. If the problem had occurred in a private residence or business the owner might also have had to pay a service charge of $20 to $30. If situations of this type occur throughout the infested area, one could easily assume that thousands of dollars are spent each year on similar "simple" problems. A recent report by one firm in Marshall, Texas estimated $1,000 to $1,500 in business attributable to IFA damage to air conditioners (H. K. Poulan personal communication).

Another consistent complaint has been the invasion of telephone pedestals by IFA (D. Carlson, personal communication). These pedestals are used for various connections or linkings of telephone cables or wires in the field. There are literally thousands of these units located throughout the south. Two problems are associated with IFA in the pedestals. First, the telephone maintenance people may be stung when they attempt to work within the pedestal, and second, the IFA may damage the equipment.

Two other isolated problems have been reported. Lofgren et al. (1975) cited a report of the inactivation of airport runway lights in a manner similar to the air conditioner incident. R. Grothaus (personal communication) noted damage to asphalt roads in and near Camp LeJeune, North Carolina, This problem was discovered on a few secondary roads where the IFA tunneled underneath the asphalt and removed sufficient dirt so that a small portion of the road collapsed.

CONTROL OF IMPORTED FIRE ANTS

Historically, control of various ant species has relied on chemicals and, more specifically, the combination of a chemical toxicant with an attractive food material to form a toxic bait. The efficacy and desirability of this approach rely on the fact that control of an ant colony ultimately depends on the death of the queen. In most cases the worker population of the colony may be reduced drastically, but the colony will survive and multiply if the colony queen survives. The prime requisite of any effective toxic bait is a chemical with a delayed toxic or physiological activity. Toxic compounds that affect the foraging ants too quickly prevent them from returning the bait to the colony and distributing it to their nest mates and queen.

While toxic baits have been the preferred approach to IFA control in recent years, control in the 1950s and the early 1960s relied heavily on the use of residual or contact insecticides such as chlordane, heptachlor, and dieldrin. These compounds were used extensively for IFA control, either as aqueous drenches for individual mounds or as granular formulations applied as residual treatments over large areas. It was the use of these chemicals in the federal-state IFA control program (promulgated by the U.S. Congress in 1957) that incited much of the controversy over IFA control in the late 1950s. During the period from 1957 to 1962 thousands of acres were treated at application rates ranging from 0.25 to 2 lb per acre (0.11 to 0.91 kg per ha). These treatments resulted in numerous reports of adverse environmental effects to a variety of organisms. Consequently, the development of mirex bait by the USDA in the early 1960s (Lofgren et al. 1964) was hailed as an outstanding development since it appeared to provide a means to control IFA over large tracts of land with minimal environmental effects. Mirex bait consisted of a granular carrier (12–30 mesh corncob grits) onto which a food attractant (soybean oil) containing the toxicant mirex was absorbed. While several variations of the formulation were tested, the most extensively used bait contained 0.3% mirex, 14.7% soybean oil, and 85% corncob grits (Banks et al. 1976). The bait was applied with ground or aerial equipment, usually at the rate of

1.25 lb per acre (1.36 kg per ha). At this rate the actual application rate of the toxicant, mirex, was 4.2 per ha.

In 1968 a study was initiated by the USDA and cooperating states to test the feasibility of eradicating IFA with mirex bait. The results of these studies (Banks et al. 1973) revealed that eradication might be technically feasible since no insurmountable technical problems were detected. Coincident with these studies other problems arose with regard to the use of mirex. It was discovered that even though extremely small quantities of the toxicant were applied (4.2 per ha), residues could still be detected in a large number of nontarget organisms (Markin et al. 1974). Eventually, residues were also detected in adipose tissue from humans (Kutz et al. 1974) and laboratory studies suggested that mirex might be a carcinogen (Ulland et al. 1977). The final result was that mirex bait registrations were cancelled by the Environmental Protection Agency in December of 1978.

With no adequate bait formulations available for control of IFA, a number of investigators began to evaluate the potential of other methods, particularly the treatment of individual mounds or colonies of IFA with chemicals (Morrill 1976, Hillman 1976, Williams and Lofgren 1983) or hot water (Tschinkel 1980). Over the next several years a variety of chemicals were tested and registered as mound drenches or fumigants (Williams and Lofgren 1983; Metcalf et al. 1982).

The primary objective in the application of mound drenches and injection or fumigant treatments is the rapid exposure of the ant population within the mound to the chemical. While this may sound simple, a variety of factors can influence the effectiveness of the treatments including soil type, soil moisture content, temperature, and mound height and shape. Of primary importance is the need to kill the colony queen. Killing thousands of worker ants will only result in a temporary abatement of the colony growth since the surviving workers will move the queen, and any remaining brood, to a new location, where they will rapidly begin development again. Another important factor is the percentage of the worker ants that are away from the nest at the time of treatment. Normally, these ants will not return to the nest when it has been contaminated with insecticide, but will establish satellite queen-less colonies. These may persist for many weeks unless they are also treated.

IFA construct the largest mounds when the soil is wet and/or cool. Thus, the colonies and queen should be easier to kill at this time. In contrast, IFA mounds are not maintained well during hot, dry weather and in sandy soil and they may become flattened, dispersed, and irregularly shaped. At this time the ants move deeper into the soil to obtain a favorable temperature and moisture environment. Obviously, it is more difficult to contact the workers and queen at this time. Probably the best advice for utilization of mound drench treatments is to follow directions on the label, use common sense, and re-treat new or satellite colonies as needed. Finally, mound treatments are labor intensive and thus have limited usefulness in alleviating IFA problems over large areas.

Efforts to develop chemical formulations that could be applied as residual area treatments for cropland similar to the granular formulations of the chlorinated hydrocarbons that were used in the 1950s have been unsuccessful (Sheppard 1982, Stringer et al. 1980, Banks et al. 1982). The USDA began an accelerated search for delayed-action toxicants in the mid 1970s to replace mirex for fire ant bait control. Williams (1983) reported the evaluation of more than 5,000 chemicals. Twenty-nine compounds were tested in baits; however, only one was developed commercially. This compound was an amidinohydrazone (AC 217,300; tetrahydro-5,5-dimethyl-2(1H)-pyrimidinone, [3-[4-(trifluoromethyl)phenyl]-1-[2-[4-(trifluoromethyl)phenyl]ethenyl]-2-propenylidene]hydrazone). The chemical was formulated at concentrations of 2.5 to 10% in a bait consisting of extruded corn pellets and soybean oil. It proved to be very effective against laboratory colonies consisting of a queen, immatures, and 10,000 to 100,000 workers. In every case either the entire colony was killed or most of the workers and the colony queen were killed (Williams et al. 1980a). Subsequently, Banks et al. (1981) reported that AC 217,300 formulated with soybean oil on an extruded corn pellet bait was quite effective in controlling natural populations of IFA in large-scale tests. Conditional registration of AC 217,300 in a bait called Amdro® was approved by the EPA in 1980 for use against IFA in pasture, range grasses, lawns, turfs, and nonagricultural land. The bait is similar to that tested by Banks et al. (1981) and contains 0.88% AC 217,300. It is registered for broadcast application at 1 to 1.5 lb per acre (1.12 to 1.68 kg per ha) or 4 to 6 g AI per acre (9.85 to 14.78 g per ha). Full registration of this bait to include application to agricultural crops has not been received to date.

Other older compounds, such as boric acid, are registered for ant control and may be useful in homes. Boric acid is known to be moderately toxic to IFA, but the necessity of using water-based sugar formulations limits its usefulness for large-scale outdoor application.

While standard toxicants provide the most effective and rapid approach to IFA control by affecting all colony members, including the queen, compounds that inhibit reproduction can also cause death of the colony. Thus, chemicals that interfere with larval development or the ability of the queen to lay eggs will ultimately kill the colony. The undesirable aspect of this approach is that the workers in the colony are not affected and remain a problem. Workers that lose their queen cannot replace her by feeding special food to some larvae, as is the case with honey bees. There is the possibility that orphaned workers will adopt a new fertilized queen following mating flights from other colonies (Tschinkel and Howard 1978) although evidence that this occurs to any extent in the field is lacking. Also, queen adoption is dependent, at least partly, on production of the queen recognition pheromone, which occurs about two weeks after their dealation (Glancey et al. 1981a).

The first actual utilization of the reproduction inhibition approach was presented by Edwards (1977) and Hrdy et al. (1977). They demonstrated control of populations of the pharoah ant, Monomorium pharaonis (L.) in buildings with applications of baits of two juvenile hormone analogs, hydroprene and methoprene. Vinson and Robeau (1974) and Banks et al. (1978) demonstrated that IFA colonies in the laboratory could be killed with Stauffer MV-678 (1-(8-methoxy-4,8-dimethyl-nonyl)-4-(1-methylethyl)benzene) and JH-25 ((E)-1-[7-ethoxy-3,7-dimethyl-2- octenyl)oxy]-4-ethyl benzene. Subsequently, Banks and Schwartz (1980) reported on field tests in which 76% of the IFA colonies were destroyed with a soybean oil bait of Stauffer MV-678. In other unpublished data 85 to 90% control of IFA has been obtained following two aerial applications of baits of Stauffer MV-678 made with soybean oil and pregel defatted corn grit.

Other IGRs that have shown promise against IFA are: Ciba-Geigy CGA-38531, 1-(3-ethoxy butoxy)-4-phenoxybenzene; Montedison JH-286, 1[(5-chloro-pent-4-ynyl)-oxy]-4-phenoxybenzene; and Maag Argochemicals RO 13-5223, ethyl[2-(p-phenoxyphenoxy)ethyl]carbamate (Banks et al. 1983). The latter compound is under commercial development and has been assigned the tradename Logic®.

Lofgren and Williams (1982) reported on a novel chemical for control of IFA (avermectin B_1a) that acts as a highly potent inhibitor of reproduction by queens. Avermectin B_1a is a natural product derived from a soil microorganism, Streptomyces avermitilis. In laboratory tests, concentrations as low as 0.0025% in soybean oil completely stopped reproduction by the queen. In field tests worker brood was found in only 8 of 928 colonies that had access to baits applied at rates ranging from 0.0077 to 7.41 g per ha. Glancey et al. (1982) found that avermectin B_1a causes irreversible cell and tissue damage to the ovaries of the queen. The damage is characterized by atrophy of the squamous epithelium that shields the ovarioles and pycnosis of the nurse cell nucluei. This physiological damage results in complete or partial reduction in the number and size of eggs laid. Commercial development of this compound is being conducted under an experimental use permit from the EPA. The bait formulation is called Affirm®.

The current status of biocontrol of IFA has been reviewed by Jouvenaz et al. (1981). Populations of the IFA in the United States appear to be disease free (Jouvenaz et al. 1977). However, two Microsporida and a neogregarine have been recovered from populations in South America. The most common species, *Thelohania solenopsae*, was found in 25% or more of the colonies examined at the end of the rainy season, but much lower infection rates (about 5%) were noted at the end of the dry season (Jouvenaz et al. 1980). An undescribed microsporidian was also detected that infected a much lower percentage of the colonies. The neogregarine, a *Mattesia* species, was found in a small number of colonies. Virus-like particles were discovered in an undescribed *Solenopsis* species from Brazil and a bacterial infection was found in one colony of *S. invicta*. The importance of all of these organisms in limiting populations of IFA in South America has never been determined.

Three groups of parasitic arthropods are known to affect IFA. *Solenopsis* (formerly *Labauchena*) *daguerri*, a workerless social parasitic ant, has been thoroughly studied (Silviera-Guido et al. 1973). This species has been found in association with *S. richteri* in Uruguay. The queens "yoke" themselves to the IFA queen permanently by grasping her between the head and thorax. The IFA workers care for the parasitic queen and her progeny preferentially, reducing the vigor of the IFA colony. While high incidences of parasitism (30 to 40%) have been found in some areas in Uruguay, the rate is about 4% in most locations. No instances of these social parasites attacking *S. invicta* has been reported (Jouvenaz et al. 1981). The other two groups of parasites associated with fire ants in South America are eucharitid wasps (*Orasema* sp.) and phorid flies of the genera *Pseudacteon* and *Apodicranea*. The life cycles of these parasites have been studied by Williams (1980), but their importance for IFA control has not been determined.

Various nonspecific predators of IFA queens, workers, and larvae have been studied in the United States. Again the true significance of any of these is undetermined, but there can be little question that newly mated queens are very vulnerable to predation. Buren et al. (1978) suggested that manipulation of predators of newly mated queens could lead to an effective method of pest management for IFA. However, this possibility appears remote when considering the changing population and agricultural trends in the southern United States and the potential for IFA to invade and populate disturbed habitats. Certainly increasing needs for food production and living space for man will continue to provide disturbed or altered habitats that will favor IFA queen survival over that of the nonspecific predators.

The greatest possibilities for biocontrol of IFA appear to remain undiscovered in South America. Hopefully, specific diseases or parasites of IFA will be detected when future research efforts are made in this direction. Behavioral chemicals may be useful for IFA control (see Chapter 7) but as yet none have proven effective.

The overall goal for IFA control should be an integrated pest management program incorporating all the preceding procedures. This goal will not be achieved until much more research is accomplished on nonchemical control methods, particularly the search for, and development of, biological control agents in the homeland of the IFA in South America.

REFERENCES

Adams. C. T. 1983. Destruction of eggplants in Marion County, Florida by red imported fire ants. *Fla. Entomol.* 66: 518–520.

Adams, C. T. and C. S. Lofgren. 1981. Red imported fire ants: Frequency of sting attacks on residents of Sumter County, Georgia. *J. Med. Entomol.* 18: 378–382.

Adams, C. T. and C. S. Lofgren. 1982. Incidence of stings or bites of the red imported fire ant and other arthropods among patients at Ft. Stewart, Georgia USA. *J. Med Entomol.* 19: 366–370.

Adams, C. T., J. K. Plumley, C. S. Lofgren, and W. A. Banks. 1976. Economic importance of the red imported fire ant, *Solenopsis invicta* Buren. I. Preliminary investigations of impact on soybean harvest. *J. Ga. Entomol. Soc.*. 11: 165–169.

Adams, C. T., J. K. Plumley, W. A. Banks, and C. S. Lofgren. 1977. Impact of the red imported fire ant, *Solenopsis invicta* Buren on harvest of soybeans in North Carolina. *J. Elisha Mitchell Sci. Soc.* 93: 150–152.

Adams, C. T., T. E. Summers, C. S. Lofgren, D. A. Focks, and J. C. Prewitt. 1981. Interrelationships of ants and the sugarcane borer in Florida sugarcane fields. *Environ. Entomol.* 10: 415–418.

Adams, C. T., W. A. Banks, C. S. Lofgren, B. J. Smittle, and D. P. Harlan. 1983. Impact of the red imported fire ant, *Solenopsis invicta*, on the growth and yield of soybeans. *J. Econ. Entomol.* 76: 1129–1132.

Anonymous. 1976. Fire ant control. Council for Agricultural Science and Technology. CAST Rep. No. 62, 2nd ed. Rep. No. 65. 24 pp. Iowa State Univ., Ames.

Apperson, C. S. and E. E. Powell. 1983. Correlation of the red imported fire ant, *Solenopsis invicta* Buren, with reduced soybean yields in North Carolina. *J. Econ. Entomol.* 76: 259–263.

Baer, H., D. T. Liu, M. Hooten, M. Blum, F. James, and W. H. Schmid. 1977. Fire ant allergy: Isolation of 3 allergenic proteins from whole venom. *Ann. Allergy* 38: 378.

Banks, W. A. and D. P. Harlan. 1982. Tests with an insect growth regulator, Ciba Geigy CGA-38531, against laboratory and field colonies of red imported fire ants. *J. Ga. Entomol. Soc.* 17: 462–466.

Banks, W. A. and M. Schwarz. 1980. The effects of insect growth regulators on laboratory and field colonies of red imported fire ants. *Proc. Tall Timbers Conf. Ecol. Anim. Control Habitat. Manag.* 7: 95–105.

Banks, W. A., B. M. Glancey, C. E. Stringer, D. P. Jouvenaz, C. S. Lofgren, and D. E. Weidhaas. 1973. Imported fire ants: Eradication trials with mirex bait. *J. Econ. Entomol.* 66: 785–789.

Banks, W. A., D. M. Hicks, J. K. Plumley, D. P. Jouvenaz, D. P. Wojcik, and C.

S. Lofgren. 1976. Imported fire ants: 10-5, an alternate formulation of mirex bait. *J. Econ. Entomol.* 69: 465–467.

Banks, W. A., C. S. Lofgren and J. K. Plumley. 1978. Red imported fire ants: Effects of insect growth regulators on caste formation and colony growth and survival. *J. Econ. Entomol.* 71: 75–78.

Banks, W. A., H. L. Collins, D. F. Williams, C. E. Stringer, C. S. Lofgren, D. P. Harlan, and C. Mangum. 1981. Field trials with AC 217,300, a new amidino-hydrazone bait toxicant for control of red imported fire ants. *Southw. Entomol.* 6: 158–164.

Banks, W. A., D. P. Harlan and C. E. Stringer. 1982. Effectiveness of emulsions and control release granules of chlorpyrifos and isofenphos on red imported fire ants in cultivated fields. *J. Ga. Entomol. Soc.* 17: 259–265.

Banks, W. A., L. R. Miles, and D. P. Harlan. 1983. The effects of insect growth regulators and their potential as control agents for imported fire ants. *Fla. Entomol.* 66: 172–181.

Bellinger, F., R. E. Dyer, R. King, and R. B. Platt. 1965. A review of the problem of the imported fire ant. *Bull. Georgia Acad. Science* 23: 1–22.

Blust, W. E., B. H. Wilson, K. L. Koonce, B. D. Nelson, and J. E. Sedberry, Jr. 1982. The red imported fire ant, *Solenopsis invicta* Buren: Cultural control and effect on hay meadows. Bull. No. 738, Louisiana State University, Baton Rouge, LA.

Brown, D. W. and D. A. Goyer. 1982. Effects of a predator complex on lepidopterous defoliators of soybean. *Environ. Entomol.* 11: 385–389.

Brown, L. L. 1972. Fire ant allergy. *South. Med. J.* 65: 273–277.

Brown, R. 1982. IFA damage to young citrus trees. Paper read at Imported Fire Ant Conference, March 25–26, 1982, Austin, Texas.

Bruce, W. A., L. D. Cline, and G. L. LeCato. 1978. Imported fire ant infestation of buildings. *Fla. Entomol.* 61: 230.

Buren, W. F., G. F. Allen, and R. N. Williams. 1978. Approaches toward possible pest management of the imported fire ants. *Bull. Entomol. Soc. Amer.* 24: 418–421.

Burns, E. C. and D. G. Melancon. 1977. Effect of imported fire ant invasion on lone star tick (Acarina: Ixodidae) populations. *J. Med. Entomol.* 14: 247–249.

Buschman, L. L., W. H. Whitcomb, R. C. Hemenway, D. L. Mays, W. Ru, N. C. Leppla, and B. J. Smittle. 1977. Predators of velvetbean caterpillar eggs in Florida soybeans. *Environ. Entomol.* 6: 403–407.

Campbell, T. E. 1974. Red imported fire ant, a predator of direct seeded longleaf pine. U.S. Forest Serv. Res. Note SO-179, pp. 1–3.

Canter, L. W. 1981. Cooperative Imported Fire Ant Program: Final Programmatic Environmental Impact Statement. U.S. Dept. of Agriculture, Hyattsville, MD. 150 pp. and appendices.

Clemmer, D. I. and R. E. Serfling. 1975. The imported fire ant: Dimensions of the urban problem. *South. Med. J.* 68: 1133–1138.

Cockerman, K. L. and E. Oertel. 1954. Control of ants in apiaries. *Glean. Bee Cult.* 82: 348–349.

Combs, R. L., Jr. 1982. The black imported fire ant, a predator of the face fly in Northeast Mississippi. *J. Ga. Entomol. Soc.* 17: 496–501.

Crance, J. H. 1965. Fish kills in Alabama ponds after swarms of the imported fire ant. *Prog. Fish Cult.* 27: 91–94.

Dutcher, J. D. and D. C. Sheppard. 1981. Predation of pecan weevil larvae by red imported fire ants. *J. Ga. Entomol. Soc.* 16: 210–213.

Edwards, J. P. 1977. Control of *Monomorium pharaonis* with an insect juvenile hormone analogue. Proc. 8th Int. Cong. IUSSI, pp. 81–82. Wageningen, The Netherlands.

Glancey, B. M., J. D. Coley, and F. Killibrew. 1979. Damage to corn by the red imported fire ant. *J. Ga. Entomol. Soc.* 14: 198–201.

Glancey, B. M., A. Glover, and C. S. Lofgren. 1981a. Pheromone production by virgin queens of *Solenopsis invicta* Buren. *Sociobiology* 6: 119–127.

Glancey, G. M., R. K. Vander Meer, A. Glover, C. S. Lofgren, and S. B. Vinson. 1981b. Filtration of microparticles from liquids ingested by the red imported fire ant, *Solenopsis invicta* Buren. *Insectes Soc.* 28: 395–401.

Glancey, B. M., C. S. Lofgren, and D. F. Williams. 1982. Avermectin B_1a: Effects on the ovaries of red imported fire ant queens. *J. Med. Entomol.* 19: 743–747.

Green, H. B. 1967. The imported fire ant in Mississippi. Miss. State Univ. Exp. Sta. Bull. 737, 23 pp.

Gross, H. R., Jr. and W. T. Spink. 1969. Responses of striped earwigs following applications of heptachlor and mirex, and predator-prey relationships between imported fire ants and striped earwigs. *J. Econ. Entomol.* 62: 686–689.

Harlan, D. P., W. A. Banks, C. E. Stringer, P. M. Bishop, L. Miles, and J. Mitchell. 1981. Red imported fire ants: Damage to corn in the field and greenhouse. *Proc. Annu. Miss. Insect Control Conf.* 27: 39.

Harris, W. G. and E. C. Burns. 1972. Predation on the lone star tick by the imported fire ant. *Environ. Entomol.* 1: 362–365.

Hays, S. B. and K. L. Hays. 1959. Food habits of *Solenopsis saevissima richteri* Forel. *J. Econ. Entomol.* 52: 445–447.

Herzog, D. C., T. E. Reagan, D. C. Sheppard, K. M. Hyde, S. S. Nilakle, M. Y. B. Hussein, M. L. McMahan, R. C. Thomas, and L. D. Newsome. 1976. *Solenopsis invicta* Buren-influence on Louisiana pasture soil chemistry (Hymenoptera: Formicidae). *Environ. Entomol.* 5: 160–162.

Hill, E. P. 1969. Observations of imported fire ant predation on nestling cottontails. Proc. 23rd Ann. Conf. Southeast Assoc. Fish Wild. Comm. pp. 171–181.

Hillman, R. C. 1976. Control of red imported fire ants, *Solenopsis invicta*, on home grounds in North Carolina, USA. *J. Elisha Mitchell Sci. Soc.* 92: 69–70.

Howard, F. W. and A. D. Oliver. 1978. Anthropod populations in permanent pastures treated and untreated with mirex for red imported fire ant control. *Environ. Entomol.* 7: 901–903.

Hrdy, I., J. Krecek, V. Rupes, J. Zdorek, J. Chmela, and J. Ledvinka. 1977. Control of the pharaoh's ant (*Monomorium pharaonis*) with juvenoids in baits. Proc. 8th Int. Congr. IUSSI, pp. 83–84. Wageningen, The Netherlands.

Hunt, T. N. 1976. Agricultural losses due to the imported fire ant as estimated by North Carolina, USA, farmers. *J. Elisha Mitchell Sci. Soc.* 92: 69.

James, F. K., Jr. 1976. Fire ant sensitivity. *J. Asthma Res.* 13: 179–183.

Johnson, A. S., III. 1962. Antagonistic relationship between ants and wildlife with special reference to imported fire ants and bobwhite quail in the southeastern

United States. M.S. Thesis, Auburn Univ., Auburn, AL. xii. 100 pp.

Jones, D. and W. L. Sterling. 1979. Manipulation of red imported fire ants in a trap crop for boll weevil suppression. *Environ. Entomol.* 8: 1073–1077.

Jouvenaz, D. P., G. E. Allen, W. A. Banks, and D. P. Wojcik. 1977. A survey for pathogens of fire ants, *Solenopsis* spp., in the southeastern United States. *Fla. Entomol.* 60: 275–279.

Jouvenaz, D. P., W. A. Banks, and J. D. Atwood. 1980. Incidence of pathogens in fire ants, *Solenopsis* spp., in Brazil. *Fla. Entomol.* 63: 345–346.

Jouvenaz, D. P., C. S. Lofgren, and W. A. Banks. 1981. Biological control of imported fire ants: A review of current knowledge. *Bull. Entomol. Soc. Amer.* 27: 203–208.

Krispyn, J. W. and J. W. Todd. 1982. The red imported fire ant as a predator of the southern green stink bug on soybean in Georgia. *J. Ga. Entomol. Soc.* 17: 19–26.

Kutz, F. W., A. R. Yobs, W. G. Johnson, and G. B. Wiersma. 1974. Mirex residues in human adipose tissue. *Environ. Entomol.* 3: 882–884.

Landers, J. L., J. A. Garner and A. McRae. 1980. Reproduction of gopher tortoises (*Gopherus polyphemus*) in southwestern Georgia. *Herpetologica.* 36: 353–361.

Lockey, R. F. 1974. Systemic reactions to stinging ants. *J. Allergy Clin. Immunol.* 54: 132–146.

Lockey, R. F. 1979. Allergic and other adverse reactions caused by the imported fire ant. *In* "Advances in Allergology and Clinical Immunology," pp. 441–448 (O. Oehling, E. Methov, I. Glazer, and C. Orbesman, eds.), Pergamon Press, New York.

Lofgren, C. S. and C. T. Adams. 1981. Reduced yield of soybeans in fields infested with the red imported fire ant, *Solenopsis invicta* Buren. *Fla. Entomol.* 64: 199–202.

Lofgren, C. S. and C. T. Adams, 1982. Economic aspects of the imported fire ant in the United States. *In* "The Biology of Social Insects," pp. 124–128 (M. D. Breed, C. D. Michener, and H. E. Evans, eds.), Westview Press, Boulder, Colo.

Lofgren, C. S. and D. F. Williams. 1982. Avermectin B_1a, a highly potent inhibitor of reproduction by queens of the red imported fire ant. *J. Econ. Entomol.* 75: 798–803.

Lofgren, C. S., F. J. Bartlett, C. E. Stringer, and W. A. Banks. 1964. Imported fire ant toxic bait studies: Further tests with granulated mirex-soybean oil bait. *J. Econ. Entomol.* 57: 695–698.

Lofgren, C. S., W. A. Banks and B. M. Glancey. 1975. Biology and control of imported fire ants. *Annu. Rev. Entomol.* 20: 1–30.

Long, W. H., E. A. Cancienne, E. J. Concienne, R. N. Dobson, and L. D. Newsom. 1958. Fire-ant eradication program increases damage by the sugarcane borer. *Sugar Bull.* 37: 62–63.

Lopez, J. D. 1982. Emergence pattern of an overwintering population of *Cardiochiles nigriceps* in central Texas. *Environ. Entomol.* 11: 838–842.

Lyle, C. and I. Fortune. 1948. Notes on an imported fire ant. *J. Econ. Entomol.* 41: 833–834.

MacConnell, J. G., M. S. Blum, and H. M. Fales. 1970. Alkaloid and fire ant venom: Identification and synthesis. *Science* 168: 840–841.

Markin, G. P., H. L. Collins, and J. Davis. 1974. Residues of the insecticide mirex in terrestrial and aquatic invertebrates following a single aerial application of mirex bait-Louisiana-1971–72. Pestic. Monit. J. 8: 131–134.

McDaniel, S. G. and W. L. Sterling. 1979. Predator determination and efficiency on Heliothis virescens eggs in cotton using ^{32}P. Environ. Entomol. 8: 1083–1087.

McDaniel, S. G. and W. L. Sterling. 1982. Predation of Heliothis virescens (F.) eggs on cotton in east Texas. Environ. Entomol. 11: 60–66.

Metcalf, R. L., D. J. Severn, E. L. Alley, M. Blum, M. K. Hinkle, H. N. Nigg, W. H. Schmid, and J. Wood. 1982. In "Environmental Toxicology, Appendix K," pp. 190–200 (S. L. Battenfield, ed.), Atlanta, Georgia, June 1982. U.S. Environ. Prot. Agency and U.S. Dept. Agric. 255 pp.

Mills, H. B., 1967. Chairman. Report of committee on imported fire ant. National Acad. Sci. to Administrator, U.S. Dept. Agri., Agri. Res. Serv. 15 pp. [Unpublished].

Morrill, W. L. 1976. Red imported fire ant control with mound drenches. J. Econ. Entomol. 69: 542–544.

Morrill, W. L. 1977. Red imported fire ant foraging in a greenhouse. Environ. Entomol. 6: 416–418.

Morrill, W. L. 1978. Red imported fire ant predation on the alfalfa weevil and pea aphid. J. Econ. Entomol. 71: 867–868.

Mount, R. H. 1981. The red imported fire ant, Solenopsis invicta, as a possible serious predator on some native southeastern vertebrates: Direct observations and subjective impressions. J. Ala. Acad. Sci. 52: 71–78.

Mount, R. H., S. E. Trauth, and W. H. Mason. 1981. Predation by the red imported fire ant. Solenopsis invicta, on eggs of the lizard Cnemidophorus sexlineatus (Squamata: Teiidae). J. Ala. Acad. Sci. 52: 66–70.

Negm, A. A. and S. D. Hensley. 1969. Effect of insecticides on ant and spider populations in Louisiana sugarcane fields. J. Econ. Entomol. 62: 948–949.

Reagan, T. E. 1981. Sugarcane borer pest management in Louisiana: Leading to a more permanent system. Proc. 2nd Inter-American Sugarcane Seminar: Insect and Rodent Pests-1981. Florida International University, Miami, Fla., Oct. pp. 100–110.

Reagan. T. E., G. Coburn, and S. D. Hensley. 1972. Effects of mirex on the arthropod fauna of a Louisiana sugarcane field. Environ. Entomol. 1: 588–591.

Rhoades, R. B., W. L. Schafer, M. Newman, R. Lockey, R. M. Dozier, P. F. Wubbena, A. W. Tower, W. H. Schmid, G. Neder, T. Brill, and H. J. Wittig. 1977. Hypersensitivity to the imported fire ant in Florida. Report of 104 cases. J. Fla. Med. Assoc. 64: 247–254.

Ridlehuber, K. T. 1982. Fire ant predation on wood duck ducklings and pipped egg. Southwestern Natur. 27: 220.

Roe, R. L. II. 1973. A biological study of Solenopsis invicta Buren, the red imported fire ant, in Arkansas, with notes on related species. M.S. Thesis, University of Arkansas, Fayetteville, AK. 135 pp.

Russell, C. E. 1981. Predation on the cowpea curculio by the red imported fire ant. J. Ga. Entomol. Soc. 16: 13–15.

Sheppard, C. 1982. Effects of broadcast diazinon sprays on populations of red imported fire ants. *J. Ga. Entomol. Soc.* 17: 177–183.

Silviera-Guido, A., J. Carbonell, and C. Crisci. 1973. Animals associated with the Solenopsis (fire ants) complex with special reference to *Labauchena daguerri. Proc. Tall Timbers Conf. Ecol. Anim. Control Habitat Manag.* 4: 41–52.

Smittle, B. J., C. T. Adams, and C. S. Lofgren. 1983. Red imported fire ants: Detection of feeding on corn, okra and soybeans with radioisotopes. *J. Ga. Entomol. Soc.* 18: 78–82.

Snodgrass, G. L. 1976. An evaluation of the black imported fire ant, *Solenopsis richteri*, as a predator in soybeans in Northeast Mississippi. M.S. Thesis. Mississippi State Univ., Mississippi State, MS. 86 pp.

Sterling, W. L. 1978. Fortuitous biological suppression of the boll weevil by the red imported fire ant. *Environ. Entomol.* 7: 564–568.

Sterling, W. L., D. Jones, and D. A. Dean. 1979. Failure of the red imported fire ant to reduce entomophagous insect and spider abundance in a cotton agroecosystem. *Environ. Entomol.* 8: 976–981.

Stringer, C. E., W. A. Banks, and J. A. Mitchell, 1980. Effects of chlorpyrifos and acephate on populations of red imported fire ants in cultivated fields. *J. Georgia Entomol. Soc.* 15: 413–417.

Summerlin, J. W. and S. E. Kunz. 1978. Predation of the red imported fire ant on stable flies. *Southw. Entomol.* 3: 260–262.

Summerlin, J. W., J. K. Olson, R. R. Blume, A. Aga, and D. E. Bay. 1977. Red imported fire ants: Effects on *Onthophagus gazella* and the horn fly. *Environ. Entomol.* 6: 440–442.

Triplett, R. F. 1973. Sensitivity to the imported fire ant: Successful treatment with immunotherapy. *South. Med. J.* 66: 477–480.

Tschinkel, W. R. 1980. A simple, non-toxic home remedy against fire ants. *J. Georgia Entomol. Soc.* 15: 102–105.

Tschinkel, W. R. 1982. History and biology of fire ants. *In* "Proceedings of the Symposium on the Imported Fire Ant," pp. 16–35 (S. L. Battenfield, ed.), Atlanta, Georgia, June 1982. U.S. Environ. Prot. Agency and U.S. Dept. Agri., 255 pp.

Tschinkel, W. R. and D. F. Howard. 1978. Queen replacement in orphaned colonies of the fire ant, *Solenopsis invicta. Behav. Ecol. Sociobiol.* 3: 297–310.

Tschirley, F. H. 1982. Executive summary. *In:* "Proceedings of the Symposium on the Imported Fire Ant," pp. 1–8 (S. L. Battenfield, ed.), Atlanta, Georgia, June 1982. U.S. Environ. Prot. Agency and U.S. Dept. Agri., 255 pp.

Ulland, B. M., N. P. Page. R. A. Squire, E. K. Weisburger and R. L. Cypher. 1977. A carcinogenicity assay of mirex in Charles River CD rats. *J. Natl. Cancer Inst.* 58: 133–140.

U.S.D.A. 1958. Observations on the biology of the imported fire ant. Rep. ARS 33–49, 21 pp.

Vinson, S. B. 1982. Effect of the IFA on ground-nesting bees. Paper read at Imported Fire Ant Conference, March 25–26, Austin, Texas.

Vinson, S. B. and R. Robeau. 1974. Insect growth regulators: Effects on colonies of the imported fire ant. *J. Econ. Entomol.* 67: 584–587.

Whitcomb, W. H., H. A. Denmark, A. P. Bhatkar, and G. L. Greene. 1972. Preliminary studies on the ants of Florida soybean fields. *Fla. Entomol.* 55: 129–142.

Whitcomb, W. H., T. D. Gowan, and W. F. Buren. 1982. Predators of *Diaprepes abbreviatus* (Coleoptera: Curculionidae) larvae. *Fla. Entomol.* 65: 150–158.

Wilkinson, R. C. and C. W. Chellman. 1979. Toumeyella scale, red imported fire ant, reduce slash pine growth. *Fla. Entomol.* 62: 71–72.

Williams, D. F. 1983. The development of toxic baits for the control of the imported fire ant. *Fla. Entomol.* 66: 162–172.

Williams, D. F. and C. S. Lofgren. 1983. Imported fire ant control: Evaluation of several chemicals for individual mound treatments. *J. Econ. Entomol.* 76: 1201–1205.

Williams, D. F., C. S. Lofgren, W. A. Banks, C. E. Stringer, and J. K. Plumley. 1980a. Laboratory studies with 9 amidinohydrazones, a promising new class of bait toxicants for control of red imported fire ants. *J. Econ. Entomol.* 73: 798–802.

Williams, D. F., C. S. Lofgren, and A. Lemire. 1980b. A simple diet for rearing laboratory colonies of the red imported fire ant. *J. Econ. Entomol.* 73: 176–177.

Williams, J. L. 1976. Status of the greater wax moth, *Galleria mellonella*, in the U.S.A. *Amer. Bee J.* 116: 524–526.

Williams, R. N. 1980. Insect natural enemies of fire ants in South America with several new records. *Proc. Tall Timbers Conf. Ecol. Anim. Control Habitat Manag.* 7: 123–134.

Wilson, E. O. and J. H. Eads. 1949. A report on the imported fire ant *Solenopsis saevissima* var. *richteri* Forel in Alabama. Ala. Dep. Conserv. Spec. Rep. 53 pp. 13 plates [Mimeographed].

Wilson, N. L. and A. D. Oliver. 1969. Food habits of the imported fire ant st areas in southern Louisiana. *J. Econ. Entomol.* 62: 1268–1271.

Wilson, N. L. and A. D. Oliver. 1970. Relationship of the imported fire ant to Nantucket pine tip moth infestations. *J. Econ. Entomol.* 63: 1250–1252.

Yeagar, W. 1978. Frequency of fire ant stinging in Lowndes County, Georgia. *J. Med. Assoc. Ga.* 67: 101–102.

Chapter 9

THE BIOLOGY, ECONOMIC IMPORTANCE, AND CONTROL OF THE PHARAOH'S ANT,
Monomorium pharaonis (L.)
J. P. Edwards

INTRODUCTION

Pharaoh's ant *Monomorium pharaonis* (Fig. 9.1) is probably the most widely distributed of all the ant pests. The monomorphic workers are about 2 mm long, yellow-brown, with a noticeable lighter area at the dorsal anterior gaster (Fig. 9.1). Queens (4mm) are larger than workers but are otherwise similar in shape and coloration. Queens are winged when they emerge from the pupal stage but the wings are lost soon after mating. The distinctive, black, winged males are only rarely present in a colony and are seldom seen in natural infestations.

Within the Formicidae, the genus *Monomorium* is placed in the subfamily Myrmicinae (tribe Solenopsidini) along with closely related genera like *Solenopsis* (Bolton and Collingwood 1975). Many species within the tribe Solenopsidini are lestobiotic (i.e., feed on the brood of other ant species) and Dumpert (1981) states that *M. pharaonis* lived originally as a thieving ant species in its natural habitat. The type-specimen described by Linnaeus (1758) was from Egypt and it is likely that the species originally came from the North African-Middle Eastern region. However, in the last 200 years, with increasing trade throughout the world, the species has been distributed worldwide and is now truly cosmopolitan. This ant is widespread in Europe (Eichler 1978) and is present in the Soviet Union, United States, Canada, South America, Australia, Africa, and Japan and is probably widely distributed throughout the tropics. In nontropical climates the species is dependent on artificial heating and thus, in temperate regions, pharaoh's ants are invariably associated with human habitation. In Europe and in North America, infestations are frequently associated with hospitals, bakeries, factories, offices, and large domestic apartment blocks. In the United Kingdom the species was first reported in London in 1828 (Donisthorpe 1927) and since that time has become distributed throughout most of the country. In a recent survey it was estimated that more than 10% of English hospitals were infested (Edwards and Baker 1981).

Until 1972, infestations of *M. pharaonis* were mainly regarded as a nuisance. However, Beatson (1972) demonstrated that the species was able to transmit a variety of pathogenic organisms including *Salmonella, Pseudomonas, Staphylococcus, Streptococcus, Klebsiella,* and *Clostridium* sp, and Alekseev et al. (1972) demonstrated the survival and mechanical transmission of the plague organism *Pasteurella pestis* by *M. pharaonis* after ants had fed on carcasses of animals that had died of the disease. Although houseflies, cockroaches and other domestic pests are able to act as vectors of pathogenic organisms, pharaoh's ants present a particularly serious infection hazard for the following reasons. The small size of foraging workers enables them to pass through the smallest gaps to invade wounds and equipment and they readily chew through paper, plastic, and rubber to gain entry into packages and sealed chambers (Fig. 9.2). Because the species has a high requirement for moisture, ants are often found feeding at sluices, drains, toilets, and other areas likely to harbor disease organisms. Foraging workers have been found in sterile supplies, in sets for giving intravenous fluid and under wound dressings on postoperative patients (Beatson 1973, Cartwright and Clifford 1973,

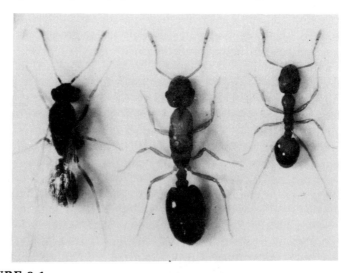

FIGURE 9.1
Adult stages of *Monomorium pharaonis*. Male (left) queen (center) and worker (right).

FIGURE 9.2
Pharaoh's ants inside a cellophane biscuit packet.

Anonymous 1974). Often, these small insects remain undetected and once a food source has been located by a foraging ant, many others will be recruited to feed at that site, thus increasing the chance of disease transmission. Moreover, when a foraging worker becomes contaminated with potential pathogens and returns to the nest, it is likely that these organisms will be passed on to other nest mates, thus greatly increasing the number of individuals capable of transmitting the pathogen. Furthermore, the conditions of warmth and high humidity in the nest, together with the presence of stored food material, could promote the survival of some pathogens (e.g., gram-negative bacteria). Thus, it is apparent that infestations of M. pharaonis in hospitals and domestic residences constitute a serious threat to public health.

BIOLOGY

Colonies of M. pharaonis are polydomic (i.e., comprised of several nests) and there is no aggression between nests. In temperate regions, nests are always situated indoors and are often inaccessibly located in foundations, wall cavities, and underground ducting systems. Nests are unstructured (i.e., not built) and occupy any suitable crevice. The nest contains the brood stages (eggs, larvae, and pupae), numerous workers, and several queens. In laboratory colonies the number of queens per nest is variable, ranging from a few individuals to several hundreds. Queens generally remain inside the nest and are solely responsible for egg laying since the workers are sterile. Workers perform the "house-keeping" duties within the nest (feeding queens and rearing brood stages) and forage outside the nest for food and water. Workers exhibit age polyethism such that young workers are mainly responsible for brood tending while older workers leave the nest to scout and forage for food (J. P. Edwards unpublished). As in the majority of social hymenoptera, eggs laid by queens may give rise to males, queens, or workers. Males are produced from haploid (unfertilized) eggs, and queens and workers are produced from diploid (ferilized) eggs. Peacock et al. (1950) have given the following developmental periods for workers: egg, 7.3 days; larva, 17 days; prepupa, 3.1 days; pupa, 9 days (total 36.4 days). Both sexual forms (males and queens) take slightly longer to reach adulthood (41.25 days), the difference being mainly associated with a longer larval period. Peacock et al. (1950) have also recorded the longevity of adults and show that queens may live up to 39 weeks, workers nine to ten weeks, and males, depending on whether or not they mate, three to eight weeks. In our experience these values are reasonable estimates, although in this laboratory, some queens have survived for more than 52 weeks (J. P. Edwards unpublished).

In the majority of ant species, new nests are founded by single fertilized queens following the "nuptial flight." In M. pharaonis, there is no nuptial flight and though both males and virgin queens are winged, neither is capable of sustained flight. Instead, fertilization of newly produced queens takes place in or near the nest and there appears to be little cross-fertilization between nests. Adult males often emerge from the pupal stage slightly before the queens. When the queens emerge, they are highly attractive to the males which follow them around the nest. Hölldobler and Wust (1973) have reported the existence of a sex pheromone produced in the Dufour's gland and bursa copulatrix of virgin females that acts as a short-range attractant and stimulates mating. Mating lasts for five to 15 minutes and males may mate up to four times. Queens probably mate only once and lose their wings after mating. In laboratory colonies, males and queens may be produced at all times of the year and the factors responsible for the initiation of sexual production are not fully understood. However, it is known that the presence of queens in a nest inhibits the rearing of new sexual forms and it has been suggested that queens inhibit sexual production by competing for "profertile" food at the expense of developing sexual larvae (Buschinger and Kloft 1973) or by producing an inhibitory pheromone (Berndt 1977). Petersen-Braun (1975) has suggested that new sexual forms are produced regularly every three to four months, coinciding with the death of extant queens. In this laboratory, however, the period between batches of sexual forms can be as much as one year (J. P. Edwards unpublished).

In M. pharaonis the foundation of new nests and the dissemination of the colony is achieved by "budding" (sociotomy). Thus, new nests are founded by groups of workers carrying brood stages (eggs, larvae, and pupae) to a new nest site. Although queens often accompany these emigrant groups, they are not essential for the successful foundation of a new colony, since workers can rear new queens and males from existing brood. Often, these emigrating groups will occupy a temporary nest while searching for a more permanent abode. Such temporary nests are highly mobile, and probably the main method by which infestations are spread is the inadvertent transportation of such nests to previously uninfested premises. Many factors can precipitate the formation of satellite nests. In domestic infestations, changes in the environment of the nest (e.g., temperature or the availability of food or water) or overcrowding may lead to partial or total migration. From the practical viewpoint it is of interest that the presence of insecticides may also induce the migration of nests to untreated areas (Green et al. 1954). The polydomic habit, coupled with a lack of aggressive behavior between occupants of different nests, enables the species to respond rapidly to changing environmental conditions. This adaptability, together with a lack of a complex nest structure, has predisposed the species to a synanthropic existence and thus, pest status.

Another factor that has contributed in this respect is the omnivorous nature of M. pharaonis. In the laboratory, these ants will feed on a wide variety of foodstuffs: fats, proteins, and carbohydrates, and will also take and kill small insects. Based on observations of pharaoh's ants in Nigeria, Sudd (1962) suggested that the most important source of food was the dead or dying bodies of insects that were attracked to lights in a house.

Once a food source is located by a foraging worker, a chemical trail is laid from the food to the nest by the returning worker who enters the nest and "excites" other workers to follow the trail to the food source (Sudd 1957, 1960). These chemical trails are often many meters long and comprise a chemical substance (faranal) produced in the Dufour's gland of workers (Ritter et al. 1977) and deposited on the ground by the sting. Workers forage by day and night and do not avoid light, even strong sunlight. When workers return to the nest with food they often feed other workers, queens and developing larvae. However, a large proportion of food is stored in the nest in discrete piles from which material is removed for feeding queens and larvae. Subjective observation suggests that the food is partly chewed by workers prior to storage.

In addition to the trail pheromone faranal, workers produce other chemicals (Ritter et al. 1975) that may act as trail substances (Edwards and Pinniger 1978) or as alarm pheromones or defense secretions. These chemicals, mostly pyrrolidines, are produced in the poison gland of workers and are also dispensed via the sting. Because the sting and associated glands have become adapted for trail laying, etc., pharaoh's ants are not able to sting, neither are their jaws capable of piercing human flesh. The inability of pharaoh's ants to inflict painful lesions on humans is in marked contrast to their close relatives, the fire ants (*Solenopsis* sp.), which are renowned for their painful sting (See Chapter 8).

Queens begin to lay eggs soon after mating. For the first four to five weeks the oviposition rate is four to six eggs per queen per day; subsequently, the rate rises to 25–35 eggs per queen per day and this rate is maintained until some four to five weeks before death, during which time the oviposition rate decreases rapidly to three to seven eggs per queen per day (J. P. Edwards unpublished). Thus during her lifetime a queen may lay more than 4,500 eggs, most of which are destined to become workers. With several hundred queens per nest, there is obvious potential for a rapid increase in the worker population, despite the fact that there is considerable mortality in the larval stages (Peacock et al. 1950). Within the nest, workers of *M. pharaonis* separate the brood into distinct sections each containing individuals at similar stages of development. Eggs tend to be placed in groups and newly hatched larvae probably obtain much of their nourishment by feeding on eggs in their immediate vicinity until they are removed to "larval" sections of the nest. Piles of eggs, larvae and pupae are constantly inspected by the nurse workers and all stages are frequently touched with the antennae and "licked" by workers. The larvae are typical apodous "grubs" resembling the larvae of many other ant species (Fig. 9.3). When larva are fully grown, the sac containing the larval excreta is removed by workers and the insects enter a short prepupal (pharate pupa) stage before pupal eclosion. Although adult workers are able to emerge from the pupal stage unaided, adult emergence is often assisted by nurse workers. Sexual larvae are not easily distinguishable from worker larvae until the former are about half-grown. At this stage the sexual larvae may be distinguished because the color of gut contents is much lighter (suggesting some nutritional differences between workers and sexuals) and the larvae are hairless. Fully grown sexual larvae are noticeably larger. Figure 9.3 shows examples of the various developmental stages of *M. pharaonis*.

Colonies of *M. pharaonis* remain active and viable over a relatively wide temperature range. Peacock et al. (1955b) reported the lower threshold for colony survival over a protracted period to be about 18°C. At lower temperatures the insects become sluggish or immobile, and when diurnal and nocturnal temperatures vary from about 6°C to 11°C, death

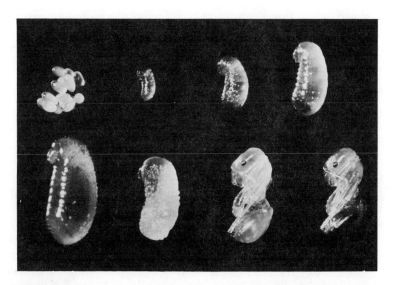

FIGURE 9.3
Developmental stages of pharaoh's ant workers. Top (left to right) eggs, very small larva, small larva, medium larva. Bottom (left to right) fully grown larva, pre-pupa, newly-formed pupa, mature pupa (pharate adult).

occurs within 7 days (Peacock et al. 1955a). However, small colonies may survive several days at such low temperatures and can subsequently be revived when restored to favorable conditions. The high temperature threshold for survival of colonies of *M. pharaonis* is not known. In the laboratory these insects will flourish at temperatures of 27°C–30°C and observations in hospital ducts suggest that they exist at higher temperatures, perhaps up to 45°C. Survival at high temperatures is probably dependent on a good supply of water since this species appears to require relatively high humidity, especially inside the nest, for survival. In infested buildings, ants are often most numerous in areas where high temperature and humidity prevail, that is, boiler houses, heating ducts, kitchens, and laundries. Away from these areas, ants are most often found near heating pipes and radiators, sinks, drains, and toilets.

CONTROL

Historically, control methods used against this species can be traced back almost 100 years. Riley (1889) and Bellevoye (1889) used sponges soaked in sugar and raw liver, respectively, to attract workers in infested premises. Subsequently, these ants were killed when the baits were dropped into benzene or hot water. It is doubtful whether this tedious method controlled the infestations despite the fact that in just six weeks Bellevoye (1889) captured and killed about 350,000 workers and over 800 queens and males in a single infested house.

The earliest report of an insecticide-based control program was made by Lintner (1895), who discussed the use of pyrethrum against M. pharaonis. More recently, a wide variety of inorganic and organic chemicals have been tested both in the laboratory and in the field against this species. Peacock et al. (1950), in a comprehensive review of the biology and control of M. pharaonis, evaluated baits containing sodium fluoride and/or thallium sulfate and reported a considerable measure of success against several natural infestations. Rogers and Herrick (1953) tested three residual organochlorine insecticides (chlordane, DDT, and benzene hexachloride) applied as sprays in kerosene and obtained the best results with chlordane. In an attempt to obtain greater persistence of residual insecticides, Morgan and Price (1954) used dieldrin formulated as a resin or lacquer and a similar technique was subsequently used with considerable success (Papworth 1958). However, as Kane (1967) pointed out, restrictions on the use of dieldrin for other purposes led to a reluctance to use this compound in some areas and other, less potentially hazardous insecticides (e.g., aprocarb) were incorporated into lacquers. Beatson (1968) evaluated baits containing 0.125% chlordecone (kepone) and, until this chemical was withdrawn from use in the mid 1970s, kepone baits were the most effective method of control against M. pharaonis. The withdrawal of chlordecone coupled with increasing concern about the use of persistent organochlorine insecticides prompted the search for safer and more effective materials, and led to the development of the control methods in current use.

Modern control methods are based on either the use of residual insecticides, usually applied as spray treatments, or on the use of baits containing slow-acting poisons or other agents (e.g., juvenile hormone analogs). Spray treatments are directed toward rapid elimination of worker ants in infested premises. Although such treatments often achieve spectacular reductions in the population of foraging workers, they seldom result in complete eradication of an infestation. There are several reasons for this. First, spray treatments with residual contact poisons do not, generally, affect the nest and, because the nest contains reserves of food material, colonies may survive for considerable periods of time without the need to forage (Kretzschmar and Berndt 1976). Second, in many infested premises there are areas that may harbor infestations but that are not suitable for the application or persistence of a residual insecticide. Finally, application of residual insecticides, particularly if such treatment is carried out on a limited basis, may simply serve to induce the migration of the infestation. However, residual spray treatments can be of benefit in areas where rapid reduction of the foraging worker population is essential (e.g., operating theaters). Of the modern contact insecticides used against M. pharaonis, bendiocarb and fenitrothion are the most effective contact poisons tested at this laboratory (J. P. Edwards unpublished). Bendiocarb has the advantage that it has no unpleasant odor and does not leave visible residues on treated surfaces. Bendiocarb is also available as a dust formulation (1% active ingredient [AI]) and this may be used in areas where spraying is not possible. However, where premises infested by pharaoh's ants harbor infestations of the cockroach Blattella germanica [strains of which may be bendiocarb-resistant (Barson and McCheyne 1978)], fenitrothion is an acceptable alternative.

The use of baits against pharaoh's ants is probably the most effective way of eliminating infestations, and the technique has several advantages over other control methods. For example, the placing of baits in infested premises is often much less disruptive to those working or living there than is a treatment involving insecticidal sprays of lacquers. Moreover, given adequate intensity of bait placement, the active ingredient in the bait will be found and transported by workers back to nests that are otherwise inaccessible to treatment by other methods. In recent years, baits containing a variety of active ingredients have been tested by several workers. Among the materials tested have been slow-acting poisons such as boric acid (Newton 1980), bacterial pathogens (Vankova et al. 1975), chemosterilants (Berndt et al. 1972), and the juvenile hormone analog methoprene (Edwards 1975b). Of these, only two compounds (boric acid and methoprene) have been generally used in practical control operations.

Edwards (1975b) evaluated the insect juvenile hormone analog methoprene (isopropyl-11-methoxy-3,7,11-trimethyl dodeca-2,4-dienoate) against this species. Food containing 0.5% methoprene was fed to laboratory colonies and resulted in the disruption of the brood stages and induced sterility in queens. Sterility in queens was associated with the degeneration of the ovaries although the physiological basis of this effect is not understood. The result of these effects was the destruction of the colony —the death of adult stages (workers and queens) occurring as a result of natural mortality. Subsequently, methoprene baits were developed for practical use and have successfully eradicated a number of infestations in a variety of situations ranging from small retail premises to large hospital complexes (Edwards 1975a, 1982, Edwards and Clarke 1978). In 1980 methoprene was approved for use against M. pharaonis by the United Kingdom Pesticides Safety and Precautions Scheme and is currently used by several commercial pest control organizations in Britain, Europe, and the United States.

Newton (1980) evaluated baits containing chlordecone, boric acid, or methoprene against laboratory colonies of *M. pharaonis* and concluded that boric acid baits (5% or 7% AI) were the most effective because they led to the most rapid reduction in the numbers of queens. However, since queens account for such a small proportion of the total adult population, and because they are unable to found new colonies in the absence of brood stages, their presence for a period of time after treatment is unimportant in practical terms. Moreover, because boric acid is toxic, adult workers may die after feeding on the baits, and, as their numbers are reduced, the effectiveness of the treatment may be impaired because there are fewer workers able to transport the bait back to the nest. Furthermore, although treatments utilizing boric acids baits can achieve eradication, it is usually necessary to use a higher intensity of baiting at more frequent intervals than is the case with methoprene baits (L. F. Baker personal communication). This is an important practical consideration because of the labor costs involved in baiting. Thus, of the methods currently available, methoprene baits are probably the most effective way of eradicating infestations of this species and therefore this technique will be covered in some detail.

Once the identity of the infesting species has been confirmed, a thorough survey should be made to determine the extent of the infestation throughout the infested site. This should be done using the liver-baiting technique, whereby small pieces of raw liver (e.g., 0.5 cm^3) are placed throughout the infested premises. These baits may be laid on small pieces of card or polythene, or wrapped in aluminum foil and held in place with adhesive tape (Fig. 9.4). The liver baits are left *in situ* for 24 hours and then examined for the presence of ants. The location of positive baits should be marked on a plan of the infested premises that can then be used to plan the treatment procedure. This preliminary survey should be done as described and not on the basis of visual inspection or on reported sightings of ants by those persons working or living in the infested areas.

It is also important that the preliminary survey of the infestation should include all areas (e.g., service ducts, roof spaces, outbuildings etc.) even if such areas are thought to be free from infestation. On the basis of the preliminary survey the areas to be treated can be identified. Treatment should be carried out in all areas where baits indicate the presence of ants and in all ajoining areas even if these appear to be free from infestation. In this way, a *cordon sanitaire* is established around the infestation to preclude the movement of nests from a treated to an untreated area. Special care should be taken to treat potential escape routes such as underground ducting and outbuildings that are connected to infested areas by such ducting. Similar care should be taken to ensure that the distribution of ants by artificial means (e.g., from movement of foodstuffs, equipment, potted plants, etc.) does not take place during treatment.

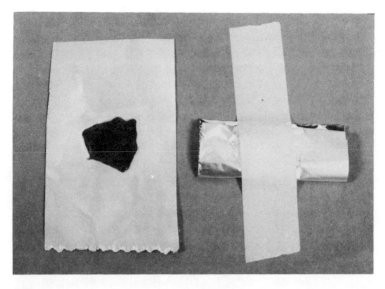

FIGURE 9.4
Liver baits for detecting Pharaoh's ant infestations. Liver on kitchen foil (left) and folded with adhesive tape for application (right).

The treatment procedure (Edwards and Clarke 1978) involves the placing of attractive, food-based bait consisting of a mixture of liver powder, honey, and cake containing the juvenile hormone analog methoprene (0.5% AI) throughout the treatment area. The intensity of baiting will vary depending on the perceived level of infestation, but an average intensity of one bait per 3 m^2 is appropriate for most infestations. Higher baiting intensities may be required in heavily infested areas or in areas where extremes of temperature humidity may reduce the stability of persistence of baits. Baits can be applied using a hand-cranked pressure applicator such as a grease gun (Fig. 9.5) or by any other convenient method. Normally, each bait (approximately one gram in weight) is applied direct to the substrate. However, in some areas the baits may need to be protected. For example, in areas that are frequently washed or cleaned (e.g., kitchens and operating theaters) or in areas with high cockroach populations (which may feed on the baits) baits can be protected by careful placement or by enclosing baits in small plastic tubes or similar containers. In other areas use should be made of natural cavities to preclude the possibility that baits may be inadvertently removed or disturbed by cleaning staff. As a general rule, baits are best placed near floor level, in corners, and along skirting boards. However, because these ants are known to nest in door and window frames, light switches, and fuse boxes, etc.,

FIGURE 9.5
Applying treatment baits with a hand-cranked applicator (grease-gun).

these locations should also be considered as potential bait sites. Because ants rely on food, heat, and moisture for survival, baits should also be placed near heating pipes or radiators, by sinks, drains, and toilets, and in or near food cupboards. However, baits should not be placed directly on hot pipes or in drains as in such situations the stability or attractiveness of baits could be adversely affected.

In control programs using methoprene baits, the baiting should be carried out as described above, and the baits should be left *in situ* for one or two weeks whereupon the treatment procedure should be repeated. Because baiting techniques rely on worker ants to transport the active ingredient back to the nest, no attempt should be made to reduce the worker population immediately prior to or during the baiting period. For this reason, the use of contact insecticides at these times is counterproductive and could seriously detract from the effectiveness of a bait treatment.

After the treatment has been completed, a period of time should be allowed for natural mortality to eliminate adult ants. In treatment programs using methoprene baits this period should be 20–25 weeks after treatment. At this time a thorough liver baiting should be undertaken to confirm the eradication of the infestation or to detect any remaining pockets of infestation. If residual infestation is detected at the post-treatment survey, the areas concerned should be retreated immediately. Although the decline in worker population is quite rapid during the post-treatment period (approximately 98% reduction after 12 weeks) there may be some areas (e.g., operating theaters) where it is necessary to achieve a more rapid elimination of worker ants. In such areas, a residual insecticide may be applied about 14 days after the baiting period.

Having eradicated an infestation of M. pharaonis care should be taken to prevent the possible reintroduction of the species. In this respect it is important to ensure that premises servicing the infested site (e.g., staff accommodation, laundries, food suppliers, etc.) are free from infestation or are treated at the same time as the main infestation. In our experience, this is particularly necessary in the case of hospital staff residences as there is considerable circumstantial evidence to suggest that the transportation of nests in personal belongings of nurses and others is the main method by which infestation is spread to previously uninfested premises.

At present it is possible, using the techniques outlined above, to achieve eradication of infestations of pharaoh's ants in domestic and institutional premises. However, it is likely that further improvements could be made to existing control techniques to improve their efficiency and effectiveness. For example, despite the predominance of baiting techniques in modern control methods, there is little detailed knowledge about the attractiveness of various food materials to foraging workers. Similarly, the potential use of pheromones to increase the attractiveness of baits remains largely unexplored. Nevertheless, as a result of the introduction of modern control methods, the pharaoh's ant is no longer "the most persistent and difficult of all our house-infesting ants to control or eradicate" (Smith 1965).

ACKNOWLEDGMENT

The author would like to thank L. F. Baker for his comments on this chapter.

REFERENCES

Alekseev, A. N., V. A. Bibiovka, L. I. Brikman, and Z. K. Kantarbaeva. 1972. The persistence of viable plague microbes on the epidermis and in the alimentary tract of Monomorium pharaonis in experimental conditions. Medskya Parazitol. 41: 237–239.

Anonymous. 1974. Pharaoh's ants in hospitals. Brit. Med. J. 2: 66.

Barson, G. and N. B. McCheyne. 1978. Resistance of the german cockroach

(*Blatella germanica*) to bendiocarb. *Ann. Appl. Biol.* 90: 147–153.

Beatson, S. H. 1968. Eradication of pharaoh's ants and crickets using chlordecone baits. *Int. Pest Control* 10: 8–10.

Beatson, S. H. 1972. Pharaoh's ants as pathogen vectors in hospitals. *Lancet* 1: 425–427.

Beatson, S. H. 1973. Pharaoh's ants enter giving sets. *Lancet* 1: 606–607.

Bellevoye, M. A. 1889. Observations on M. *pharaonis* (Latr.). *Insect Life* 2: 230–233.

Berndt, K. P. 1977. Physiology of reproduction in the pharaoh's ant (*Monomorium pharaonis* L.). 1. Pheromone mediated cyclic production of sexuals. *Wiad. Parazytol.* 23:163–166.

Berndt, K. P., K. Kretzschmar, and Mitschmann. 1972. Koniginnentechnik zur becampfung von sozialen Insecten speziell von *Monomorium pharaonis*. *Angew Parasitol.* 13: 226–244.

Bolton, B. and C. A. Collingwood. 1975. Handbooks for the identification of British Insects, Volume 6 part 3 (c) Hymenoptera, Formicidae. Royal Entomological Society of London, London. 34 pp.

Buschinger, A. and W. Kloft. (1973). Zur Funktion der Konigin im sozialen Nahrungshaushalt der Pharaoameise *Monomorium pharaonis* (L.) Forschber. Landes Nrhein–Westf. No. 2306. 34 pp.

Cartwright, R. Y. and C. M. Clifford. 1973. Pharaoh's ants. *Lancet* (ii) 1455–1456.

Donisthorpe, H. 1927. British ants–Their life history and classification. 2nd Edit. Routledge and Sons, London. 436 pp.

Dumpert, K. 1981. The social biology of ants. Pitman, London. 298 pp.

Edwards, J. P. 1975a. The effects of a juvenile hormone analogue on laboratory colonies of pharaoh's ant, *Monomorium pharaonis* (L.) *Bull. Entomol. Res.* 75–80.

Edwards, J. P. 1975b. The use of juvenile hormone analogues for the control of some domestic insect pests. *Proc. 8th Brit. Insectic. Fungi. Conf.* 1: 267–275.

Edwards, J. P. 1982. Control of *Monomorium pharaonis* (L.) with methoprene baits: Implications for the control of other pest species. *In* "The biology of social insects" (Breed, M. D., Mitchener, C. D. and Evans, H. E., eds.). Westview Press, Boulder, Colo. 419 pp.

Edwards, J. P. and L. F. Baker. 1981. Distribution and importance of the Pharaoh's ant *Monomorium pharaonis* (L.) in National Health Service Hospitals in England. *J. Hosp. Infect.* 2: 249–254.

Edwards, J. P. and B. Clark. 1978. Eradication of pharaoh's ants with baits containing the insect juvenile hormone analogue methoprene. *Int. Pest Control* 20: 5–10.

Edwards, J. P. and D. B. Pinniger. 1978. Evaluation of four isomers of 3-butyl-5-methyloctahydroindolizine, a component of the trail pheromone of pharaoh's ant, *Monomorium pharaonis*. *Ann. Appl. Biol.* 89: 395–399.

Eichler, W. 1978. Die verbreitung der pharaoameise in Europa. *Memorabilia Zool.* 29: 31–40.

Green, A. A., M. J. Kane, P. S. Tyler, and D. G. Halstead. 1954. The control of pharaoh's ants in hospitals. *Pest Infest. Res.* (1953): 24.

Hölldobler, B. and M. Wust. 1973. Ein Sexual pheromon bei der Pharaoameise *Monomorium pharaonis* (L.) *Zeit. Tierpsychol.* 32: 1–9.

Kane, M. J. 1967. The control of *Monomorium pharaonis* L. with arprocarb. *Public*

Health Inspector 75: 2 pp.

Kretzschmar, K. H. and K. J. Berndt. 1976. Zur hungeranfalligkeit von kolonien der pharaoameise im himblick auf die bedampfung. *Zeit. Gest. Hyg.* 22: 653–657.

Linnaeus, C. 1758. Systema Naturae. Vol. 1 (10th Edition) facsimile edition published by trustees of the British Museum (Natural History), London (1956). 824 pp.

Lintner, J. A. 1895. *Monomorium pharaonis* (Linn.). *49th Ann. Report. N. Y. State Museum* 109–114.

Morgan, M. T. and M. D. Price. 1954. Insect proofing of hospitals with the new insecticidal resins. *The Hospital* 767–771.

Newton, J. 1980. Alternatives to chlordecone for pharaoh's ant control. *Int. Pest Control.* 22: 112–135.

Papworth, D. S. 1958. Practical experience with the control of ants in Britain. *Ann. Appl. Biol.* 46: 106–111.

Peacock, A. D., D. W. Hall, I. C. Smith, and A. Goodfellow. 1950. The biology and control of the ant pest *Monomorium pharaonis* (L.) *Misc. Publ. Dept. Agric. Scot.* 17: 51 pp.

Peacock, A. D., J. H. Sudd and A. T. Baxter. 1955a. Studies in pharaoh's ant *Monomorium pharaonis* (L.) 11. Colony foundation. *Entomol. Mon. Mag.* 91: 125–129.

Peacock, A. D., F. L. Waterhouse, and A. T. Baxter. 1955b. Studies in pharaoh's ant, *Monomorium pharaonis* (L.) 10. Viability in regard to temperature and humidity. *Entomol. Mon. Mag.* 91: 37–42.

Petersen-Braun, M. 1975. Untersuchungen zur sozialen organization der Pharao-ameise *Monomorium pharaonis* (L.) 1. Brutzyklus und seine steuerung durch populationseigene faktoren. *Insectes Soc.* 22: 269–292.

Riley, C. V. 1889. The little red ant. *Insect Life* 2: 106–108.

Ritter, F. J., I. E. Bruggeman-Rotgans, E. Verkuil, and C. J. Persoons. 1975. The trail pheromone of the pharaoh's ant *Monomorium pharaonis*: Components of the odour trail and their origin. *In* "Pheromones and defensive secretions on social insects" (Noirot, C. H., P. E. Howse, and G. Le Masne, eds.). University of Dijon. 248 pp.

Ritter, F. J., I. Bruggemann-Rotgans, P. Verwiel, E. Talman, F. Stein, J. LaBrijn, and C. J. Persoons. 1977. Faranal trail pheromone from the Dufour's gland of the pharaoh's ant, structurally related to juvenile hormone. *Proc. 8th Int. Cong. IUSSI Pudoc.* 325 pp.

Rogers, T. H. and G. W. Herrick. 1953. Fieldwork against pharaoh's ants. *The Sanitarian* 61: 399–402.

Smith, M. R. 1965. House-infesting ants of the eastern United States. USDA Tech. Bull. No. 1326.

Sudd, J. H. 1957. Communication and recruitment in pharaoh's ant. *Brit. J. Anim. Behav.* 5: 104–109.

Sudd, J. H. 1960. The foraging method of pharaoh's ant. *Anim. Behav.* 8: 67–75.

Sudd, J. H. 1962. The natural history of *Monomorium Pharaonis* (L.) infesting houses in Nigeria. *Entomol. Mon. Mag.* 98: 164–166.

Vankova, J., E. Vobrazkova, and K. Samsinak. 1975. The control of *Monomorium pharaonis* with *Bacillus thuringiensis*. *J. Invertebr. Pathol.* 26: 159–163.

Chapter 10

BIOLOGY, ECONOMICS, AND CONTROL OF CARPENTER ANTS

Harold G. Fowler

INTRODUCTION

Only a few members of the Holarctic nominate subgenus of the cosmopolitan formicine genus *Camponotus* regularly nest in wood, and collectively are known as carpenter ants. Carpenter ants are coevolved components of forested regions, yet their role in the mineral and nutrient cycling, and their influence on other forest organisms remains largely unknown. Both the New World and Old World faunas contain eight recognized species, but only one species, *C. herculeanus*, is found in both faunas. Because of their nesting habits, these ants are of some economic importance, and are apparently more problematical in the New World than in the Old World. *Camponotus herculeanus* and *C. pennsylvanicus* are undoubtedly the most pestiferous. Here I briefly discuss what little we know of their biology, ecology, and behavior, and also their economic impact and current control practices.

COLONY LIFE CYCLES

A carpenter ant colony is a highly integrated biological entity, regulated by the queen and her endocrine cycles, the brood, the workers, the reproductives, and many as yet unidentified feed-back loops (Fig. 10.1).

Mature colonies produce both female and male reproductives, which overwinter in the nest (Fig. 10.1). Colonies may produce up to five separate mating swarms (Hölldobler and Maschwitz 1965) over a period of three months or more (Fowler and Roberts 1982a). At the time of mating, the males' fat bodies and digestive systems begin to degenerate (Forbes

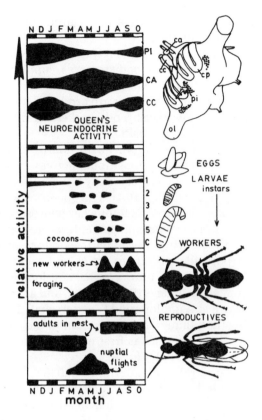

FIGURE 10.1

Schematic diagram of yearly cycles in a carpenter ant colony. Featured are brood development, worker and reproductive production, and queen oviposition and neuroendocrine cycles (pi = pars interacerebralis; ca = corpora allata; cc = corpora cardiaca; cp = corpora penduculata; ol = optic lobes). Also given is the yearly foraging intensity. Data taken from Pricer (1908), Palma-Valli and Deyle (1981), Sanders (1964), Sanders and Baldwin (1972), Fowler and Roberts (1982a), and Fowler and Roberts (1980).

1956, Hölldobler 1966). However, the mandibular gland secretions of the male induce the females to take flight (Hölldobler and Maschwitz 1965), but do not produce a behavioral response in the workers (Payne et al. 1975). Swarming typically occurs in late morning or early afternoon, and is modulated by temperature (Hölldobler and Maschwitz 1965). Mating probably occurs in the air, although Goetsch and Kathner (1937) reported pairs copulating on the ground. Soon after mating, the males die, and the females alight, remove their wings, and search out small cavities, to begin oviposition, typically under tree bark.

Queens normally found colonies singly and claustrally (Pricer 1908), depending on flight muscle and fatbody reserves, supplemented by occasionally eating an egg or larva (Wheeler 1933). During the first year of colony existence only five to 25 workers are produced (Pricer 1908). With the maturation of the first worker brood, workers assume all tasks, and the queen's only function is oviposition.

From the second year onward, the queen characteristically demonstrates two distinct periods of ovipositional activity, resulting in a bimodal production of workers. Occasionally, an additional midsummer brood may also be produced (Fig. 10.1). The queen's oocytes do not develop during the winter, during which time her neurosecretory products are stored in the corpora cardiaca and pars interacerebralis (Palma-Valli and Deyle 1981). In the spring, these neurosecretory products are released, and the corpora allata becomes active and oocyte maturation is resumed (Palma-Valli and Deyle 1981). Normal developmental times from egg to adult are from 45 to 70 days (Pricer 1908), and are apparently controlled by temperature and photoperiod (Vanderschaff 1970). Reproductives mature during the summer and overwinter in the nest, while a portion of the fall brood overwinters as first instar larvae (Eidmann 1943, Hölldobler 1961, Benois 1972a) (Fig. 10.1).

Mature colonies generally contain from 3,000 to 6,000 workers (Fowler 1982), with an upper limit of approximately 15,000 (Pricer 1908, Sanders 1970). Satellite nests may be present in adjacent trees or cavities around large central nests containing the queen (Sanders 1964, 1970). Moreover, some large colonies may exist with their separate queens as part of a large superstructure, yet each colony maintains its reproductive independence (Hölldobler 1962). There is no evidence to suggest a polygynous organization in any species of carpenter ants, and we may assume that colonies are functionally monogynous. Both brood and reproductive populations in a colony are highly correlated with the standing worker population (Fowler 1982), although sexuals are probably not produced until a colony is 3 to 5 years old (Pricer 1908). Moreover, a strong correlation exists between the reproductive output of a queen and the amount of neurosecretory products in storage, both of which increase yearly (Palma-Valli and Deyle 1981). It is likely that the colony can assess the reproductive condition of their queen through the quantity of attractive pheromone she produces (Fowler and Roberts 1982b). Although major workers will oviposit when placed in groups of 40 or more (Vanderschaff 1970), it is likely that these eggs are trophic, since no neurosecretory products that are necessary for oocyte maturation have been found in workers (Gawande 1968).

POLYMORPHISM AND POLYETHISM

Carpenter ants are among the most polymorphic ants found in the Holarctic. Workers range in head width from less than 1 mm to more than 4 mm. Within a mature colony, worker size variation follows a simple monophasic allometric pattern (Weyrauch 1933, Kill 1934, Smith 1942a, Fowler 1982), skewed toward the smaller-size classes. Over the life span of a colony, as worker populations increase, mean worker head width also increases (Leutert 1962, Fowler 1982).

Both larval food quantity and quality influence final worker size. Reduced food supply, or food high in protein content, results in the production of fewer workers of smaller size than when larvae are fed more food of a lower protein content (Smith 1942b, 1944). Presumably, worker-queen differences are determined through similar trophic pathways (Ezhikov 1934).

Size-biased division of labor occurs among the workers (Buckingham 1911, Lee 1938, Gotwald 1968, Fowler 1982). While foraging, for example, smaller workers tend aphids, while larger workers receive honeydew from the tending workers and transport it to the nest (Pricer 1908, Fowler and Roberts 1980), allowing them to effectively exploit the allometry of lead carriage (Nielsen et al. 1982). Size-biased division of labor also occurs for tasks performed within the nest (Lee 1938, Buckingham 1911, Fowler 1982). The topography of the antennal sensilla and peripheral receptor capacity is size-dependent in the worker caste, and consequently the ability of workers to discriminate and respond to a variety of compounds is size-dependent (Masson and Friggi 1974, Fowler 1982, Payne et al. 1975, Brand et al. 1973).

Traniello (1977) found that the oldest workers with the greatest degree of ovarian degeneration were most likely to be foragers. This suggests that a temporal division of labor is superimposed on the size-biased division. Among foraging ants, both size and task constancy have been documented (Kill 1934, Fowler and Roberts 1980, Ebbers and Barrows 1980). Thus, individual specializations are also important components of carpenter ant colony polyethism.

NUTRITION AND FEEDING

The feeding relationships of carpenter ants are still not well known. Honeydew is apparently the major source of nutrition (Pricer 1908, Gotwald 1968, Fowler and Roberts 1980). The high respiratory quotient of workers (Jensen and Holm-Jensen 1980) indirectly supports their dependence on this food source. Colonies have been reported tending a variety of aphid and membracid species (Fowler 1982) and feeding on plant sap (Gotwald 1968).

Although the carbohydrate requirements of the colony may be met by honeydew, it is unlikely that this resource alone can meet the protein demands of the colony. Ayre (1963a, 1967) suggested that crop enzymes may contribute to the digestion of food within the crop, or when regurgitated. In particular, fungi may be major elements of diet of carpenter ants, due to the presence of crop amylase (Ayre 1963a, 1967). The presence and survivorship of the chestnut blight, *Endothia parasitica*, in the digestive system of carpenter ants clearly implicates these ants as potential vectors (Anagnostakis 1982), and suggests that fungi may be a regular component of their diet.

The importance of insect prey in the nutrition of colonies is unknown. Ayre (1963b) found that *C. herculeanus* killed 98% of all live insects offered, and consumed 87% of all dead insects offered. Carpenter ants have been cited as important predators of a variety of forest insect pests (Meyers and Campbell 1976, Green and Sullivan 1950, Oilon 1965, Hammer 1910, Brooks 1918, Gosling 1978). Although carpenter ants have a well-developed venom apparatus, the sting is not functional, but rather, venom is released from the body via the acidopore, a nozzle-like structure formed by an inrolling of the posterior portion of sternum VII. The acidopore is ringed with fine setae that direct the venom away from the ant's body (Hung and Brown 1964). When attacking prey or defending the colony, ants first bite and then spray the contents of their poison gland at the site of the bite. Formic acid, which is synthesized and stored in the poison gland (Hefetz and Blum 1978), additionally functions as an attractant at low concentrations, but as an alarm pheromone at high concentrations (Ayre and Blum 1971). The associated Dufour's gland contains an array of alkanes and ketones, principally N-undecane or tridecane (Ayre and Blum 1971, Brophy et al. 1973, Bergstrom and Lofqvist 1972). The contents of the Dufour's gland may act as wetting agents, assisting the spread of formic acid over the cuticle of the prey or enemy. N-undecane also functions as a synergist of formic acid, attracting and calming workers (Ayre and Blum 1971).

Proteases are absent in the digestive glands (Ayre 1963a), and foragers are apparently unable to ingest solid tissue, due to a filtering apparatus in the digestive system. The nitrogeneous demands are apparently met entirely in liquid form, that is, the hemolymph and water-soluble protein of the prey, possibly aided by extraoral digestion (Ayre 1963a). Forbes and McFarlane (1961) suggested that the postpharyngeal glands were important in protein digestion. Although lipases are present in this gland, its function is more closely associated with trophallaxis and larval nutrition (Ayre 1963a). The elevated presence of granular hemocytes found in all stages suggests that the hemolymph is the storage site of lipids (Costa Leonardo and Cruz Landim 1978).

The proventriculus of the carpenter ant is perhaps the most specialized part of its digestive system. Food particles in excess of 200 μm are not ingested, while particles greater than 150 μm in diameter are filtered and condensed into infrabuccal pellets, which are later voided (Eisner and Happ 1962). Only particles smaller than 100 μm in diameter are swallowed (Eisner and Happ 1962).

Carpenter ant larvae are distinguished by single-hooked hairs, the presence of chiloscleres, or a pair of thickened and hardened portion of the cuticula lying on either side of the labrum, and by the presence of the praesepium, or shallow depression on the ventral side of the anterior somites. The praesepium or food basket into which workers deposit food for the larvae is poorly developed in young and prepupal larvae (Wheeler and Wheeler 1953). The reduction of the praesepium may be an adaptation to a nearly complete liquid diet.

RECRUITMENT, ORIENTATION, AND FORAGING

Camponotus pennsylvanicus, and probably the majority of carpenter ants, rely on both pheromonal and motor display for recruitment. Trail pheromone is elaborated by the hindgut (Blum and Wilson 1964) and is deposited on the substrate by dragging the tip of the abdomen (Hartwick et al. 1977, Traniello 1977), probably as a ritualization of defecation behavior (Hölldobler 1978). A single gas chromatographic peak of hindgut extract is highly active in eliciting trail-following (Barlin et al. 1976) in *C. pennsylvanicus*. Both *C. pennsylvanicus* and *C. americanus* follow the hindgut extract of the other species, but not those of other *Camponotus* subgenera. The addition of poison gland materials to the hindgut materials greatly enhances the intensity of trail-following (Traniello 1977). Motor displays, nevertheless, form an integral part of the recruitment system of *C. pennsylvanicus*. If a returning ant is able to stimulate workers by motor display, the number of ants exhibiting trail-following increases dramatically (Traniello 1977). Similar mechanisms are employed during colony migrations. The removal of those workers involved in motor displays greatly diminishes recruitment to a food source or a new nest site and recruitment decreases exponentially in intensity with distance from the nest. The low probability of discovery of ephemeral resources and the low rate of recruitment of discovered resources greatly hinder the development of toxic baits as an alternative control measure.

Orientation to the nest site by foragers in the field may be mediated by trail pheromones. The hindgut material may serve as a long-term olfactory orientation cue to food resources (Traniello 1977), and may persist for up to several weeks in the field (David and Wood 1980). At stable resources, such as aphid colonies, landmark and astrotactic orientation are employed. Traniello (1977) suggested that landmark orientation was employed by laboratory colonies, and consequently confirmed in the laboratory and in the field (Hartwick et al. 1977, David and Wood 1980). Trunk-trails (David and Wood 1980) and subterranean foraging galleries (H. G. Fowler unpublished) serve to provide potent long-term orientation cues. As workers are long lived, the possibility exists that topographic traditions can be transmitted to younger foragers by older foragers through association. Topographic traditions could also be transmitted directly through the site constancy exhibited by certain foragers or indirectly through foraging trunk trails and subterranean foraging galleries (Fowler and Roberts 1980).

ENEMIES AND ASSOCIATIONS

A wide variety of animals prey on carpenter ants, ranging from toads, frogs, a variety of birds, bears, other ants, beetles, spiders, and rats through man (Fowler 1982). It is impossible to estimate the importance of predators in regulating carpenter ant populations. Parasites and pathogens may be important, especially phorid flies (*Apocephalus* spp.), eucharid wasps (*Schizaspidia* spp.), and a variety of fungi.

The relationship of most inquilines and myrmecophiles with their host remains cloudy (Fowler 1982). A complex of Müllerian mimics may be associated with North American carpenter ants. These include longhorn beetles (Wheeler 1910), and possibly at least three species of predatory mirids (Wheeler and Henry 1980), but experimental data are lacking.

NESTING BEHAVIOR

This facet of carpenter ant biology, more than any other, is responsible for their economic status. Carpenter ant workers gnaw out diagnostic laminar galleries in the heartwood of standing trees or in seasoned wood. The lamellae do not vary greatly among species (Hölldobler 1944, Benois 1972b, Graham 1918, Eidmann 1929, Pricer 1908). Wood is not consumed, but rather deposited outside the nest. Consequently, as the colony grows and more space is needed to house an increasing worker and brood population, excavation must increase proportionally. In early spring it is not uncommon to find large accumulations of wood dust outside these rapidly expanding nests (McCook 1906).

Fielde and Parker (1904) first reported on the response of carpenter ants to substrate vibrations. The function of substrate vibrational responses, at least in the nest, is one of communication. The lamellae that characterize carpenter ant nests are carriers of substrate-borne vibrations. Heads and gasters are drummed against the lamellae (Markl and Fuchs 1972, Fuchs 1976a, 1976b), and substrate vibrations are transmitted through the lamellae and are received by the subgenual organs (Autrum and Schneider 1948). Vibrational communication is apparently used to communicate alarm, causing ants to either rapidly engage in defense behaviors or to escape, depending on worker size (Fowler 1982).

Maschwitz (1974) suggested that the metapleural gland secretions of most species of ants were important bactericidal and bacteristatic agents. As such, these glands are important in permitting ants to maintain an antiseptic environment within the nest. It is therefore noteworthy that carpenter ants either lack metapleural glands, or these glands are greatly atrophied (Ayre and Blum 1971). Where, if any, antiseptic compounds are produced and stored in carpenter ants remains unknown.

In urban areas, up to 75% of all standing trees in central New Jersey may house carpenter ant colonies (Fowler and Parrish 1982). In forests, up to 70% of cedar trees in Minnesota (Graham 1918) and 10% in Ontario (Sanders 1964) contain carpenter ant colonies. Infestation rates ranging from 0.4% to 4.2% have been recorded for balsam fir and spruce stands over varying time intervals for New England and the Canadian Maritime provinces (Sanders 1964).

Radioisotope tagging has been employed to locate nests, estimate their size and foraging territories, and to study colony activity (Kloft et al. 1960, Sanders and Baldwin 1972). Such methods have indicated that large colonies may have satellite nests in nearby trees. No suitable or economical method has yet been found to locate colonies within human structures.

McComb and Noble (1982) documented a relatively high utilization rate of vertebrate nest boxes and natural tree cavities as nest sites by carpenter ants. Cavities afford a readily inhabitable site for a large emigrating colony and minimize the energitic and temporal costs of gnawing-out suitable living space. Usage of other convenient cavities, such as wall voids, by emigrating colonies is thus to be expected.

In the Northeastern United States, current development practices of bulldozing trees and leaving them on site for future removal while simultaneously beginning construction of residences in suburban developments may account for part of the growth of carpenter ant problems (Fowler 1983). As the nest site begins to deteriorate in the felled tree, the colony may emigrate into partially constructed structures, usually wall voids, and the structure is subsequently raised around an already large

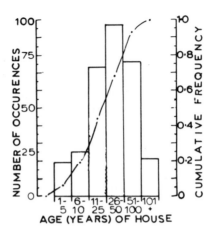

FIGURE 10.2
The distribution of carpenter ant nests in Swedish homes as a function of years since construction. The age distribution of homes in Sweden is not known, and thus it is impossible to examine electivity on the part of the ants.
(*Source:* Butovitsch 1976)

colony. Once occupied the owners of the structure may find a serious carpenter ant problem. Consequently, structures of all ages are utilized as nesting sites by carpenter ants (Fig. 10.2). By placing suburban developments in wooded areas, the risks of carpenter ant invasion are greatly increased. Butovitsch (1976) found that the closer a structure is to woodland, the higher the probability of invasion by emigrating carpenter ant colonies.

Structural design as well as location may also influence the risks of invasion by carpenter ants. Butovitsch (1976) found houses lacking basements to be more prone to carpenter ant invasions. Other structural features undoubtedly play a role in the nesting choices of carpenter ant colonies, but quantitative data are unavailable.

DETERMINANTS OF PEST STATUS

Although carpenter ants may produce direct quantitative economic damages to trees and seasoned wood, the pest status of carpenter ants is nevertheless dependent on their perception by urban dwellers. Previous attempts have failed to clarify if carpenter ants are structurally damaging, or merely nuisance pests (McCook 1906, Anonymous 1967, Francoeur 1977). Are carpenter ants considered pests by the public because of the damage they cause or might cause, or are there other factors responsible for their pest status? Moreover, throughout the Northeastern United States, the number of carpenter ant complaints is increasing yearly (Fowler 1983), and in Finno-Scandinavia, *C. herculeanus* is becoming more and more of a problem (Butovitsch 1976), possibly through range expansion (Jensen and Nielsen 1982).

Carpenter ant complaints demonstrate significant annual cycles in climatically harsh northern states (Fowler and Roberts 1983, Fowler 1983). This suggests that weather may be a major factor which triggers public complaint. Rainfall, for instance, has been thought to greatly affect the nesting success of carpenter ants (Weber 1942). It is well known that carpenter ants tend to be more problematic in the more northern states (Levi and Moore 1978).

What determines the patterning of public complaints? It is unlikely that the public frequently inspects their structures and become worried about the colonies they find. Rather, it is probable that the public responds by complaining to sightings in the home. A comparison of public complaints and the number of pest control jobs handled to the swarming and foraging activity of carpenter ant colonies suggests that complaints are probably a function of swarming activity (Fig. 10.3). In the more northern states, swarming is probably under stricter climatic controls and more colonies are also probably located within human structures resulting in a greater number of complaints (Fowler 1983). Not surprisingly, pest control companies are busier during periods of swarming activity, not during peak foraging, which is almost always nocturnal and conducted outside a structure (Fig. 10.3). The annual swarming tends to be brief, lasting only a few hours. Perhaps for this reason, pest control companies generally guarantee a treatment for one year or less. When swarming occurs indoors, a large colony must be present in the structure, and the homeowner generally seeks remedial action without regard to structural damage, if any.

ECONOMICS OF CARPENTER ANTS

Because of lack of detailed field studies, it is impossible to estimate the savings in forests produced by carpenter ant predation on defoliating insect pests, but the impact of carpenter ants is probably substantial. Likewise, in those situations in which carpenter ants are pestiferous, data are lacking to evaluate the magnitude of their impact. This lack of information is partially due to the difficulties in assessing the true incidence of

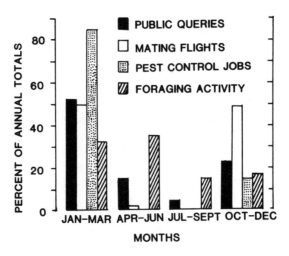

FIGURE 10.3
The seasonal distribution of public queries concerning carpenter ants in
New Jersey (Fowler 1983), swarming flights of carpenter ants (Fowler and
Roberts 1982a), pest control company carpenter-ant jobs (Whitworth un-
published), and the foraging activity of field colonies of carpenter ants
(Fowler and Roberts 1980, Fowler 1982a).

carpenter ants in structures and trees, and of relating incidence data into
economic damage. Also, carpenter ants are principally urban pests, and
no quantitative tools have yet been satisfactorily developed for translat-
ing human perception into action thresholds. However, some basic costs
can be estimated and shed some light on the magnitude of the problem
in the Northeastern United States.

In 1980, L. Hagman (personal communication) conducted a survey
of pest control companies in central New Jersey. He found that each company handled an average of 200 carpenter ant jobs a year. A standard
partial treatment runs upward from $50 U.S. dollars (all dollar values U.S.
in this chapter), while a standard treatment can cost $500 or more (J.
Graydon personal communication). The 1979 Bell Telephone Directory
Yellow Pages for Middlesex County, NJ, lists 124 pest control companies
or branches serving this county. Rough extrapolations suggest that approximately 24,800 carpenter ant jobs are thus handled annually in Middlesex County. Given the costs listed earlier, between $1.2 and $12 million
were probably spent during 1980 for carpenter ant control in Middlesex
County alone. A figure of $5 to $6 million probably would represent a
realistic outlay for direct expenses to Middlesex County residents. Given
these initial approximations, it is not unreasonable to expect that at least

$25 million is spent by the residents of New Jersey for carpenter ant control yearly, or that over $25 million is spent annually in the Northeastern United States in direct costs to pest control companies.

Even these estimates do not represent the magnitude of the economic impact of carpenter ants in urban areas. For example, if carpenter ants are responsible for only a small percentage of the enormous household insecticide market, several million dollars can be added to the figures given above. Other direct costs include the replacement of shade trees weakened by carpenter ants and felled by winds (Fowler and Parrish 1982, Fowler and Roberts 1982c), as well as the costs of repairing structural damage to houses, or replacing electrical cable (Cooke 1956). Indirect costs, such as the effects on ornamentals by plant-sucking insects protected by carpenter ant colonies, and the psychological effects of sighting ants in a household by a suburban resident, may greatly amplify these initial costs.

Nevertheless, carpenter ants probably rarely cause serious structural damage to buildings, as they prefer to nest in preexisting voids. For example, Wane and Homer (1969) found that only 63 (2.6%) cases of wood damage were reported in 2,432 carpenter ant jobs handled by Californian pest control operators. This suggests that programs should be developed to educate the public on carpenter ants and their damage. The public should be made aware of the fact that even if their houses harbor carpenter ant colonies, structural damage is probably a rare occurrence.

In forests, the economic impact of carpenter ants has likewise not been adequately quantified. Up to 102 ft^3 of wood volume per acre, or 10% of the merchantable pulp wood volume, is lost to the nesting activities of carpenter ants in Ontario (Sanders 1964). Damage to forest products has been documented (e.g., Graham 1918) yet no comprehensive economic costs are available. Additionally, structural weakening of trees through carpenter ant nesting activities greatly increases the risk of wind breakage (Sanders 1964, Fowler and Roberts 1982c). The possible involvement of carpenter ants in the spread of tree pathogens, such as chestnut blight (Anagnostakis 1982) may further magnify their pest potential to forestry.

CONTROL AND PROPHYLACTIC MEASURES

A survey of 23 pest control companies in the northeastern United States indicated that 69.6% of all companies provide complete treatments for carpenter ant problems, while the remainder provide only partial or spot treatments (T. Whitworth personal communication). Complete treatments usually include locating the nest, or if the nest cannot be found, drilling and dusting all suspected wall voids. Partial treatments generally involve only local applications of residual insecticides. The key to the successful control of carpenter ant colonies, however, depends on locating the nest. Killing only a part of the foraging force, as partial treatments or home-owner treatments tend to do, does not greatly affect mature colonies, and the problem will tend to recur. Locating a nest is a tedious and some-times impossible task, if soundings and probings fail to reveal its loca-tion. It is not uncommon for a structure to harbor more than one colony, and thus even a well-conducted inspection may not identify all potential problems. Not surprisingly, 96% of all pest control companies offer guar-antees for one year or less, and 40% of all jobs must be re-treated (T. Whit-worth personal communication). This attests to the difficulty of carpenter ant control, and may also reflect the frequency of emigration of colonies into buildings after treatment (Anonymous 1967), and the possibility of having multiple colonies nesting within a single structure.

Pesticides presently employed for carpenter ant control by profes-sional pest control companies are, in descending order of usage: Ficam® (39%), diazinon dust (26%), Dursban® (13%), Drione® (9%), and chlordane (4%) (T. Whitworth personal communication). Additionally, a large mar-ket for household insecticides exists, most of which incorporate some formulation of organophosphate or pyrethroid.

In shade trees and firewood, carpenter ant colonies may be treated by drilling and dusting, or by the application of a 4%–5% formulation of malathion, a 1% formulation of rotenone, or a 0.2% pyrethrum formu-lation. Unless risks of possible wind damage are high, shade trees gener-ally do not need to be treated (Fowler and Parrish 1982).

Prevention of carpenter ant nests is the key to their management in urban areas. The association of carpenter ant nesting with mechanical damage to shade trees underscores the importance of sanitation in man-aging carpenter ants. In structures, prevention is best achieved by fre-quent inspection. Many carpenter ant colonies are introduced into resi-dences in firewood, and then emigrate into wall voids. If firewood must be kept indoors, it should be thoroughly inspected beforehand and treat-ed as discussed earlier.

Contrary to commonly held conceptions, carpenter ants, especially C. herculeanus and C. pennsylvanicus, produce the most damage in sound wood, in which previous fungal decay is minimal or absent (Butovitsch 1976). Wood deterioration is not a necessary first step in producing car-penter ant problems. Special attention should be given to frequently re-ported nesting sites: around clothes dryer vents, rain gutters, window and door frames, porches, exterior columns, sills, joists, and subfloors (Levi and Moore 1978). Although carpenter ants are not usually found in rotten wood, wood decay may provide them with channels for invasion. Also, the increasing use of insulation in wall voids provides a convenient nesting substrate for carpenter ant colonies, and also increases problems related to their treatments, due to the fact that dusts are less likely to achieve uniform coverage.

EPILOGUE

I have tried to present our present state of knowledge concerning carpenter ants in a concise form. Other pertinent information has not been included due to space limitations, and the reader will undoubtedly find this presentation somewhat slanted toward the New World. However, there is a tremendous gap in our knowledge of this important group of ants, and I hope that serious investigators will begin to bridge this gap.

ACKNOWLEDGMENTS

This chapter would not have been possible without a grant from the New Jersey Pest Control Association. C. Collingwood, T. Whitworth, J. Grayden, and L. Hagman provided me with useful information, and allowed me to use some of it. For their help, I am extremely thankful.

REFERENCES

Anagnostakis, S. L. 1982. Carpenter ants as carriers of *Endothia parasitica*. Paper presented at the USDA Chestnut Conference, Jan. 1982.

Anonymous, 1967. Carpenter ants annoy many operators: PC survey uncovers treatments hints. *Pest Control.* 35: 42, 44.

Autrum, H. and W. Schneider. 1948. Vergleichende Untersuchungen uber den Erschutterungssinn von Insekten. *Z. Vergl. Physiol.* 31: 77–88.

Ayre, G. L. 1963a. Feeding behavior and digestion in *Camponotus herculeanus* (L.) *Entomol. Exp. Appl.* 6: 165–170.

Ayre, G. L. 1963b. Laboratory studies on the feeding habits of seven species of ants in Ontario. *Can. Entomol.* 95: 712–715.

Ayre, G. L. 1967. The relationship between food and digestive enzymes in five species of ants. *Can. Entomol.* 99: 408–411.

Ayre, G. L. and M. S. Blum. 1971. Attraction and alarm of ants (*Camponotus* spp.) by pheromones. *Physiol. Zool.* 44: 77–83.

Barlin, M. R., M. S. Blum, and J. M. Brand. 1976. Species-specificity studies on the trail pheromone of the carpenter ant, *Camponotus pennsylvanicus*. *J. Ga. Entomol. Soc.* 11: 162–164.

Benois, A. 1972a. Evolution du couvain et cycle annuel de *Camponotus vagus*

Scop. (= *pubescens* Fabr.) dans la region d'Antibes. *Ann. Zool. Ecol. Anim.* 4: 325–351.

Benois, A. 1972b. Etude ecologique de *Camponotus vagus* Scop. (= *pubescens* Fabr.) dans la region d'Antibes: nidificacion et architecture des nids. *Insectes Soc.* 19: 111–129.

Bergstrom, G. and J. Lofqvist. 1972. Similarities between the Dufour's gland secretions of the ants *Camponotus ligniperda* (Latr.) and *Camponotus herculeanus* (L.). *Entomol. Scand.* 3: 225–238.

Blum, M. S. and E. O. Wilson. 1964. The anatomical source of trail-substances in formicine ants. *Psyche* 71: 28–31.

Brand, J. M., R. M. Duffield, J. G. MacConnell, M. S. Blum, and M. H. Forbes. 1973. Caste specific compounds in male carpenter ants. *Science* 179: 388–389.

Brooks, F. E. 1918. Papers on deciduous fruit insects. I. The grape curculio. U.S. Dept. Agri. Bull. No. 730.

Brophy, J. J., G. W. K. Cavill, and J. J. Shannon. 1973. Venoms and Dufour's gland secretions in an Australian specis of *Camponotus*. *J. Insect Physiol.* 19: 791–798.

Buckingham, E. N. 1911. Division of labor among ants. *Proc. Amer. Acad. Arts. Sci.* 46: 425–507.

Butovitsch, V. 1976. Uber Vorkommen und Schadwirkung der Rossameisen *Camponotus herculeanus* und *C. ligniperda* in Gebauden in Schweden. *Mater. Organ.* 11: 162–170.

Cooke, E. 1956. Enemies of electric cables. *Telecon House Mag.* 32: 1–14.

Costa Leonardo, A. M. and C. da Cruz Landim. 1978. Estudo das celulas do sangue em *Camponotus refipes* (Fabricius). *Studia Entomol.* 20: 235–251.

David, C. T. and D. L. Wood. 1980. Orientation to trails by a carpenter ant, *Camponotus modoc* in a giant sequoia forest. *Can. Entomol.* 112: 993–1000.

Ebbers, B. C. and E. M. Barrows. 1980. Individual ants specialize on particular aphid herds. *Proc. Entomol. Soc. Wash.* 82: 405–407.

Eidmann, H. 1929. Zur Kenntnis der Biologie der Rossameise (*Camponotus herculeanus* [L]). *Z. Angew. Entomol.* 14: 229–253.

Eidmann, H. 1943. Die Uberwinterung der Ameisen. *Z. Morphol. Okol. Tiere* 39:217–275.

Eisner, T. and G. M. Happ. 1962. The infrabuccal pocket of a formicine ant: A social filtration device. *Psyche* 69: 107–118.

Ezhikov, T. 1934. Individual variability and dimorphism in social insects. *Amer. Nat.* 68: 333–334.

Fielde, A. M. and G. H. Parker. 1904. The reactions of ants to material vibrations. Proc. Acad. Nat. Sci., Phila. 56: 642–649.

Forbes, J. 1956. Observations on the gastral digestive tract in the male carpenter ant, *Camponotus pennsylvanicus* DeGeer. *Insectes Soc.* 3: 506–511.

Fowler, H. G. 1982. Caste and behavior in the carpenter ant, *Camponotus pennsylvanicus* (DeGeer). Ph.D. Thesis, Rutgers University, New Brunswick, NJ. 271 p.

Fowler, H. G. 1983. Urban structural pests: Carpenter ants displacing subterranean termites in the public concern. *Environ. Entomol.* 12: 997–1002.

Fowler, H. G. and M. D. Parrish. 1982. Urban shade trees and carpenter ants. *J. Arboricult.* 8: 281–284.

Fowler, H. G. and R. B. Roberts. 1980. Foraging behavior of the carpenter ant, *Camponotus pennsylvanicus*, in New Jersey. *J. Kan. Entomol. Soc.* 53: 295–304.

Fowler, H. G. and R. B. Roberts. 1982a. Seasonal occurrence of founding queens and the sex ratio of *Camponotus pennsylvanicus* in New Jersey. *J. N.Y. Entomol. Soc.* 90: 247–251.

Fowler, H. G. and R. B. Roberts. 1982b. Entourage pheromone in carpenter ant, *Camponotus pennsylvanicus* (DeGeer), queens. *J. Kan. Entomol. Soc.* 55: 568–570.

Fowler, H. G. and R. B. Roberts. 1982c. Carpenter ant induced wind breakage in New Jersey shade trees. *Can. Entomol.* 114: 649–650.

Fowler, H. G. and R. B. Roberts. 1983. Activity cycles of carpenter ants (*Camponotus*) and subterranean termites (*Reticulitermes*): inference from synanthropic records. *Insectes Soc.* 30: 323–331.

Francoeur, A. 1977. Synopsis taxonomique et economique des fouirmis du Quebec. *Ann. Soc. Entomol. Quebec* 22: 205–212.

Fuchs, S. 1976a. An informational analysis of the alarm communication by drumming behavior in nests of carpenter ants (*Camponotus*). *Behav. Ecol. Sociobiol.* 1: 315–336.

Fuchs, S. 1976b. The response to vibration of the substrate and reactions to the specific drumming in colonies of carpenter ants (*Camponotus*). *Behav. Ecol. Sociobiol.* 1: 155–184.

Gawande, R. B. 1968. A histological study of neurosecretion in ants (Formicoidea). *Acta Entomol. Bohem.* 65: 349–363.

Goetsch, W. and B. Kathner. 1937. Die Koloniegrundung der Formicen und ihre experimentelle Beeinflussung. *Z. Morphol. Okol. Tiere* 33: 201–260.

Gosling, D. C. L. 1978. Observations on the biology of the twig pruner, *Ehaphidionoides parallelus* in Michigan, USA. *Great Lakes Entomol.* 11: 1–10.

Gotwald, W. H., Jr. 1968. Food gathering behavior of the ant *Camponotus noveboracensis* (Fitch). *J. N.Y. Entomol. Soc.* 76: 278–296.

Graham, S. A. 1918. The carpenter ant as a destroyer of sound wood. *Minn. State Entomol.* 17th Ann. Report., Minn. Agric. Exp. Sta.

Green, G. W. and C. R. Sullivan. 1950. Ants attacking the larvae of the forest tent caterpillar, *Malacosoma disstria*. *Can. Entomol.* 82: 194–195.

Hammer, A. G. 1910. Life history of the codling moth in northwestern Pennsylvania. *U.S. Dept. Agri. Bur. Entomol.*, Bull. 80: 71–111.

Hartwick, E. B., W. G. Friend, and C. E. Atwood. 1977. Trail-laying behavior of the carpenter ant, *Camponotus pennsylvanicus*. *Can. Entomol.* 109: 129–136.

Hefetz, A. and M. S. Blum. 1978. Biosynthesis and accumulation of formic acid in the poison gland of the carpenter ant, *Camponotus pennsylvanicus*. *Can. Entomol.* 109: 129–136.

Hölldobler, B. 1961. Temperaturunabhaenigige rhythmishe ercheinungun bei Rossameisenkolonien (*Camponotus ligniperda* Latr. und *Camponotus herculeanus* L.). *Insectes Soc.* 2: 175–189.

Hölldobler, B. 1962. Zur Frage der Oligogynie bei *Camponotus ligniperda* und *Camponotus herculeanus*. *Z. Angew. Entomol.* 49: 337–352.

Hölldobler, B. 1966. Futterverteilung durch Mannchen der Rossameisen. *Waldygiene* 4: 228–250.

Hölldobler, B. 1978. Ethological aspects of chemical communication in ants. *Adv. Stud. Behav.* 8: 75–115.

Hölldobler, B. and U. Maschwitz. 1965. Der Hochzeitsschwarm der Rossameise *Camponotus herculeanus* L. *Z. Vergl. Physiol.* 50: 551–568.

Hölldobler, K. 1944. Ueber die fortslich wichtigen Ameisen des nordoskarelischen Uralis. *Tiel. I. Z. Angew. Entomol.* 30: 606–622.

Hung, A. C. F. and W. L. Brown, Jr. 1964. Structure of the gastric apex as a subfamily character of the Formicinae. *J. N.Y. Entomol. Soc.* 74: 198–200.

Jensen, T. P. and I. Holm-Jenson. 1980. Energetic cost of running in workers of three ant species, *Formica fusca* L., *Formica rufa* L., and *Camponotus herculeanus* L. *J. Comp. Physiol.* 137: 151–156.

Jensen, T. P. and M. G. Nielson. 1982. En status over ubdredelsen af myreslaegten *Camponotus* i Danmark. *Entomol. Meddr.* 49: 113–116.

Kill, V. 1934. Untersuchungen uber Arbeitsteilung bei Ameisen (*Formica rufa* L., *Camponotus herculeanus* L. und *C. ligniperda* Latr.). *Biol. Zentr.* 54: 114–146.

Kloft, W., B. Hölldobler, and A. Haisch. 1960. Traceruntersuchungen zur Abgrenzung von *Camponotus herculeanus* (L.) und *C. ligniperda* (Latr.) *Entomol. Exp. Appl.* 8: 20–36.

Lee, J. 1938. Division of labor among the workers of the Asiotic carpenter ant (*Camponotus japonicus* var. *aterreamus*). *Peking Nat. Hist. Bull.* 13: 137–145.

Leutert, W. 1962. Beitrage zur Kenntnis des Polymorphismus bei *Camponotus ligniperda* Latr. *Mitt. Schweiz Entomol. Gesellsch.* 35: 146–154.

Levi, M. P. and H. B. Moore. 1978. The occurrence of wood-inhabiting organisms in the United States. *Pest Control* 46: 14–16, 18, 20, 45, 49.

Markl, H. and S. Fuchs. 1972. Klopfsignale mit Alarmfunktin bei Rossameisen (*Camponotus*). *Z. Vergl. Physiol.* 76: 204–225.

Maschwitz, U. 1974. Vergleichende Untersuchungen zur Funktion der Ameisenmetathorakaldruse. *Oecologia* 16: 303–310.

Masson, C. and A. Friggi. 1974. Codage de l'information par les cellules des recepteurs olfactifs de l'antenne de *Camponotus vagus*. *J. Insect Physiol.* 20: 763–782.

McComb, W. C. and R. E. Noble. 1982. Invertebrate use of tree cavities and vertebrate nestboxes. *Am. Midl. Nat.* 107: 163–179.

McCook, H. C. 1906. A guild of carpenter ants. *Harper's Mag.* 113: 293–300.

Meyers, J. H. and B. J. Campbell. 1976. Predation by carpenter ants as a deterrent to the spread of the cinnabar moth. *J. Entomol. Soc. Brit. Col.* 73: 7–9.

Nielsen, M. G., T. F. Jensen, and I. Holm-Jenson. 1982. Effect of load carriage on the respiratory metabolism of running workers of *Camponotus herculeanus* (Formicidae). *Oikos* 39: 137–142.

Oilon, J. H. 1965. Bionomics of the spruce budmoth, *Zeirapheira ratzburgiana* (Ratz.) (Lepidoptera: Olethreutidae). *Pytoprotext.* 46: 5–13.

Palma-Valli, G. and G. Deyle. 1981. Controle neuro-endocrine de la ponte chez les reines de *Camponotus lateralis* Oliver. *Insectes Soc.*, 28: 167–181.

Payne, T. L., M. S. Blum, and R. M. Duffield. 1975. Chemoreceptor responses of all castes of a carpenter ant to male-derived pheromones. *Ann. Entomol. Soc. Amer.* 68: 385–386.

Pricer, J. L. 1908. The life history of the carpenter ant. *Biol. Bull.* 14: 177–218.

Sanders, C. J. 1964. The biology of carpenter ants in New Brunswick. *Can. Entomol.* 96: 894–909.

Sanders, C. J. 1970. The distribution of carpenter ant colonies in the spruce-fir forests of northwestern Ontario. *Ecology* 51: 865–873.

Sanders, C. J. and W. F. Baldwin. 1972. Iridium-192 as a tag for carpenter ants of the genus *Camponotus*. *Can. Entomol.* 101: 416–418.

Smith, F. 1942a. Polymorphism in *Camponotus*. *J. Tenn. Acad. Sci.* 17: 367–373.

Smith, F. 1942b. Effect of reduced food supply upon the stature of *Camponotus* ants. *Entomol. News* 53: 133–135.

Smith, F. 1944. The nutritional requirements of *Camponotus* ants. *Ann. Entomol. Soc. Amer.* 38: 401–408.

Traniello, J. F. A. 1977. Recruitment behavior, orientation, and the organization of foraging in the carpenter ant, *Camponotus pennsylvanicus* (DeGeer). *Behav. Ecol. Sociobiol.* 2: 61–79.

Vanderschaff, P. 1970. Polymorphism, oviposition by workers, and arrested larval development in the carpenter ant, *Camponotus pennsylvanicus* DeGeer. Ph.D. Thesis, University of Kansas, Lawrence.

Wane, E. and C. Homer. 1969. Survey on carpenter ants. *Pest Control* 37: 19–21.

Weber, N. A. 1942. On ant nesting habits in North Dakota in 1941 compared with drouth years. *Can. Entomol.* 74: 61–62.

Weyrauch, W. K. 1933. Ueber unterscheidende Geschlechtsmerkmale. 2. Beitrag. Die Variabilitat der Koperlange bei den Camponotinen. *Z. Morphol. Okol. Tiere* 27: 384–400.

Wheeler, A. G. and T. J. Henry. 1980. Seasonal history and host plants of the ant-mimic *Barberiella formicorides* Poppius, with description of the fifth instar (Hemiptera: Miridae). *Proc. Entomol. Soc. Wash.* 82: 269–275.

Wheeler, W. M. 1910. Ants, their structure, development and behavior. Columbia Univ. Press, New York.

Wheeler, W. M. 1933. Colony-founding among ants, with an account of some primitive Australian species. Harvard Univ. Press, Cambridge, Mass.

Chapter 11

THE BENEFICIAL ECONOMIC ROLE OF ANTS
William H. Gotwald, Jr.

INTRODUCTION

There are compelling and incontrovertible reasons to rank the ants as the "premier social insects" (Wilson 1971). Certainly they are the most widely distributed and are numerically the most abundant of the eusocial insects. It has been estimated that at any given moment there are at least 10^{15} living ants perambulating about the earth (Wilson 1971). The Formicidae, the single family to which the ants belong, is a thoroughly diversified group that trophically competes with other terrestrial life forms for nutrient molecules that are biologically packaged in a myriad, and sometimes in seemingly inaccessible, ways. Evidence of their abundance is most impressive in the tropics. In biomass studies of central Amazonia, ants were found not only to be the predominate insect form (Fig. 11.1), but also to be, along with the termites, among the most numerous members of the entire fauna (Fittkau and Klinge 1973). Taking into consideration their abundance alone, it should not be surprising to discover that ants and humans compete and interact with one another in a variety of ways. A significant number of these interactions may be regarded as economically beneficial.

The ants have not gone unnoticed, even by our ancient progenitors, many of whom paid special attention to these redoubtable creatures. The cohesive behavior and industriousness of ants have inspired poets and philosophers alike to muse about the lessons that ants might teach mankind. We are advised, for example, by the Book of Proverbs (6:6–8) of the Old Testament to follow the ant's example:

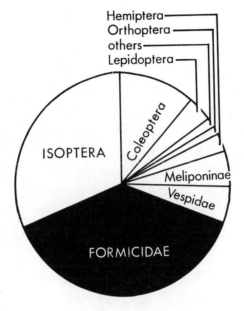

FIGURE 11.1
Composition of the insect biomass of a central Amazonian rain forest.
(*Source*: Fittkau and Klinge 1973)

> Go to the ant, thou sluggard; consider her ways
>   and be wise:
> Which having no guide, overseer or ruler,
> Provideth her meat in the summer, and gathereth
>   her food in the harvest.

The Greek writer Aesop fabled the same theme in "The Grasshopper and the Ant". He wrote:

> The thrifty ant says truly neighbor,
> I get my living by hard labor;
> But you, that in this storm came hither,
> What have you done when 'twas fair weather?
> I've sung, replies the grasshopper.

In a more practical frame of mind, ancient Chinese fruit growers apparently observed the red tree ant, *Oecophylla smaragdina* F., devour insects harmful to their citrus trees (Fig. 11.2a). It was their distinction to add an ingenious twist to the tradition of the ant-inspired homily; they

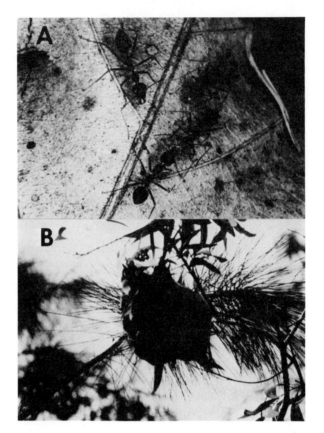

FIGURE 11.2
Oecophylla smaragdina.
A: Foraging workers on the forest floor in Malaysia.
B: Arboreal *Oecophylla* nest consisting of leaves bound together with
larval-produced silk. Photos by the author.

began to culture the ant, to deliberately encourage the ant to reside in
their orchards. Culturing of the red tree ant is still in practice (DeBach
1974). *Oecophylla* nests, consisting of leaves bound together with larval-
produced silk, are often picked in the wild during colder months of the
year when these normally pugnacious ants are relatively docile (Fig.
11.2b). The nests are then sold to the fruit growers for placement in the
citrus trees. During the early months of the growing season, the ants were

traditionally fed dog intestines and silkworm larvae in order to encourage them to refrain from foraging and thus to remain in the trees (Groff and Howard 1924). The potential for ant manipulation in biological control schemes is examined in Chapter 12.

Clearly, in reviewing the beneficial economic effects of ants, the word "beneficial" is necessarily defined in an anthropocentric context. That is, beneficial implies an improvement in the well-being of the human species. These beneficial economic effects of ants can be derived at several levels in a hierarchy of relationships that ranges in result from rather general, pandemic benefits along a continuum of increasingly specific direct benefits. The pandemic benefits originate from the roles ants play in ecosystems, roles that help create environments hospitable to human endeavors. For instance, ants are both primary and secondary consumers in the food chain and are, as well, fed on by other secondary consumers. In an ecological sense then, they function with all other organisms to produce an evolving but essentially predictable stability in ecosystems. This kind of benefit, that is ecosystem stability, is far too general to fall within the purview of this chapter. However, other pandemic benefits can be included: Ants move and aerate the soil and facilitate the cycling of nutrients. They therefore may have a profound effect on agriculture. Ants are also pollinators of flowering plants and are predators of innumerable pest invertebrates. Among the more direct economic benefits, two are of immediate interest: (1) ants are consumed as human food and (2) they are utilized in the practice of medicine. It is on this hierarchy of benefits that this chapter is organized.

ANTS AND THE SOIL

Charles Darwin (1882) was the first scientist to conclusively demonstrate that animal life played a vital role in the conditioning and genesis of soil. Specifically, he studied the role of earthworms in the formation of the so-called "vegetable mould," the soil's upper layer. He presented his analysis of the humble earthworm's prodigious contribution to the earth's humid soils in the book, *The Formation of Vegetable Mould Through the Action of Worms with Observations on Their Habits*, a tome published shortly before his death.

Darwin concluded that, "It may be doubted whether there are many other animals which have played so important a part in the history of the world, as have these lowly organized creatures." He had not considered the ants.

It is now well established that ants rival the earthworms in their capacity to modify and transform the soil. This they accomplish through the construction of subterranean nests, labyrinthic networks of galleries and chambers, and by the act of depositing above the ground vast quantities of excavated soil. In this way they subject the soil to conditions that alter the soil's physical and chemical properties (Petal 1978).

The physical changes brought about by ants manifest themselves primarily in the soil profile. Ants bring to the surface soil from the deeper layers of the profile and at the same time carry organic matter into their nests. Their subterranean chambers and galleries increase soil porosity and thus promote aeration and water infiltration. The increased downward movement of organic matter improves the water-holding capacity of the soil (Petal 1978).

The chemical alterations that occur also derive from the accumulation of organic matter in the nest and from the subsequent processes of decomposition. Depending on the soil type, the pH of an ant nest may be higher or lower than the pH of the surrounding soils. The amassment of exchangeable cations, for instance, will produce an increase in pH, while an influx of organic matter will cause a decrease. Even the potassium and phosphorus content of ant nests may be higher than in the neighboring soil (Petal 1978).

Soil Movement

It was Shaler (1891) who first noted that "the work of ants in the sandy soils resembles that of earthworms in the clayey ground." In New England he observed that certain untilled fields were covered to a depth of several centimeters with a fine sand. Pebbles were absent from the covering: He attributed this surface layer of uniformly sized particles to the soil-carrying activities of ants. As a result of studies of a field in Massachusetts during two summer seasons, Shaler was able to conclude that the "average transfer of soil matter from the depths to the surface was sufficient to form a layer each year having a thickness of at least one-fifth of an inch over the area" observed.

Numerous scientists have since described the impact of ants on the soil. Branner (1909) pointed to the shared role in the tropics of ants and termites in moving and modifying soil. In the forest region of southwest Nigeria, Nye (1955) found that, of the soil fauna of the region, the termites and ants have the most acute influence on the chemical and physical characteristics of the soil. Jacot (1936) proposed that "ants are the most active of soil animals." Thorp (1949) estimated that ant mounds in some regions of the Western plains represented about 3,400 lb of soil per acre that had been moved to the surface from the subsoil horizons. In the dry sandy heath areas of Denmark, Nielsen (1982) showed that the ant *Lasius alienus* (Forst), through its nest systems, improves soil aeration. Furthermore, he noted this ant's digging activity is the main factor in the turnover of soil.

Even ants that produce only small mounds of excavated soil are important to the process of soil genesis. Such ants, for instance, play a significant role in the development of Brown Podzolic soils in New England (Lyford 1963). In conventional studies of podzolic soils, emphasis has always been placed on the downward movement of materials from the upper few centimeters of soil. Lyford (1963), in a departure from traditional approaches, focused his attention on the upward movement of subsoil material. He discovered that ants are responsible for this upward movement and estimated that "if two new mounds per square yard are constructed each year an amount of about 50 grams per square yard each year is returned to the surface." He calculated that this amounts to about an inch of soil every 250 years and "conceivably a thickness of as much as 10 to 18 inches of soil material could be returned to the surface in a period of 3-4000 years." The return to the surface by ants of B horizon material consisting of mineral particles coated with sesquioxide-humus substances figures importantly in the eluviation-illuviation podzolization cycle. This fact has been generally overlooked by pedologists (Lyford 1963).

Nutrient Cycling

Ants are obviously involved in the decomposition of organic matter and nutrient cycling, at least to the extent that they accelerate the downward movement of organic materials. Earthworms, as Darwin (1882) noted, perform a similar service by covering with their castings all manner of dead plant and animal remains and by dragging "an infinite number of dead leaves and other parts of plants into their burrows, partly for the sake of plugging them up and partly as food."

The most stunning example of this downward movement is found in the attines, the fungus-growing ants of the New World. Weber (1982) pointed out that "through the formation of their fungus gardens, the attines are perhaps the animal species chiefly responsible for introducing organic matter into the soil in tropical America." Dead ants and exhausted garden substrate are relegated to refuse disposal areas, some of which are located in subterranean chambers but others of which are situated above ground (Haines 1978).

The magnitudes of energy and nutrient flow though the attine colony in relation to the flow through the forest leaf litter were calculated by Haines (1978). The annual flow rates of 13 elements found in plant parts foraged by the ants, expressed per square meter of ant refuse dump, ranged from 16 to 98 times the flow in leaf litter. This increase in element flow through the refuse dumps accounted for the fourfold increase in the quantity of fine roots in the dump compared to the forest floor. Trees apparently recycle the elements directly from the dumps rather than from lower positions in the soil profile. Haines (1978) concluded that "when calculated per area of an ant nest, energy flow in leaf material funneled through nests by the ant was 11 times the energy flow in leaf litter falling on the nests." When calculated per area of forest, annual energy flow through the nests was 1.7% of the energy flow in leaf litter fall. It was further calculated that the ant colonies, including their fungus gardens, assimilate about 22% of the energy that flowed through them.

Changes in the soil of ant nests, as occur with the accumulation of nutrients, are followed by a characteristic plant succession (Petal 1978). Briese (1982) found that because ants take plant and animal materials into their nests, often mixing these materials with excavated soil, there is an increase in carbon, nitrogen, and phosphorus in the surface soil around the nest sites. These elevated mineral levels are similar to those found in the soil beneath *Atriplex vesicaria* shrubs that result from litter deposition. By concentrating nutrients at other sites, the ants contribute to the mineral mosaic typical of the *A. vesicaria* shrub-dominated steppe in southeastern Australia and "hence may influence overall community nutrition" (Briese 1982). Local disturbances caused in sand prairie in Illinois by the ant *Formica obscuripes* Forel increase heterogeneity of the plant community. This increase in community complexity is most obvious at early successional sites (Beattie and Culver 1977).

Importance of Ants to Soil Genesis

The importance of the soil fauna in producing "desirable" soil conditions should not be underestimated. Although humification and mineralization of the soil rests largely on the microflora, animals, especially earthworms, termites, and ants, are necessary for preparation of the litter before microbial activity can begin (Kevan 1962). Quite conclusively, ants are responsible in a significant way in some regions for assisting in bringing about chemical and physical changes in the soil (Fig. 11.3).

Though it can be argued that, in light of the advanced state of contemporary agricultural technology, soil fertility can be maintained and enhanced without the activities of ants (and earthworms), the continued stability of undisturbed soils in forests, plantations, and certain types of orchards may require the activities of ants and other members of the soil fauna (Kevan 1962). Forest management schemes should be hesitant to disturb the biotic equilibrium of the soil, if soil fertility is to be maintained. The widespread use of pesticides on tree crops, for example,

Plant Succession

AIR WATER SOIL ORGANIC MATERIALS

Humification
Mineralization

FIGURE 11.3
Ants rival the earthworms in their capacity to modify and transform soil. By nesting in the soil, they increase aeration and water infiltration of the soil; they deposit above ground vast quantities of excavated soil; and they accelerate the downward movement of organic materials in the soil, which in turn, through the processes of humidification and mineralization, influences plant succession.

ultimately may be counterproductive, if, in altering the soil fauna as a result, soil fertility is depressed.

ANTS AS POLLINATORS

Pollination Biology

A flower is an assemblage of structures that includes a "space barrier" between the stamens and the pistil. This gap, regardless of whether it exists within a single flower or between male and female flowers, makes necessary the process of pollination. Pollination may result from the intercession of abiotic agents, most notably wind and water, or biotic agents, second organisms that serve as pollination agents or pollen vectors. Biotic agents of pollination include insects, birds, and small mammals such as bats. Insects usually visit flowers in search of pollen or nectar and most belong to the orders Diptera, Lepidoptera, and Hymenoptera (Percival 1965).

Importance of Ant Pollination

The relative importance of ants as pollinators has yet to be determined. Critical studies of the role of ants in pollination biology are rare in the literature. Surely the attention given bees (solitary bees, bumblebees, and honeybees), the foremost pollen vectors, is distracting, and some authors almost entirely ignore other biotic agents, consigning them to the status of inconsequential. In fact, some biologists are careful to note that ants may actually avoid the role of pollinator. Faegri and van der Pijl (1971) declared, "If there was ever a scoundrel in the pollination drama, that role has been assigned to the ants." They pointed out that ants are notoriously fond of sugar (and may eat pollen as well) and frequently visit floral nectaries to satisfy their needs. These authors observed, however, that (1) ants are so small that they can "sneak" in and out of blossoms without ever coming in contact with either anthers or stigma; (2) the bodies of ants are hard and smooth and not well adapted to transporting pollen; and (3) ants are ignominiously "bellicose" and may successfully defend blossoms against nectar or pollen feeding insects that may indeed be pollinators. Ants therefore may be considered the "prototypes of nectar thieves."

Many plants have evolved ant-deterring structures that prevent ants from entering the flowers. The sticky belts on the stems of *Viscaria vulgaris* (= *Lynchnis viscaria* or German catchfly) are a striking example (Faegri and van der Pijl 1971). The petals of some plants have a repellent, no doubt chemical, effect on ants (van der Pijl 1955), although repellent or toxic tissues may be directed more specifically at insects that chew through flower structures in their quest for nectar. Janzen (1977) observed that in lowland tropical habitats ants are conspicuous sugar collectors at extrafloral nectaries, honeydew-producing homopterans, and broken fruits but are evidently absent from the most abundant sugar source of all: floral nectaries. He hypothesized that ants avoided floral nectar, because it contains chemicals that are "powerfully repugnant, indigestible, or toxic" to the ants.

Janzen's hypothesis stimulated a flurry of responses that challenged not only the accuracy of his conclusion but also the veracity of the premise on which he based his hypothesis, that is, that ants do not visit flowers in the lowland tropics. Ants do indeed visit such flowers. For instance, six species of ants were found by Barrows (1977) to visit flowers of the tropical vine, *Pachyptera hymenaea*, presumably "to rob nectar." Rico-Gray (1980) witnessed ants visiting the flowers of the black mangrove (*Avicennia germinans*). In challenging Janzen's hypothesis, Schubart and Anderson (1978) suggested that "a more valid observation is that ants do not visit flowers as frequently or as systematically as do winged insects and birds."

Ants do visit floral nectaries, but the question remains: do ants act as pollinators? Obviously, when ants crawl about flowers or inflorescences they may serve inadvertently as the agents of autogamous pollination (Faegri and van der Pijl 1971), but it is also true that genuine cases of ant pollination, although few in number, have been documented. In the United States, Kincaid (1963) showed that ants pollinate *Orthocarpus pusillus*, and Hagerup (as cited in Faegri and van der Pijl 1971) concluded that if pollination by means of biotic agents is to take in such areas as the Sahara desert, ants must be involved. The most conclusively documented case of ant pollination involves a specialized mutualistic system in which the small annual plant *Polygonum cascadense* (Polygonaceae) is pollinated. On the hot, dry slopes of the western Cascades of Oregon, Hickman (1974) found that cross-pollination of *P. cascadense* is mediated by the ant *Formica argentea* Wheeler and that the ants are essential for normal seed set. He described ten traits that adapt *P. cascadense* for pollination by the ants. Among these are the following: (1) the plants must grow in hot and dry habitats where ants are abundant and active; (2) the nectaries must be morphologically accessible to the ants; (3) the plants must be short and prostrate; (4) pollen volume per flower must be small in order not to stimulate self-grooming in the ant through excessive loading of pollen; and (5) seeds must be few per flower, thus requiring fewer pollen transfers. Hickman (1974) offered that these traits might be of predictive value in discovering other plants that are ant-pollinated. Perhaps most importantly, he concluded that this ant-flower system of pollination minimizes the interaction energy outputs of both plant and pollinator. Additional cases of ant pollination have been described. Three species of plants (of the families Cruciferae, Boraginaceae, and Umbelliferaceae) that blossom in early June on the Colorado alpine tundra are predominantly cross-pollinated by ants of the genera *Formica* and *Leptothorax* (Petersen 1977). Because conditions on the tundra (e.g., high wind velocity, cold temperature, and food shortage) limit the role of flying pollinators, Petersen (1977) presumed that the niche normally occupied by flying pollinators in other ecosystems is filled by high-altitude ants. In another case, *Diamorpha smallii* Britton (Crassulaceae), a self-incompatible annual plant endemic to granite outcrops in the southeastern United States, is pollinated by the ant *Formica schauffusi* Mayr (Wyatt 1981 a,b). The plant and its habitat appear to fit the "ant-pollination syndrome" first proposed by Hickman (1974). Contrary to the assertion that ants are not efficient pollen carriers, Wyatt (1981a) found large numbers of sticky pollen grains adhering primarily to the thoracic setae and sculpturing on flower-visiting workers. No doubt the discovery of other ant-pollination systems can be anticipated.

But do ants pollinate plants of economic value? It is possible, but evidence in support of such a conclusion is anecdotal. Ants have been observed as possible pollinators of the following crop plants (Free 1970): Cashew nut (*Anacardium occidentale* L.), beet (*Beta vulgaris* L.), bird chili (*Capsicum frutescens* L.), coconut (*Cocos nucifera* L.), lychee (*Litchi chinensis* Sonn.), and cocoa (*Theobroma cacao* L.). Only in the case of cocoa has the involvement of ants as occasional pollinators been assumed, but the pollination biology of cocoa is still poorly understood. In Brazil, Vella and Magalhaes (1971) noted that there were a greater number of pollinations in cocoa trees occupied by the ant *Azteca chartifex spiriti* Forel. The ants did not themselves pollinate the flowers. Rather they apparently secreted a substance that attracted pollinating insects to the trees. In all cases above but *C. frutescens*, ants most likely serve in self-pollination.

Therefore, the relative role of ants as pollinators is not clearly understood. Under certain conditions, ants may serve as important pollinators for specific plants, but as yet, they have not been identified as major pollinators of any economically valuable crop.

ANTS AS PREDATORS

Trophic Biology of Predatory Ants

Ants are primitively carnivorous, most likely inheriting their predatory habits from solitary wasp ancestors (Wheeler 1910, Brian 1965, Sudd 1967, Wilson 1971), and although most ants engage in predation to some extent, it is most exclusively practiced in the subfamilies Ponerinae, Dorylinae, and Ecitoninae. Ants utilize a wide variety of arthropods as nutrient sources. They demonstrate a remarkable range in diets from those of trophic specialists, like some species of the genera *Proceratium* and *Discothyrea* that prey on arthropod eggs (Brown 1957), to trophic generalists, epitomized in the army ant genus *Dorylus*, some species of which consume not only an abundant array of invertebrates but also any vertebrate that cannot elude their attacks (Gotwald 1982). Ants are intriguing because they have evolved a trophic preference for other social insects, in fact, to such an extent that they are regarded as the principal predators of social species (Wilson 1971).

Predator Ants in Pest Control: Historical Background

The impact of ant predation on prey populations can be substantial. Gosswald (1958) estimated that large colonies of Formica rufa L. could collect 65,000 to 100,000 caterpillars per day. Kajak et al. (1972) measured the predatory impact of the ant genus Myrmica on a meadow ecosystem and found that these ants removed 32% of the newly emerged Diptera, 43% of the leafhoppers, and 49% of the lycosid spiders. It is readily apparent that predator ants could significantly depress the size of pest populations.

The practical application of predatory ants in the control of agricultural pests was first recorded by the Chinese, as discussed earlier, in the regulation of citrus pests. This early excursion into biological control is described in the Ninth and Tenth century book, "Ling Pio Lu Yi" or "Wonders from South China," by Liu Shun. It contains the following passage (Liu 1939):

> There are many kinds of ants in Lingnan (South China). Sometimes one finds that the ants are carried in a bag and sold on the market. The ants are yellow with long legs, larger than ordinary ants and live in a nest made from leaves and twigs. They are bought for the protection of the orange for it is said that without these ants most of these fruits would be wormy.

Predacious ants were employed in a similar fashion by date growers of Yemen on the southern Arabian Peninsula. The growers transported colonies of the predatory ants from nearby mountains to oases where they were released to control phytophagous ants that attacked the date palm (Doutt 1964, van den Bosch et al. 1982).

American Indians made their own use of predaceous ants by placing their pest-infested blankets and furs on the nests of the occident ant, Pogonomyrmex occidentalis (Cresson). These ants harvested the lice and other unwanted guests that annoyed people of the American plains (McCook 1882). The same practice was adopted by "old pioneers" and campers and by laborers who prepared the prairie sod for grain. These laborers were provided with barracks "and by reason both of personal uncleanliness and abundance of certain objectionable insects in the prairie grass, soon become infested with parasites" (McCook 1884). It was their habit to place their infested garments on the mounds of Formica rufa. North American entomologists were sensitized to the possible economic value of predatory ants in agroecosystems by McCook (1882) who learned from missionaries in China of the use of Oecophylla smaragdina.

He pondered the feasibility of using predatory ants to control coccid pests of citrus in Florida. After considerable deliberation, he was forced to conclude that "we have no indigenous species upon which to experiment, either to utilize or develop a bait that will make ants so highly beneficial as insecticides as to justify any dependence upon them as protectors of fruits." But McCook went on to contemplate the possibility of importing O. smaragdina to be cultured in the southern United States. As early as 1879, the United States Department of Agriculture was considering the "friendly offices of ants in destroying the eggs and larva of the cotton-worm" (McCook 1882). Their prime candidate at the time was Solenopsis xyloni McCook.

Army Ants and the Ponerinae in Pest Control

Curiously, the ponerines, dorylines, and ecitonines, all of whom have earned considerable notoriety as predators, are not the source of many species of known economic benefit. Shortly after the turn of this century, the United States Department of Agriculture experimented with the importation of the ponerine Ectatomma tuberculatum (Oliver) to control the cotton boll weevil. The association of this ant with cotton, whose extrafloral nectaries the ants visit, was well known to the Indians of Guatemala, who referred to the ant as the "kelep." In 1904, about 4,000 of these ants, in 89 colonies, were collected and transported to Texas where they were released (Cook 1904, 1905). Little was heard of the success of the initial release as a bitter controversy developed between O. F. Cook, who sponsored the importation, and W. M. Wheeler, the world-renowned myrmecologist. Wheeler (1904) remarked that "the attempt to establish the ant in Texas will prove to be about as successful and profitable as an attempt to acclimatize in the same state some rare Central American orchid, the South African secretary bird or the Australian wombat." This kind of rhetoric no doubt clouded the real issue: the empirical determination of the ant's value to cotton agriculture.

Weber (1946) suggested that E. tuberculatum be reevaluated and that E. ruidum Roger be examined for their beneficial economic potential. While these and other ponerine ants are predacious on numerous insect pest species, they also tend and protect homopterous insects that are deleterious to plant growth. Thus the merit of using such ants to combat prey species must be evaluated in the context of their total role in the agroecosystem.

The Dorylinae and Ecitoninae comprise the Old and New World army ants, respectively. They have not been ignored by entomologists in the search for economically beneficial ants. The fate of screw-worm larvae, Cochliomyia hominivorax, was investigated in Texas by Linquist

(1942). When mature these carcass-infesting larvae leave the host animal either when the host is alive, whereupon the larvae drop to the substrate and are dispersed over a wide area, or when the host is dead and subsequently concentrate beneath and about the host carcass. Linquist (1942) found, when observing the emergence of the fly larvae from their hosts, that a "great number of ants. . . . quickly appeared and began to attack and kill the larvae." Among these predator ants was the ecitonine *Labidus coecus* (Latreille) which was the most frequently observed species. In controlled experiments. 4.1% emergence rate of adult flies occurred when host carcasses were exposed to ants, while a 93.1% emergence rate was recorded from carcasses protected against ant attack. Ants therefore constituted an important natural control of *C. homnivorax*, and *Labidus* figured most importantly among these ants.

Army ants of the Old World tropics have been accorded some credit for preying on pest species, but so little quantified data exist to support such claims that their true economic potential is unknown. Wellman (1908) considered the African driver ant, *Dorylus (Anomma) nigricans* Illiger, to be economically important because it rids "bungalows" of such unwelcomed guests as bugs, beetles, cockroaches, mice, rats, and snakes. Dutt (1912) believed that *D. (Alaopone) orientalis* Westwood was beneficial in India because it attacked the nuisance ant, *Pheidole indica*. Burgeon (1924) regarded the driver ant, *D. (Anomma) wilverthi* Emery, to be of general value since it destroyed agricultural pests. Both Alibert (1951) and Strickland (1951) implied that driver ants were valuable in cocoa farms, although they offered no evidence. Gotwald (1974) investigated the behavior of these ants in cocoa farms in Ghana to determine the extent to which these predators affected the faunal composition of the ground and canopy microhabitats. The driver ants were never observed foraging in cocoa trees, although in other habitats they commonly forage in trees to heights of 9 m or more. Dominant canopy ant species may successfully repel foraging driver ants, thus it was concluded that "driver ants are not of direct economic importance in determining the composition" of the cocoa canopy fauna (Gotwald 1974).

Predatory Ants in Agroecosystems

Pheidole megacephala (Fabricius) is credited with destroying nematodes, larvae, and adults of *Carpophilus* beetles, larvae of bud moths, root maggots, root grubs, and fruit maggots. The overall economic effect of this ant in pineapple is rated as positive, in spite of the fact that it also attacks beneficial insects and tends coccid pests (Phillips 1934). Additionally, field experiments in Hawaii indicate that *P. megacephala* is effective in controlling houseflies and blowflies (Phillips 1934).

In the Solomon Islands, the incidence of immature nutfall of coconuts caused by the coreid *Amblypelta* "depends indirectly upon certain species of ants, some of which protect the palms against *Amblypelta*, whilst others do not" (Brown 1959, Greenslade 1971). Reduction in nutfall, for instance, is associated with replacement of *Pheidole megacephala* by *Oecophylla smaragdina* as the dominant ant. Coreids of the genus *Theraptus* in East Africa severely damage developing coconut fruits but can be controlled through predation by *Oecophylla longinoda* (Latreille) (Way 1953). However, three other species of ants, none of which prey on *Theraptus*, may replace *O. longinoda*, resulting in an increase in coreid-produced damage. Way (1953) proposed methods by which the undesirable ant species could be controlled or manipulated so that *O. longinoda* would continue, unchallenged, in its pursuit of *Theraptus*.

Ants are impressive predators of fruit flies of the family Tephritidae. Most susceptible to attack are those fruit fly life stages associated with the soil, that is, the mature larvae that have left the host fruit, the pupae, and the emerging adults (Bateman 1972). Of the soil predators the ants are most crucial. For instance, in Switzerland, heavy mortality is inflicted on mature larvae of *Rhagoletis cerasi*, a pest of cherries, by the ant *Myrmica laevinodis* Nylander (Bateman 1972, Boller and Prokopy 1976).

In Central and South America, larvae of lepidopterous stalk borers of the genera *Castniomera* and *Castinia* damage banana plants by burrowing into the pseudostem, thus lowering the resistance of the plants to wind. One species, *Castniomera humboldti*, endemic in banana plantations from Costa Rica through northern South America, is controlled by ants that prey on its eggs and larvae. Thus far, all epidemics of the *Castniomera* stalk borer in Central America have resulted from the destruction of the predator ants with insecticides such as dieldrin. Ostmark (1974) cautions that "where this stalk borer exists ground-inhabiting ants should be carefully protected."

In the Neotropics, leaf-cutting ants (tribe Attini) are important defoliant pests of citrus. In Trinidad, ants of the genus *Azteca* have been found to protect citrus trees against defoliation by the leaf-cutter *Atta cephalotes* L. Trees without *Azteca* average 59% defoliation compared with 27% in trees occupied by *Azteca* (Jutsum et al. 1981). *Azteca* was experimentally removed from some trees, after which 80% of the trees were attacked by *A. cephalotes*. Although the suppression of *A. cephalotes* by *Azteca* is clearly an economic advantage, it occurs only on larger trees and thus the protection provided is incomplete. *Azteca* has the disadvantage of protecting homoptera that can damage citrus. The relative merits of an *Azteca* presence in citrus must be carefully evaluated before a strategy can be devised to utilize *Azteca* in pest management.

Predatory Ants in Forest Ecosystems

The value of predacious ants in the natural control of forest pests has been investigated in Europe for over 50 years. Researchers including Bruns (1959), Kloft (1960), Lange (1958), Otto (1960), and Pavan (1959) have provided details on the biology of red wood ants, known as the *Formica rufa* group, and have developed techniques for collecting and rearing these ants. Although the value of these ants to forest pest control is not fully elucidated, evidence that they make a positive contribution does exist. Gosswald and Horstmann (1966), for example, studied the effects of *Formica polyctena* (Forst) on populations of the green oak leaf roller, *Tortrix viridana*, and concluded that these ants do indeed prevent defoliation of trees on which they forage. They determined, in a cost-benefit analysis, that it would be profitable to artificially colonize forests with *F. polyctena*.

Only recently has similar attention been given to ant predators of forest insects in North America. Finnegan (1971) established criteria for evaluating the efficaciousness of predacious ants as limiting agents of forest pests and applied the criteria to the indigenous ants of Quebec Province, Canada. He concluded that "there is no clear prospect of using them successfully in biological control work on a large scale."

Red wood ants constitute the best candidates for natural control of pests in temperate coniferous-deciduous forests. These ants, according to Finnegan (1974), combine certain qualities, not found in other forest predators, that make them ideal to the task of control. These characteristics include the fact that the ants: (1) attain a high population density, requiring a large, continuous supply of food selected from a wide array of prey; (2) forage at all levels in the forest, from in the soil to the canopy; (3) have an extended period of activity, foraging for as many as 200 days in central Europe; (4) are polygynous, assuring a prolonged nest life since dead queens are easily replaced; (5) form colonial nests, that is, groups of colonies that exchange brood, worker ants, and queens, thereby promoting greater stability and permanence in the ant population over large areas; (6) forage for and attack all stages in the life cycle of prey; and (7) specialize in foraging for those prey that explosively increase in total numbers of individuals.

With these characteristics in mind and given the fact that the indigenous ants of eastern Canada show little promise in forest pest control, it was decided to introduce *Formica lugubris* Zett, from northern Italy. Since that original introduction in 1971 (and a subsequent introduction), it now appears that *F. lugubris* will become established in eastern Canada (Finnegan 1975). McNeil et al. (1978) identified the composition of the prey gathered by these ants. They found *F. lugubris* to be an "opportunistic" predator, its prey consisting entirely of invertebrates of which 95% are insects. Lepidoptera represented 58% of the total prey, the Tortricidae being the most abundant family represented. At least 80% of all tortricids collected were spruce budworms. Clearly, *F. lugubris* can remove large numbers of budworm larvae from the population, and this, in turn, reduces the level of defoliation. In 1975, for instance, defoliation was estimated to be 42.8% where the ants were present and 63.1% where they were not (Finnegan 1977, 1978).

Importance of Predatory Ants in Pest Control

Ants unequivocally play a valuable role in the control of pests. The benefit of their destruction of pest species must be weighed against the damage they may cause indirectly as a consequence of protecting harmful species of Homoptera. Thus far, most researchers rate the overall contribution of predacious ants as positive and many have suggested methods of manipulating predatory ant populations to the disadvantage of pest species (see Chapter 12). Unquestionably, predacious ants in resource forests and agroecosystems play a vital role in the regulation of certain insect populations and cannot be treated capriciously when planning and employing chemical control programs. Their demise may ultimately do more harm than good.

ANTS AS HUMAN FOOD

Nutritional Value of Insects

Cultural conditioning in industrialized societies has produced a strong bias against the consumption of insects by humans. Entomophagy is frequently regarded by such societies as a bizarre curiosity or a behavioral vestige of the distant, evolutionary past. The human species is extraordinarily adaptable, and there are a few behavioral characteristics in nonindustrialized societies, that, when methodically scrutinized, are not convincingly adaptive. The Yukpa Indians of Venezuela and Columbia, for instance, consume insects, but not as mere delicacies or opportunistic diet supplement. Indeed, research "indicates that the food use of insects by the tribesmen represents a simple response to the physiological need for proteins, fats, and other substances of animal origin" (Ruddle 1973). Rather than dietary novelties, insects are an important avenue to those nutrients lacking in the basic diets of some peoples. Insects are highly nutritious, being rich in protein, fat, vitamins, salts, and minerals (Bodenheimer 1951).

Even within the historical framework of Western religious tradition, insects were recognized as appropriate food sources. The dietary laws prescribed in the Old Testament (Leviticus 11:22) state: "Even these of them ye may eat; the locust after his kind; and the bald locust after his kind; and the beetle after his kind; and the grasshopper after his kind."

Ants Taken as Human Food

In the New World certain North American Indians commonly ate ants. This was especially true of tribes that practiced agriculture sparingly. The Digger Indians of California consumed the larvae of carpenter ants (*Camponotus*), and Indians of the American Southwest relished the honey pot ants (*Myrmecocystus*) (Bodenheimer 1951). The largest workers in the honeypot ant colony store sweet liquids in their crops and are referred to as repletes. The repletes were collected by the Indians and the honey-filled abdomens bitten off and consumed. The "syrup" extracted from these ants is said to have an "agreeable sweet taste and an odour like that of syrup of squills (a Mediterranean bulbous herb of the lily family), but slightly acid" (Bodenheimer 1951). The honeypot ant was also eaten by the Indians of Mexico (Bodenheimer 1951).

The Yukpa of South America consume ants of the genus *Atta*. These are collected during May at the beginning of the rainy season. Those individuals specifically selected are the females "bloated with eggs" and only the abdomen is eaten. The ants are wrapped in leaves and "placed on log close to the fire to roast" (Ruddle 1973). Apparently, *Atta cephalotes* Latreille females are valued throughout Amazonia, and when they emerge from their parent colony, they are collected in baskets and calabashes (Bodenheimer 1951).

In India, Burma, Malaysia, and Thailand, *Oecophylla smaragdina* is made into a paste and eaten as a condiment with curry. Reportedly, this paste is considered a delicacy. The ants, both adults and immatures, are collected from the nests and beaten into a pulpy mass with a stone. They are then enclosed in a leaf packet and sold at market. The crushed ants may be mixed with salt, tumeric, and chilies and ground between stones. The resulting paste is eaten raw with boiled rice (Bodenheimer 1951, Mathur 1954). A "brown ant of medium size" is collected from its nest in Thailand, and the adults and immatures are then pickled in salt water, tamarind juice, ginger, onion, sugar, and the leaf of "Bai Makfut" (*Citrus hystrix*) (Bristowe 1932).

Travelers and explorers among the aboriginals of Australia were quick to note the extensive food use of ants by these people. In 1889, Lumholtz observed the following:

> Several times we saw some small black ants which lay their eggs in trees. Morbora struck the trunk of the tree with my tomahawk while I held my hands out below to receive them. Several handfuls came down, and I winnowed them in the same manner as my companion did—that is, by throwing them up in the air and at the same time blowing at them. In this manner the fragments of bark were separated from the eggs, which remained in my hands, and were refreshing and tasted like nuts.

Honeypot ants of the genus *Melophorus* are a favorite food of the aboriginals, so much so that a special ceremony may be performed to promote the supply of these ants. As with the American Indians, the honey-distended abdomens of the repletes are bitten off and swallowed. The green tree ant, *O. smaragdina*, is also eaten and sometimes used in preparing a refreshing drink.

ANTS AND THE PRACTICE OF MEDICINE

Few references exist to the use of ants in medical and medicinal practice. The most celebrated of all applications, perhaps because it seems so curiously clever, is the use of ants as sutures. In medieval Spain, France, and Italy ants were employed in suturing wounds of the small intestine. Today this practice of suturing cuts still exists in some Mediterranean countries, Turkey, India, Africa and South America (Gudger 1925). This procedure in both South America and Africa requires the soldier caste of *Eciton* or *Dorylus* army ants. The soldiers are held in such a way that they are encouraged to bite on each side of the incision to be sutured. Their ice tong-like mandibles sink deeply into the skin, whereupon their bodies are snapped from their heads. A series of heads placed in this way will close an incision (Gotwald 1982).

In Mexico, the "honey" extracted from the honeypot ants was applied by Indians to bruised and swollen limbs. Supposedly, this topically applied liquid had great healing powers. The Aborigines in northern Queensland crush the adults and larvae of *O. smaragdina* in water, forming a concoction reputed to cure stomach troubles, headaches, coughs, and colds (Bodenheimer 1951).

Recent research on the venom of the tropical American ant genus *Pseudomyrmex* has isolated a polysaccharide that appears to have potential in the treatment of rheumatoid arthritis, especially its ability to curb synovial inflammation. Although the mode of action of this polysaccharide has eluded researchers, current investigations are focusing on pathways of humoral immunity (Byrne 1982).

CONCLUDING REMARKS

Ants play a vital role in determining the quality of human existence by contributing positively to environments favorable to human activities. They share this role with other organisms but have a special impact because of their diversity and prodigious numbers. They rival the earthworms—perhaps surpass them—as instruments of soil genesis. While their role as pollinators of flowering plants is not well understood, they may have a significant effect on the reproductive biology of at least some plants. No doubt as predators they make their most conspicuous contribution to the well-being of the human species. They prey on numerous arthropods that compete with *Homo sapiens* for food and secure shelter. And in some societies ants, themselves, are food. Finally, ant venoms may be a notable source for new drugs in the medical arsenal of the human species.

There is still another benefit—not economic—that can be credited to the ants. They challenge us as complex social animals to understand ourselves, to recognize the genetic limits of human social behavior but to discover, at the same time, the human potential for adaptability in an endlessly changing environment.

ACKNOWLEDGMENTS

Some of the research cited in this chapter was supported by several grants from the National Science Foundation, the latest being DEB-8113274 (W. H. Gotwald, Jr., Principal Investigator). I am grateful to Mr. B. Lorber, Institut de Biologie Moleculaire et Cellulaire du C.N.R.S., Strasbourg, France, for supplying information on the use of ants in medical practice and to Mrs. Virginia Marsicane for typing the manuscript.

REFERENCES

Alibert, H. 1951. Les insectes vivant sur les cacaoyers en Afrique occidentale. *Mem. Inst. Fr. Agri. Noire* 15: 1–174.

Barrows, E. M. 1977. Floral maturation and insect visitors of *Pachyptera hymenaea* (Bignoniaceae). *Biotropica* 9: 133–134.

Bateman, M. A. 1972. The ecology of fruit flies. *Annu. Rev. Entomol.* 17: 493–518.

Beattie, A. J., and D. C. Culver. 1977. Effects of the mound nests of the ant, *Formica obscuripes*, on the surrounding vegetation. *Amer. Midl. Nat.* 97: 390–399.

Bodenheimer, F. S. 1951. Insects as human food. Junk Publishers, The Hague. 352 pp.

Boller, E. F., and R. J. Prokopy. 1976. Bionomics and management of *Rhagoletis*. *Annu. Rev. Entomol.* 21: 223–246.

Branner, J. C. 1909. Geologic work of ants in tropical America. *Bull. Geol. Soc. Amer.* 21: 449–496.

Brian, M. V. 1965. Social insect populations. Academic Press, New York. 135 pp.

Briese, D. T. 1982. The effect of ants on the soil of a semi-arid saltbush habitat. *Insectes Soc.* 29: 375–382.

Bristowe, W. S. 1932. Insects and other invertebrates for human consumption in Siam. *Trans. Entomol. Soc. Lond.* 80: 387–404.

Brown, E. S. 1959. Immature nutfall of coconuts in the Solomon Islands. II. Changes in ant populations, and their relation to vegetation. *Bull. Entomol. Res.* 50: 523–558.

Brown, W. L., Jr. 1957. Predation of arthropod eggs by the ant genera *Proceratium* and *Discothyrea*. *Psyche* 64: 115.

Bruns, H. 1959. Siedlungsbiologische Untersuchungen in einformigen Kiefernwaldern. *Biolo. Abhandl.* 22/23: 3–52.

Burgeon, L. 1924. Les fourmis "siafu" du Congo. *Rev. Zool. Afr.* 12: 63–65.

Byrne, M. 1982. Venom of stinging jungle ant may hold a key to arthritis Rx. *Medical Tribune* 28 July: 3, 22.

Cook, O. F. 1904. Report on the kelep, or Guatemalan cotton-boll-weevil ant. U.S. Dept. Agri., Bur. Entomol. Bull. 49. 15 pp.

Cook, O. F. 1905. The social organization and breeding habits of the cotton-protecting kelep of Guatemala. U.S. Dept. Agri., Bur. Entomol., Tech. Ser. 10. 55 pp.

Darwin, C. 1882. The formation of vegetable mold, through the action of worms, with observation on their habits. D. Appleton & Company, New York. 326 pp.

DeBach, P. 1974. Biological control by natural enemies. Cambridge University Press, London. 323 pp.

Doutt, R. L. 1964. The historical development of biological control. *In* "Biological control of insect pests and weeds" (P. DeBach ed.), pp. 21–42. Reinhold Publishing Corporation, New York.

Dutt, G. R. 1912. Life histories of Indian insects. *Mem. Dept. Agri., India. Entomol.* (Ser. 4): 183–267.

Faegri, K., and L. van der Pijl. 1971. The principles of pollination ecology. Pergamon Press, Oxford. 291 pp.

Finnegan, R. J. 1971. An appraisal of indigenous ants as limiting agents of forest pests in Quebec. *Can. Entomol.* 103: 1489–1493.

Finnegan, R. J. 1974. Ants as predators of forest pests. *Entomophaga Mem. Hors. Ser.* 7: 53–59.

Finnegan, R. J. 1975. Introduction of a predacious red wood ant, *Formica lugubris* (Hymenoptera: Formidicae), from Italy to eastern Canada. *Can. Entomol.* 107: 1271–1274.

Finnegan, R. J. 1977. Predation de *Choristoneura fumiferana* (Lepidoptere: Tortricides) par *Formica lugubris* (Hymenoptere: Formicides). *Rev. Bimes. Rech.* 33: 1.

Finnegan, R. J. 1978. Predation by *Formica lugubris* on *Choristoneura fumifera* (Lepidoptera: Tortricidae). *Bi-Mon. Res. Notes* 34: 3–4.

Fittkau, E. J., and H. Klinge. 1973. On biomass and trophic structure of the central Amazonian rain forest ecosystem. *Biotropica* 5: 2–14.

Free, J. B. 1970. Insect pollination of crops. Academic Press, New York, 544 pp.

Gosswald, K. 1958. Neve Erfahrungen uber Einwirkung der rote Waldameise auf den Massenwechsel von Schadinsekten sowie einige methodische Verbesserungen bei ihrem praktischen Einsatz. *Proc. 10th Int. Cong. Entomol. (Montreal)* 4: 567–571.

Gosswald, K., and K. Horstmann. 1966. Untersuchungen uber den Einfluss der Kleinen Roten Waldameise (*Formica polyctena* Foerster) auf den Massenwechsel des Grunen Eichenwicklers (*Tortrix viridana* L.). *Waldhygiene* 6: 230–255.

Gotwald, W. H., Jr. 1974. Foraging behavior of *Anomma* driver ants in Ghana cocoa farms. *Bull. Inst. Fondam. Afr. Noire* (Ser. A) 36: 705–713.

Gotwald, W. H., Jr. 1982. Army ants. In "Social insects" (H. R. Hermann ed.), Vol. IV, pp. 157–254. Academic Press, New York.

Greenslade, P. J. M. 1971. Interspecific competition and frequency changes among ants in Solomon Islands coconut plantations. *J. Appl. Ecol.* 8: 323–349.

Groff, G. W., and C. W. Howard. 1924. The cultured citrus ant of South China. *Lingnaam Agric. Rev.* 2: 108–114.

Gudger, E. W. 1925. Stitching wounds with the mandibles of ants and beetles. A minor contribution to the history of surgery. *J. Amer. Med. Assoc.* 84: 1861–1864.

Haines, B. L. 1978. Element and energy flows through colonies of the leaf-cutting ant, *Atta colombica*, in Panama. *Biotropica* 10: 270–277.

Hickman, J. C. 1974. Pollination by ants: A low-energy system. *Science* 184: 1290–1292.

Jacot, A. P. 1936. Soil structure and soil biology. *Ecology* 17: 359–379.

Janzen, D. H. 1977. Why don't ants visit flowers. *Biotropica* 9: 252.

Jutsum, A. R., J. M. Cherrett, and M. Fisher. 1981. Interactions between the fauna of citrus trees in Trinidad and the ants *Atta cephalotes* and *Azteca* sp. *J. Appl. Ecol.* 18: 187–195.

Kajak, A., A. Breymeyer, J. Petal, and E. Olechowicz. 1972. The influence of ants on the meadow invertebrates. *Ekol. Pol.* 20: 163–171.

Kevan, D. K. McE. 1962. Soil animals. Philosophical Library. 237 pp.

Kincaid, T. 1963. The ant plant *Orthocarpus pusillus*. *Trans. Am. Micr. Soc.* 83: 101–105.

Kloft, W. 1960. Die Trophobiose zwischen Waldameisen und Pflanzanlausen mit Untersuchunger uber die Wechselwirkungen zwischen Pflanzenlausen und Pflanzengeweben. *Entomophaga* 5: 43–54.

Lange, R. 1958. Die deutschen arten der *Formica rufa*–Gruppe. *Zool. Anz.* 161: 238–243.

Linquist, A. W. 1942. Ants as predators of *Cochliomyia americana* C & P. *J. Econ. Entomol.* 35: 850–852.

Liu, G. 1939. Some extracts from the history of entomology in China. *Psyche* 46: 23–28.

Lumholtz, C. 1889. Among cannibals: An account of four years' travels in Australia and of camp life with the aborigines of Queensland. Charles Scribner's Sons, New York. 395 pp.

Lyford, W. H. 1963. Importance of ants to Brown Podzolic soil genesis in New England. *Harv. For. Pap.* 7: 1–18.

Mathur, R. N. 1954. Insects and other wild animals as human food. *Indian For.* 80: 427–432.

McCook, H. C. 1882. Ants as beneficial insecticides. *Proc. Acad. Nat. Sci. Phila.* 263–271 pp.

McCook, H. C. 1884. The rufous or thatching ant of Dakota and Colorado. *Proc. Acad. Nat. Sci. Phila.*, pp. 57–65.

McNeil, J. N., J. Delisle, and R. J. Finnegan. 1978. Seasonal predatory activity of the introduced red wood ant. *Formica lugubris* (Hymenoptera: Formicidae) at Valcartier, Quebec, in 1976. *Can. Entomol.* 110: 85–90.

Nielsen, M. G. 1982. The influence of *Lasius* species on some ecosystems in Denmark. *In* "The biology of social insects" (M. D. Breed, C. D. Michener, and H. E. Evans, eds.). p. 105. Proc. 9th Congress, IUSSI, Westview Press, Boulder, Colo.

Nye, P. H. 1955. Some soil-forming processes in the humid tropics IV. The action of the soil fauna. *J. Soil Sci.* 6: 73–83.

Ostmark, H. E. 1974. Economic insect pests of bananas. *Annu. Rev. Entomol.* 19: 161–176.

Otto, D. 1960. Statistiche untersuchunger uber die Bezeihungen zwischen Koniginnenzahl und Arbeiterinnengrosse bei den roten Waldameisen (engere *F. rufa* L. Gruppe). *Biol. Z.* 79: 719–739.

Pavan, M. 1959. Attivita per la lotta biologica con formiche del gruppo *Formica rufa* contro gli insetti dannosi alle foreste. *Min Agr. For. (Collana Verde)* 4: 1–80.

Percival, M. 1965. Floral biology. Pergamon Press, Oxford. 243 pp.

Petal, J. 1978. The role of ants in ecosystems. *In* "Production ecology of ants and termites" (M. V. Brian ed.), pp. 293–325. Cambridge University Press, Cambridge.

Petersen, B. 1977. Pollination by ants in the alpine tundra of Colorado. *Trans. Ill. State Acad. Sci.* 70: 349–355.

Phillips, J. S. 1934. The biology and distribution of ants in Hawaiian pineapple fields. *Bull. Pineapple Prod. Coop. Assoc. Exper. Stn.* 15. 57 pp.

Pijl, L. van der. 1955. Some remarks on myrmecophytes. *Phytomorphology* 5: 190–200.

Rico-Gray, V. 1980. Ants and tropical flowers. *Biotropica* 12: 223–224.

Ruddle, K. 1973. The human use of insects: Examples from the Yukpa. *Biotropica* 5: 94–101.

Schubart, H. O. R., and A. B. Anderson. 1978. Why don't ants visit flowers? A reply to D. H. Janzen. *Biotropica* 10: 310–311.

Shaler, N. S. 1891. The origin and nature of soil. In "12th Annual Report of the United States Geological Survey to the Secretary of the Interior," pp. 213–345. 1890–1891. Part I–Geology. Superintendent of Documents, Washington, D.C.

Strickland, A. H. 1951. The entomology of swollen shoot of cacao. I. The insect species involved, with notes on their biology. Bull. Entomol. Res. 41: 725–748.

Sudd, J. H. 1967. An introduction to the behaviour of ants. St. Martin's Press, New York. 200 pp.

Thorp, J. 1949. Effects of certain animals that live in soils. Sci. Monthly 68: 180–191.

van den Bosch, R., P. S. Messenger, and A. P. Gutierrez. 1982. An introduction to biological control. Plenum Press, New York. 247 pp.

Vella, F., and W. S. Magalhaes. 1971. Estudos sobre a participacao da formiga caccarema (Azteca chartifex spiriti Forel) na polinizacao do cacaueiro na Bahia. Rev. Theobroma 1: 29–42.

Way, M. J. 1953. The relationship between certain ant species with particular reference to biological control of the coreid Theraptus sp. Bull. Entomol. Res. 44: 669–691.

Weber, N. A. 1946. Two common ponerine ants of possible economic significance, Ectatomma tuberculatum (Olivier) and E. ruidum Roger Proc. Entomol. Soc. Wash. 48: 1–16.

Weber, N. 1982. Fungus ants. In "Social insects" (H. R. Hermann ed.), pp. 255–303. Academic Press, New York.

Wellman, F. C. 1908. Notes on some Angolan insects of economic or pathologic importance. Entomol. News 19: 224–230.

Wheeler, W. M. 1904. On the pupation of ants and the feasibility of establishing the Guatemalan kelep or cotton-weevil ant in the United States. Science 20: 437–440.

Wheeler, W. M. 1910. Ants: Their structure, development and behavior. Columbia University Press, New York. 663 pp.

Wilson, E. O. 1971. The insect societies. Belknap Press of Harvard University Press, Cambridge, Massachusetts. 548 pp.

Wyatt, R. 1981a. Ant-pollination of the granite outcrop endemic Diamorpha smallii (Crassulaceae). Amer. J. Bot. 68: 1212–1217.

Wyatt, R. 1981b. Patterns of ant-mediated pollen dispersal in Diamorpha smalli (Crassulaceae). Syst. Bot. 6: 1–7.

Chapter 12

UTILIZING ECONOMICALLY BENEFICIAL ANTS
Jonathan D. Majer

INTRODUCTION

The beneficial effects of ants in agriculture and forestry have already been reviewed in Chapter 11. In the following pages I discuss the methods by which ants have been utilized in crops to limit maladies and highlight the strengths and weaknesses of the various approaches. Arising from this review a checklist of points to consider when promoting beneficial is presented so that future ant manipulation attempts can draw on these experiences. This account distinguishes between species that are physically "introduced" into an area from outside and resident species that are "encouraged" to spread their range or utilize an enemy species more effectively. The term "promoted" collectively refers to both methods.

SELECTION AND DEVELOPMENT OF AN ANT AS A PEST CONTROL AGENT

The attributes possessed by ants that render them potentially useful biological control agents have been summarized by Risch and Carroll (1982a). These are: (1) Ants are extremely diverse and abundant in most tropical and some temperate ecosystems and most are predacious. (2) Ants are extremely responsive to spatial variations in the density of their food. (3) Ants can persist as effective predators in spite of temporal fluctuations in food supply. (4) Predator satiation is not likely to limit the effectiveness of ants. (5) The negative effect of ants on enemy species may exceed that resulting from direct predation alone. (6) The foraging patterns of ants can be manipulated and managed in order to maximize their contact with pests.

Surveys in Ghanaian cocoa, to take one example, have revealed almost 300 species of ants (Leston 1973). Of these, only two are of significant economic benefit to the crop (Leston 1973, Majer 1974). Clearly, the beneficial attributes mentioned are not common to all ants and one problem is to screen out the potentially beneficial species.

Following a search of the Canadian ant fauna for forest pest limiting agents, Finnegan (1971) listed the criteria used for evaluating predacious ant species. These were: (1) *Size.* Large size of individual ants is usually considered a desirable quality, particularly when the pest is large or well protected. If the ant has large colonies the size of individuals is not necessarily important. Leston's (1973) review of ants that limit tropical tree crop pests lists a number of species with small workers, for example, *Wasmannia auropunctata*, whose workers are 1.4 mm long. (2) *Food requirements and nest populations.* Species with large colonies—and hence, food requirements—are considered to be most desirable. (3) *Colonial nests.* Species with multidomous colonies or with the ability to produce numerous additional colonies over adjacent areas are particularly useful. By this attribute, ants such as *Anoplolepis longipes* are able to occupy considerable areas at high densities. (4) *Queens.* Species with more than one queen or with a queen replacement system are considered to have longer-lived colonies. (5) *Ant-Homoptera relationships.* Ants which tend and encourage undesirable Homoptera often bring about results that offset any benefits. This has thwarted the use of potentially beneficial *Formica* spp. in European conifer forests (Adlung 1966).

To Finnegan's criteria I add the following: (6) *Species habitat range.* Species with broad habitat ranges are potentially more amenable to being promoted in areas where they are not already present. Important characteristics include the ranges of temperature, moisture, and insolation that can be tolerated and also the ability to tolerate seasonal fluctuations in these factors and in food resources. (7) *Dominance hierarchy.* Species that are unlikely to be outcompeted by other ants in the crop are desirable.

Room (1973) has drawn up a formalized scheme for selecting and developing ants as pest control agents. His scheme is shown in Figure 12.1. Stage 1 involves definition of the pest or the pest complex. The biology of the pest is evaluated and their interaction with the crop is elucidated. Stage 2 involves identification of potentially useful ant species. Stage 3 is concerned with demonstrating whether or not any ant species has controlling effects on the pest complex. This may be evaluated by a sampling program designed to show how ant and pest species are distributed with respect to each other. Researchers are warned of spurious ant control

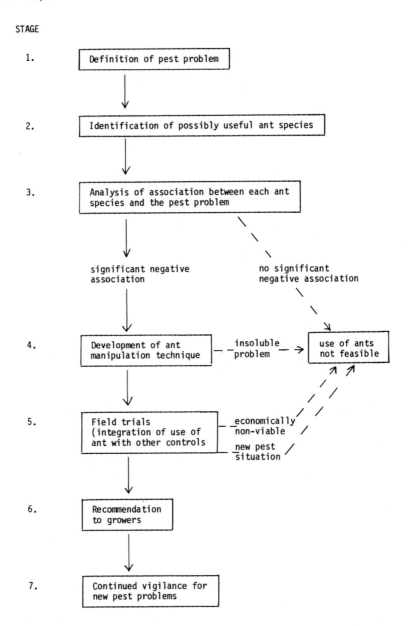

FIGURE 12.1
Suggested scheme for the selection and development of an ant as a pest control agent. (Redrawn from Room 1973)

effects resulting from ant and pest having different habitat preferences rather than the ant having any direct negative influence on the pest.

If a beneficial ant is revealed the scheme enters stage 4. Here techniques are investigated to manipulate the ant distribution in order to favor its spread. This is the most difficult stage to carry out and the remainder of this chapter explores this problem. Stage 5 is reached when field trials are performed to evaluate the practicality of ant manipulation and the efficacy of the control. The ant-based biological control method may now become incorporated into a wider integrated pest control scheme. Finally, recommendations are made to growers on how to apply the control scheme and vigilance for new pest problems that may require modifications of the scheme.

THE ANT MOSAIC AND ITS MAINTENANCE

The complexity of the interactions that exist in ant communities are usually not fully understood. This has necessitated an empirical approach to ant manipulation. The mechanisms that influence and maintain the spatial pattern of ant communities will be outlined so that workers may take them into account when attempting to manipulate ant distribution.

Of the 300 species of ants in Ghanaian cocoa, about 14 sometimes numerically predominate over other ant species (Leston 1973). Majer (1972) divided the tropical ant fauna into four status groups; dominants — which predominate numerically to the exclusion of all other dominants; codominants — dominants, which, for various reasons, are able to coexist; subdominants — which may reach dominant status if a dominant ant is removed; and nondominants — species with generally small colony size, which occur within or between the territories of dominant ants. Under certain circumstances, the dominants exhibit a dominance hierarchy based on their ability to replace each other (Greenslade 1971). This hierarchy exhibits flexibility under changing conditions. Evidence suggests that this ant status system is common to other parts of the tropics (Room 1975, Leston 1978, Taylor and Adedoyin 1978).

From the definition of dominance, ants are distributed in a three-dimensional mosaic. This has been verified in West Africa by various surveys (see references in Taylor 1977), in tropical Asia (Greenslade 1971, Room 1975) and in the New World neotropics (Leston 1978). Wilson's (1958) work in New Guinea suggests that there is a geographic patchiness in distribution superimposed over this local distribution pattern.

In order to understand how ants might be manipulated some general points on the maintenance of the mosaic need to be described. Majer's (1976a) findings are drawn on here although they are not at variance with the findings from elsewhere in the tropics.

The heterogeneity of the environment is one factor contributing to patchy ant distribution. For instance, different species of ants occur under particular cocoa canopy density regimes. Ants such as *Tetramorium aculeatum*, which are associated with dense cocoa, exhibit niche flexibility by moving into thinner canopy when adjacent dominants are removed, suggesting that the interspecific ant mosaic is maintained by a combination of competition and habitat requirements. Interspecific competition may be for food, nesting, and foraging sites or it may take the form of aggressive competition between adjacent colonies or mature colonies and founding queens.

Aggressive behavior is particularly intense between adjacent blocks of the mosaic. *Tetramorium aculeatum* is able to reduce competition with certain *Crematogaster* spp. within its territory by spacing out its foraging time while *Crematogaster castanea* may coexist with *Oecophylla longinoda* by adopting a similar colony odor.

New colony establishment is rare in mature cocoa since changes in the ant mosaic are usually compensated for by lateral spread of colonies. The role of dispersing queens is probably more important in developing cocoa farms. *Oecophylla longinoda* may aid species segregation by having queens which select their habitats when dispersing.

Climate and weather influence the structure of the mosaic by directly influencing the features of the habitat or the availability of certain types of food or by physically weakening colonies. As each dominant has a suite of nondominants associated with it (Room 1975), these are also distributed in a mosiac fashion.

EXPERIENCES WITH ANT MANIPULATION

Practical attempts at ant manipulation have included the reduction of pest-ant populations (Haines and Haines 1979), the reduction of inefficient predator populations in order to encourage more effective species (Stapley 1971), and the direct encouragement of beneficial species. All approaches can be considered as part of the same problem since the encouragement of a new species will lead to displacement of residents and species removal will lead to lateral spread of adjacent dominants; one approach is simply the compliment of the other.

Introduction of Ants from Other Regions

The introduction of ants into new geographic areas is generally considered to be risky (Room 1973) although Finnegan (1975) successfully introduced the Italian wood ant, *Formica lugubris* into Canadian forests.

Establishment was generally successful although competition with native ants and predation by birds interfered with the program's success. These problems were ameliorated by chemically treating native ants in the area and by excluding predators from nests with wire mesh. Other examples have involved accidental introductions that turned out to be beneficial. *Wasmannia auropunctata*, accidentally introduced into Cameroon from the neotropics, has been used to limit cocoa pests. Farmers successfully encourage its spread by distributing artificial nests formed in bundles of raffia leaves (Bruneau de Mire 1969). The cosmopolitan tramp species, *Anoplolepis longipes*, has been artificially introduced in some New Guinea cocoa plantations and proved to be of benefit (Room 1973), but the same species has become a pest in the Seychelles (Lewis et al. 1976).

Introduction of Ants from Within the Region

This category describes the transfer of ants from one mosaic block to another area which is held by other dominants. In Indonesia, nests of *Dolichoderus bituberculatus* were artificially established in bamboo or leaf bundles and then hung in cocoa trees to protect them against pest attack (Meer-Mohr 1927). Maintenance of the nests depended on their being placed in shady conditions and the provision of a suitable honeydew supply; without this the ant was outcompeted by adjacent dominants. The transfer of *O. smaragdina* nests to southeast Asian citrus was only temporarily successful due to competition from resident species and the absence of queens in transferred nests. Brown's (1959) work in coconuts indicated that even if queens were included, nest transfer only met with temporary success.

Selective Chemical Treatment of Ants

Mosaic blocks of unfavored species may be sprayed or poisoned to encourage adjacent beneficial residents. Phillip's (1956) trials inside Solomon Islands coconut plantations showed that *Pheidole megacephala* and *Iridomyrmex cordatus* populations were reduced by organochloride application to the lower tree while the beneficial *O. smaragdina* and *A. longipes* were less affected. The beneficial species were further encouraged by spraying the areas nested by the other two species. Similar observations were made in Tanzanian coconut plantations by Vanderplank (1960), where repeated DDT application resulted in a drastic reduction of the predominantly ground-nesting *A. longipes* and the spread of tree-nesting *O. longinoda* and *Pheidole punctulata*. Many trees remained without dominant ants following spraying, indicating an inadequate capacity of ants to rapidly fill lacunae.

The examples from coconut plantations generally involved the spread of arboreal nesters, which were untouched by the pesticides, and the decline of soil or tree-base nesters. A similar trend was observed in Ghanaian cocoa farms when blanket spraying caused the spread of tall, shade-tree nesting *Crematogaster* sp. into the lower cocoa territory originally held by *O. longinoda* and *T. aculeutum*. Selective spraying of all trees lacking *O. longinoda* led to a localized replacement of *Crematogaster* sp. by more beneficial ants (Majer 1978).

Selective Mechanical Removal of Ants

Majer (1976a) ran a field trial in which nests of the dominant species were mechanically removed from separate, replicated plots. Ant elimination was invariably followed by the spread of adjacent residents over distances as great as 50 m (e.g., *Crematogaster striatula* in Fig. 12.2). Species varied in their capacity to spread and, in one case, large gaps remained without dominants for at least 16 months. This was attributed to the unsuitability of the habitat to adjacent residents. Incidence of incipient females increased in proportion to the area from which dominant ants were removed (Fig. 12.3) although very few colonies were established in the 16 months following ant elimination. Therefore, the greatest changes in ant distribution were accounted for by lateral spread of existing colonies.

Alteration of Ant Segregation

The early Solomon Islands coconut work suggested that *O. smaragdina* was encouraged by a dense understory, since it was then able to forage independently of ground-living antagonists (Phillips 1956). Removal of understory or artificial introduction of vertical segregation by providing bridges between trees did not produce the predicted outcomes so other factors may be involved. Majer (1972) observed a natural experiment in cocoa in which a shade tree supporting a *Crematogaster depressa* colony fell into *O. longinoda* territory in cocoa. A combat immediately resulted and *C. depressa* was eliminated.

Habitat Alteration

The most extensively studied example is in Solomon Island coconuts following World War II. During the war dense undergrowth built up that was subsequently removed. Postwar vegetational changes were probably responsible for the series of replacements; *P. megacephala* > *O. smaragdina* > *A. longipes* > *O. smaragdina* and other species. Greenslade (1971) summarized the reason for this succession in terms of the complex changes

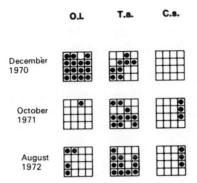

FIGURE 12.2

Changes in the distribution of *Oecophylla longinoda* (O. l.), *Tetramorium aculeatum* (T. a.), and *Crematogaster striatula* (C. s.) in a 40 × 40 m plot following selective mechanical removal of *O. longinoda* in late December 1970. The December 1970 maps represent the preremoval situation. (Redrawn from Majer 1976a)

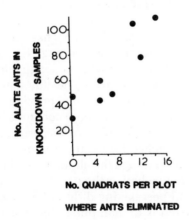

FIGURE 12.3

Relationship between area of dominant ants removed per 40 × 40 m plot (maximum of sixteen 10 × 10 m quadrats) and the incidence of alate ants in subsequent pyrethrum knockdown samples of the canopy fauna. (Redrawn from Majer 1976a)

in vegetation, its floristic composition and structural complexity, the climatic regime, and nest production. The work on Western Australian bauxite mine rehabilitation (Majer et al. 1984) provides an interesting comparison. Here the diversity and species composition of the colonizing ant fauna was strongly influenced by plant species richness and diversity, the vegetational cover in particular strata, the thickness and patchiness of litter, and the availability of dead wood. Rehabilitated areas acquired quite different ant species depending on the type of rehabilitation that was performed.

Crop Interplanting

Way (1954) noted that where coconut was interplanted with clover or citrus the queen of O. longinoda was invariably situated within the latter species. This was possibly associated with the favored Homoptera population on these plants and suggested a new option for encouraging O. longinoda. Stapley (1971) extended this observation in the Solomon Islands by interplanting with sowersop. It was possible to establish O. smaragdina on the sowersop provided that sufficient and appropriate Homoptera were present. Leston (1973) and Majer (1974) have suggested using coconuts rich in scale insects in cocoa plantations to encourage the establishment of O. longinoda.

DOES ANT MANIPULATION REALLY INFLUENCE PESTS?

In most instances the beneficial role of ants has been elucidated by direct behavioral observations, from observed negative correlations between the distribution of ants and pests (e.g., O. longinoda and certain mirids (Leston 1973)) or by noting the generally healthy state of farms where particular ants are abundant. An example is provided by Way (1963) who noted that if 70% of Zanzibar (Tanzania) coconuts in a farm were occupied by O. longinoda then the whole plot was protected from the coreids.

None of these approaches demonstrates unambiguously whether artificial spread of ants will change the abundance of pests and the second approach does not show whether the negative correlations were causal or result from ant and pest having different habitat requirements. The only infallible approach is to manipulate the distribution of beneficial ants and observe whether the abundance or distribution of pests also changes.

Stapley's (1972) extension of Phillip's (1956) work in Solomon Island coconuts provides an example. The ant P. megacephala was controlled by pesticide application to tree bases in 1969. The resident O. smarag-

dina colonies were subsequently able to spread and eventually occupy 70% of the trees. Figure 12.4 shows the changes in ant tree occupancy and coconut yield. The improved yield, which paralleled the increase in *O. smaragdina* on trees, resulted because this ant repels the coreid *Amblypelta cocophaga* and the pentatomid *Axigastus campbelli* and suggests that ant manipulation can actually have a beneficial effect on pest levels and hence yield.

Risch and Carroll (1982b), working in eastern Mexico, took the opposite approach of removing the suspected beneficial ant and observing the changes in insect abundance. They selectively removed the dominant *Solenopsis geminata* using mirex baits from half the set of paired plots of corn and squash. Ants were retained in the remaining plots. Five weeks later there were nine times as many arthropods and four times as many morphospecies in the ant removal plots as controls. These workers used the results to indicate that this ant played a substantial role as a biological control agent.

Elucidation of the beneficial role of ants is more difficult in cases where a pest complex is present in the crop. For instance, Majer's (1976a, b, c) work in cocoa revealed just over 1,000 species of insects and other

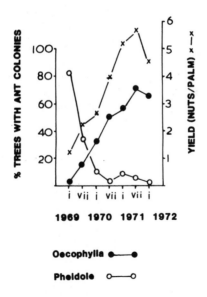

Oecophylla ●——●

Pheidole ○——○

FIGURE 12.4
The relationship between *Pheidole magacephala*, *Oecophylla smaragdina* and coconut yield in the Solomon Islands. (Redrawn from Leston 1973, using data from Stapley 1972)

invertebrates (Majer 1974), several of which directly or indirectly contributed to the degradation of cocoa. In order to investigate the influence of dominant ants on the fauna, a field trial was designed in which four species of dominant ants were selectively, mechanically removed from different subplots of cocoa with two control subplots left untouched for comparison. Monthly pyrethrum knockdown samples were taken from the control and treatment subplots in order to monitor the changes (Majer 1976c).

As mentioned, removal of a dominant ant allows the spread of adjacent dominant ants into the lacuna created. Thus, the complex fate of each subplot made the data unsuitable for interpretation by analysis of variance. A multivariate technique known as principal components analysis ordination was therefore utilized to analyze the faunal similarities of the pyrethrum knockdown samples. By plotting environmental factors on the corresponding samples displayed on the ordination diagrams it was possible to determine the relative importance of the factors that influenced the composition of the cocoa fauna. Seasonal factors accounted for axis 1 and the samples were separated according to which species of dominant ant was present on axes 2 and 3 of the ordination (Fig. 12.5).

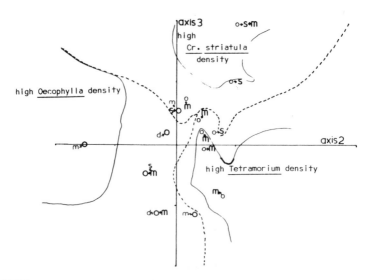

FIGURE 12.5
Dominant ants in the 144 pyrethrum knockdown samples of cocoa canopy fauna plotted on axes 2 vs. 3 of the principal components analysis sample ordinations. The separation of samples occupied by different dominant ants is depicted by the dashed lines. (Redrawn from Majer 1976c)

Although cocoa canopy density and the proximity of cocoa to forest were also involved, this work demonstrated the importance of ants in determining the makeup of the cocoa fauna. Samples dominated by exceptionally high ant densities occurred toward the extremes of the ant groupings since the fauna became even more characteristic of the dominant species when ant density was high (Fig. 12.5). In areas where ant A had been replaced by ant B as a result of ant elimination, the ordination usually placed the sample in the B group indicating that ant replacement had been accompanied by a partial change in the associated fauna. The implications of this are clear if one considers ant manipulation as a means of controlling the distribution of cocoa pests and diseases.

An additional outcome of this experiment was that an area of great damage caused by the mirid *Distanliella theobroma* occurred when *O. longinoda* had previously been eliminated, suggesting that predation by the ant was the real reason for the previously observed negative correlation between these species (Williams 1954).

The examples of research given in this section indicate that ant manipulation can have a clearly demonstrable and beneficial effect on crop pests or pest complexes. In addition, the examples give an indication of the experimental approaches that may be used to perform Stage 3 of Room's (1973) ant-mediated pest control scheme (Fig. 12.1).

PRACTICAL AND THEORETICAL CONSIDERATIONS

This section takes into account the fact that promotion of a beneficial ant should also be accompanied by an artificially induced, or natural decline in nonbeneficial ant species. Thus some of the following comments are concerned with the reduction in nonbeneficial species.

Ecological Constraints to Ant Manipulation

1. Is the mosaic as densely packed in other parts of the world as it is in Ghanaian cocoa or, are undominated (unprotected) lacunae more common elsewhere? The densely packed nature of the cocoa ant mosaic implies that the entire crop may potentially be protected by ants. The ratio of dominant ant blocks to lacunae needs to be considered where ant control is envisaged. In temperate situations there are certainly instances where dominant, or numerically dense, ants are absent from large tracts of land.

2. Can a crop or forest support large monospecific expanses of one dominant or is the small-block mosaic necessary for adequate food provision? Room (1971) has produced evidence that suggests that it may not be possible to maintain large continuous areas of colonies of the same species. This may be

because a species removes its prey spectrum so efficiently that it relies on immigration of prey from nearby territories of different ant species that do not utilize this food source. The supplementation of food in beneficial ant territories or the promotion of two or more beneficial species may overcome this problem.

3. When a lacuna is created in order to encourage a resident, will the desired adjacent species occupy the area? In a mosaic of few dominants, such as coconuts (Greenslade 1971) the outcome of non-beneficial ant removal or decline may be reasonably predicted. However, if a larger number of dominants are present and the mosaic blocks are small, such as in Ghanaian cocoa, the probability of a particular adjacent ant colony filling an artificially created lacuna is often quite low (see Fig. 12.2 and Majer 1976c).

4. How rapidly can adjacent dominants or introduced species fill artificially created lacunae? If the lapse between the decline in the nonbeneficial species and the colonization by beneficial species is not sufficiently rapid, the crop may be vulnerable to pest attack. At such times the ant lacunae in the crop may need to be protected by a control method that does not jeopardize beneficial ant colonization.

5. Is a particular beneficial species amenable to artificial introduction? Species with easily transportable colonies or that are able to be established in artificial nests (e.g., W. auropunctata and A. longipes) are particularly useful. Those with single-queen colonies or queens that are difficult to collect, and those without queen replacement systems are less suitable for introducing into crops.

6. Is the ant that is to be introduced limited by its geographical range? The geographical range of a dominant may restrict the potential for its promotion over the entire range of a crop. For instance, Taylor and Adedoyin (1978) found that although O. longinoda was widely distributed in Nigerian cocoa, T. aculeatum was found at higher levels in the wetter southern and eastern localities. The potential of the latter species as a biological control agent may therefore be lower in the drier cocoa growing areas.

7. Does the size of an ant colony limit its capacity for range extension? Some ants are only capable of producing colonies of small size. To promote such species over large expanses would therefore require the formation of an intraspecific mosaic of abutting colonies. The ecological constraints to ant manipulations and the possibility of intense intraspecific aggression between adjacent colonies may prevent such an intraspecific mosaic from being formed.

8. Does a promoted ant have sufficient niche flexibility to colonize the new area? The shade regime, nest site availability, resident Homoptera species or other factors may be unsuitable for the promoted ant. For instance, Majer (1976a) found areas, where T. aculeatum was mechanically eliminated that were not always colonized by new species; he attributed this to the dense shade being unsuitable for the adjacent dominant ants.

9. How does intrinsic growth and decline of ant colonies fit in with the desired ant distribution and abundance pattern? Greenslade (1971) has noted that territory occupied by O. smaragdina colonies expands and declines over approximately eight-year periods. This may be related to the longevity of the queen

and the implication is that short colony longevity may confound ant manipulation attempts.

Sustainability of Ant Manipulation

1. Is it necessary to supplement food in areas where a new species has been promoted, especially if they have recently been sprayed? An area colonized by a recently promoted ant may not have suitable Homoptera present or, if pesticides have been used to eliminate non-beneficial ants, prey may be reduced. Homoptera and proteinaceous food may therefore require introduction into the area to be colonized.
2. What other aftercare needs to be performed when the mosaic has been manipulated? An appropriate weeding regime may be maintained in order to provide vegetation bridges between trees to reduce contact between arboreal and ground ants, or perhaps to provide a particular soil insolation level. In addition, a farm hygiene program may need to be implemented to reduce dead wood or litter nesting sites of nonbeneficial species.
3. A mosaic of beneficial ants may represent a dysclimax—how can it be maintained at this stage? The ant mosaic in a crop at any one time may represent a stage in a succession (Greenslade 1971; Majer 1972). The desired mosaic composition may represent a transient stage in this succession and it may be necessary to manage the crop, and the ants, in order to prolong this serial stage.
4. Will the promotion of a species fail, due to its position in the dominance hierarchy in relation to that of the adjacent dominants? The order of competitive ability influences the ant's ability to colonize areas and also to replace other species. This needs to be considered when promoting a particular species. Furthermore, a beneficial species that is low down in the competitive ability rankings will tend to be in danger of being replaced by a higher ranking species.
5. Will incipient females preselect areas known to be suitable nest sites and hence reverse attempts to change ant distribution? In the selective mechanical ant removal experiment described by Majer (1976a, c) it was found that considerably more O. longinoda alates were retrieved from areas where O. longinoda had been eliminated than where other species had been removed. This suggests that species are able to select favorable nesting sites and that ant removal may be confounded by alates establishing new colonies in the areas where they have been eliminated.
6. Is the climate suitable for maintaining a promoted species throughout the year? Contrary to popular belief, the physical and biological environment of some tropical ecosystems is subject to marked seasonal fluctuations (e.g., Gibbs and Leston 1970). Thus a species introduced at one time of the year may not be able to survive the adverse climate or the lack of suitable food in another season. This point should also be considered when deciding the optimal time to artificially introduce nests of beneficial species.
7. Will a newly colonized ant bring about environmental changes detrimental to its own survival? Room (1973) has already raised this possibility; a badly damaged crop may be extremely hospitable to a sun-loving beneficial species

but when the canopy recovers it may be shaded out. Changes in food supply with crop recovery could also influence the fate of the promoted ant.

Side Effects of Ant Manipulation

1. Where a pest complex involved, will the promotion of a beneficial ant bring desirable changes in all members of the complex? The problem here is self evident and the effect of ant elimination on the pest complex would need to be assessed in crops where such programs are envisaged.
2. Will the promotion of a seemingly beneficial dominant encourage undesirable Homoptera? The majority of tropical dominant ants are associated with Homoptera of some type. Some, such as *O. longinoda* in cocoa, are associated with Homoptera of relatively low economic importance. Some *Crematogaster* spp. tend mealybugs, which are vectors of cocoa swollen shoot virus disease (SSV) and other homoptera vectors of cocoa blackpod disease (*Phythophthora palmivora*). Clearly, an ant species can only be considered as beneficial if it is not associated with undesirable Homoptera.
3. Will promotion of a beneficial ant be accompanied by that of a less desirable co-dominant? A number of dominants, in cocoa at least, are consistently associated with particular co-dominant ants. Taylor and Adedoyin (1978) noted that although one ant of the pair may be beneficial, the other may have a detrimental effect. They pointed out the manipulation of the cocoa habitat in Nigeria to favor *O. longinoda* and *T. aculeatum* might lead to an increase in *P. megacephala*, which tends the mealybug vectors of SSV.

The Role of Pesticides in Ant Manipulation

1. How permanent is the effect of a pesticide that is used to remove undesirable species? The tenacity of some ant species may require that they be exposed to pesticides for an extended period of time in order to be eliminated. The benefits of residual versus short-term pesticides therefore need to be considered if pesticides are involved in the ant manipulation program.
2. Is time of spraying nonbeneficial ants compatible with alate flight times of beneficial species? Light-trap data from Ghanaian cocoa-growing areas (J. D. Majer, unpublished data) has shown that most ants have specific periods of alate release. If these periods are known, then removal of nonbeneficial ants by spraying should be carried out prior to, or during the period of beneficial ant alate release. This would maximize the chances of alates of beneficial species establishing colonies in the lacunae created.

Other Options for Ant Manipulation

1. Is it possible to capitalize on climatically induced changes in ant abundance to promote or reduce a particular species? In Ghananian cocoa the territory of leaf-nesting ant species contracts during the dry season leaf fall and that of tree trunk nesters expands (Majer 1976a). Although it has never been evalu-

ated, it seems likely that one could capitalize on these seasonal range fluctuations by eliminating species at a time when their range is declining.

2. If the habit alteration approach is used to encourage species, how predictable is the change in dominants present? A number of workers have advocated habitat alteration by methods such as crop interplanting or farm sanitation as a means of promoting certain species of ants (Leston 1973). Although this method undoubtedly influences the ants present, the predictability of change in ant distribution is not necessarily high. Workers attempting this approach would be advised to perform adequate trials and also to quantify the ants present in differing habitat composition in order to be more certain of the outcome.

3. Is the promotion of particular ants more likely to be successful if new plantations are cultivated in order to encourage initial colonization of beneficial species? Majer (1974) devised an integrated pest control scheme for cocoa that commenced at the farm establishment phase. The scheme involved a particular land clearing and interplanting and also a crop buffer zone adjacent to the forest. This scheme has never been evaluated although the results of the bauxite mine ant succession work (Majer et al. 1984) indicates that the way in which an ecosystem is established can radically affect the type of ant community which develops. In the long term there is probably tremendous scope for tailoring the ant mosaic using particular farm establishment techniques.

Compatibility of Ant Manipulation Schemes
With Farming Procedures

1. When interplanting is employed, how economically compatible is the intercrop? A plant species planted with the major crop must either provide an economic yield or bring about sufficient improvement in the main crop's yield to compensate for the land and the possible competitive effects of its presence.

2. If the crop habitat is modified to encourage an ant, is the changed habitat compatible with the farming technique? Any cultural regime must be compatible with the agronomic and economic requirements of the crop and would need to be acceptable from the sociological point of view. For instance, it would be pointless to encourage a scheme that was too complex to implement or that involved an ant that was unduly injurious to farmers.

CONCLUDING REMARKS

This review has indicated the range of methods available to manipulate the ant fauna and some of the problems that must be considered. Many of the considerations presented above are posed as questions rather than solutions. This results from the lack of published case history data and highlights the need for more information on control attempts. Although research into the use of ants in biological control has been in virtual limbo over the past decade there is now an upsurge in interest in areas such as the New World tropics and in annual crops of the neotropics. It is hoped that the experiences reported in this chapter will contribute to the development of ant based control schemes where these seem profitable.

Finally, I would like to dedicate this chapter to the late Dennis Leston who did so much to promote the awareness of the role of ants in the tropics and elsewhere.

REFERENCES

Adlung, K. G. 1966. A critical evaluation of the European research on the use of red wood ants (Formica rufa group) for the protection of forests against harmful insects. Z. Angew. Entomol. 57: 167–189.

Brown, E. S. 1959. Immature nutfall in coconuts in the Solomon Islands. I. Distribution of nutfall in relation to that of Amblypelta and of certain species of ants. Bull. Entomol. Res. 50: 97–133.

Bruneau De Mire, P. 1969. Une fourmi utilisee au Cameroun dans la lutte contra des mirides du cacaoyer Wasmannia auropunctata Roger. Cafe Cacao 13: 209–212.

Finnegan, R. J., 1971. An appraisal of indigenous ants as limiting agents of forest pests in Quebec. Can. Entomol. 103: 1489–1493.

Finnegan, R. J. 1975. Introduction of a predacious red wood ant, Formica lugubris, from Italy to eastern Canada. Can. Entomol. 107: 1271–1274.

Gibbs, D. G. and D. Leston. 1970. Insect phenology in a forest cocoa farm locality in West Africa. J. Appl. Ecol. 7: 519–548.

Greenslade, P. J. M. 1971. Interspecific competition and frequency changes among ants in Solomon Island coconut plantations. J. Appl. Ecol. 8: 323–352.

Haines, I. H. and J. B. Haines. 1979. Toxic bait for the control of Anoplolepis longipes (Jerdon) in the Seychelles. III. Selection of toxicants. Bull. Entomol. Res. 9: 203–211.

Leston, D. 1973. The ant mosaic—Tropical tree crops and the limiting of pests and diseases. Pest. Art. News Summary 19: 311–341.

Leston, D. 1978. A neotropical ant mosaic. Ann. Entomol. Soc. Amer. 71: 649–653.

Lewis, T., J. M. Cherrett, I. Haines, J. B. Haines, and P. L. Mathias. 1976. The crazy ant Anoplolepis longipes (Jerd.) in Seychelles, and its chemical control. Bull. Entomol. Res. 66: 97–111.

Majer, J. D. 1972. The ant mosaic in Ghana cocoa farms. Bull. Entomol. Res. 62: 151–160.

Majer, J. D. 1974. The use of ants in an integrated control scheme for cocoa. Proc. Fourth Conf. West African Cocoa Entomol., pp. 181–190.

Majer, J. D. 1976a. The maintenance of the ant mosaic in Ghana cocoa farms. J. Appl. Ecol. 13: 123–144.

Majer, J. D. 1976b. The ant mosaic in Ghana cocoa farms: Further structural considerations. J. Appl. Ecol. 13: 145–155.

Majer, J. D. 1976c. The influence of ants and ant manipulation on the cocoa farm fauna. *J. Appl. Ecol.* 13: 157–175.

Majer, J. D. 1978. The influence of blanket and selective spraying on ant distribution in a West African cocoa farm. *Ref. Theobroma (Brasil)* 8: 87–93.

Majer, J. D., J. E. Day, E. D. Kabay, and W. S. Perriman. 1984. Recolonisation by ants in bauxite mines rehabilitated by a number of different methods. *J. Appl. Ecol.* 21: 355–376.

Meer-Mohr, J. C. 1927. Au sujet du role de certaines fourmis dans les plantations coloniales. *Bull. Agri. Congo Belge* 31: 97–106.

Phillips, J. S. 1956. Immature nutfall of coconuts in the British Solomon Islands Protectorate. *Bull. Entomol. Res.* 47: 575–595.

Risch, S. J., and C. R. Carroll. 1982a. The ecological role of ants in two Mexican agroecosystems. *Oecologia* 55: 114–119.

Risch, S. J. and C. R. Carroll. 1982b. Effect of a keystone predacious ant, *Solenopsis geminata*, in a nonequilibrium community. *Ecology* 63: 1979–1983.

Room, P. M. 1971. The relative distribution of ant species in Ghana's cocoa farms. *J. Anim. Ecol.* 40: 735–751.

Room, P. M. 1973. Control by ants of pest situations in tropical tree crops: A strategy for research and development. *Papua New Guin. Agri. J.* 24: 98–103.

Room, P. M. 1975. Relative distributions of ant species in cocoa plantations in Papua New Guinea. *J. Appl. Ecol.* 12: 47–61.

Stapley, J. H. 1971. Field studies on the ant complex in relation to premature nutfall of coconuts in the Solomon Islands. pp. 345–354. *Proc. Conf. Cocoa Coconuts Malaysia.*

Stapley, J. H. 1972. Premature nutfall in coconuts: Progress report to February 1972. Honiara, Dept. Agri. (mimeographed).

Taylor, B. 1977. The ant mosaic on cocoa and other tree crops in western Nigeria. *Ecol. Entomol.* 2: 245–255.

Taylor, B. and S. F. Adedoyin. 1978. The abundance and interspecific relations of common ant species on cocoa farms in western Nigeria. *Bull. Entomol. Res.* 68: 105–121.

Vanderplank, F. L. 1960. The bionomics and ecology the red tree ant, *Oecophylla* sp., and its relationship to the coconut bug *Pseudotheraptus wayi* Brown (Coreidae). *J. Anim. Ecol.* 29: 15–33.

Way, M. J. 1954. Studies on the life history and ecology of the ant *Oecophylla longinoda* Latreille. *Bull. Entomol. Res.* 45: 93–112.

Way, M. J. 1963. Mutualism between ants and honeydew-producing Homoptera. *Annu. Rev. Entomol.* 8: 307–344.

Williams, G. 1954. Field observations on the cacao mirids, *Sahlbergella singularis* Hagl. and *Distantiella theobroma* (Dist.), in the Gold Coast. III. Population fluctuations. *Bull. Entomol. Res.* 45: 723–744.

Wilson, E. O. 1958. Patchy distribution of ant species in New Guinea rain forests. *Psyche* 65: 26–38.

Chapter 13

Polistes WASPS: BIOLOGY AND IMPACT ON MAN
James E. Gillaspy

INTRODUCTION

Paper wasps (genus *Polistes*) of the social wasp family Vespidae, comprise about 200 species (Snelling 1981) distributed essentially worldwide in tropical and temperate regions. *Polistes* are prevalent, persistent, and unappreciated synanthropes or associates of man, attracted by the caterpillars flourishing on his cultivated plants, nesting on structures he erects, and making paper nests of fibers from weathered boards he provides. Remarkable biotic potential is evidenced by their common occurrence around human habitation despite man's active efforts to eliminate them (Fowler 1983). The ability of *Polistes* to seek out and modify environmental resources rests on capabilities such as those by which honey bees achieve phenomenal productivity from diffuse supplies of nectar and pollen.

In south Texas, hibernating wasps have been noted in protected locations a meter or so from the ground, to 61 m in an electric generating plant tower. Hibernating wasps can be found in corners and ceiling/wall angles of attics and barns, in hollow logs or under the protective skirt of dead and deflected leaves of palm trees. In some years at Kingsville numerous *Polistes* fall out of palm trees when a subfreezing temperature is accompanied by wind. It is believed the larger accumulations may occur at places marked by occupancy the previous year, rather than being annually independent aggregations around early-forming clusters at random favorable sites. Males exhibit territoriality by guarding narrow accesses and attempt to mate with females seeking hibernation sites. While males have not been noted in hibernating clusters in south Texas, they may live beyond the onset of winter if given food and protection (J. E. Gillaspy, personal observation).

As early as February in south Texas, females singly or together begin initiating nests. Single foundresses are often joined by others. There is much variation in choice of nesting sites. While some degree of shading and protection from wind and weather is the rule, thriving colonies of *P. exclamans* Viereck are occasionally seen on window panes, walls, attached beneath metal or translucent plastic sheet roofing, and in other exposed or unlikely places. However, confined places appear to be favored if available as evidenced by nests in tin cans, meter boxes, and other electrical equipment, clothesline pipes, large light sockets, window casements, behind screens and shutters, and narrow interstices in artificial field shelters (Gillaspy 1979b). While artificial structures appear favored (Reed and Vinson 1979), some nests can be found in adjacent shrubbery or trees. The initial nest may be placed on a smooth or painted surface to which it adheres poorly or breaks away if the paint has become flaky. Projections such as nails are often used when present. At nest initiation the stem or pedicel is most commonly about a centimeter in length from the base of attachment and terminates as the first cell. Where nests are placed on substrates subject to development of high temperatures such as sheet metal roofing, the pedicel appears to be a preadaptation that prevents transmission of heat to the brood.

Fiber for nest construction may be procured from a variety of sources, including hairs from plants (Duncan 1928) but weathered wood seems favored. When gathering fibers, the wasp moves backward, using her mandibles to rasp the surface, often with an audible sound, accumulating a ball of pulp-like material. The pulp is moist and dark as it is applied to the nest, but usually dries to light shades of gray paper-like material. The fibers are held together by a clear sizing that occasionally forms transparent "windows" between widely spaced fibers and gives the nest a lustrous, plastic-like quality. The sizing resists solution in water and strong alkali, suggesting a fibroin-like material such as silk, and is undoubtedly important in preventing wetted nests from falling apart. Janet, cited by Rau (1928), suggests the liquid offered by larvae to adults during trophallaxis is a secretion of the salivary glands, flowing from an opening at the base of the labium, hence sizing may not be entirely an adult product. The sizing is probably incorporated in the pulp as it is laid down, but may be supplemented later, since females often "lick" the surface of the nest (West-Eberhard 1969).

Although cells are rounded at their base, they become hexagonal with lengthening. West-Eberhard (1969) indicates that the straight sides and angles of the cell are determined by the wasp using its antennae, which extend into the cell on either side as the wasp straddles the partition being added to. This enables the wasp to compare the cell under construction to previously formed walls, and the hexagons formed, even though resulting from the efforts of several wasps, are sufficiently regular to form linear arrays across combs that may be 35 or more cells wide. Natural bee honeycomb is less regular, and is, according to West-Eberhard (1969), formed without the bees straddling the partition. New cells are added to the periphery of the wasp nest by the queen or single foundress as a rounded cup into which an egg is deposited. The walls are then extended as the larva develops. The pedicel is likewise enlarged progressively in diameter as the nest grows, and large nests may have more than one pedicel.

Polistes nests consist of a single comb without an outer covering or envelope, of a shape that has caused Ebeling (1978) to call them "umbrella wasps." Known tropical species have relatively short nesting cycles, usually more than one per year, with overlapping cycles throughout the year (West-Eberhard 1969). In temperate species colony activities are suspended in the fall followed by a period of winter hibernation.

Polistes tongues are short, not adapted to obtaining nectar from flowers with a long corolla, and their bodies are relatively bare of the hairs. Dependence on animal food precludes extensive stores of protein reserves, such as pollen. To some extent larvae constitute a food and moisture reserve, furnished to adults through trophallaxis (Hunt et al. 1982) and through accessibility for cannibalization. Adult wasps offer larvae chewed-up prey pellets from which the adult may also feed as the prey pellet is passed from larva to larva. Mature larvae spin a thin silken lining and conspicuous white cap for their cell and then pupate. Adults emerge about three weeks later as dark-eyed callows. Dominance interactions determine the hierarchy of egg laying and other activities among fertile females of a colony before the emergence of workers, after which fertile subordinates often leave to initiate nests on their own (Richards 1971).

Smearing of the pedicel and back of the nest from glands at the base of the fifth and sixth abdominal sternites has been noted by a number of investigators, causing repellency to ants, which has been confirmed (Hermann and Blum 1981). Of 53 pedicel-bearing nests available for laboratory study at Kingsville one *P. exclamans* Viereck and four *P. instabilis* Saussure had a pedicel and adjacent nest surface that were shiny black and lacquer-like (Rau 1931). Distribution of the black lacquer suggests its origin is from the wasp's sternal gland. Both the nest and pedicel were gray in color in 48 other nests of the above and six other species. Most were large, late-season nests, and it may be possible that smearing often diminishes when numerous wasps are available for colony defense, or perhaps where ants are not perceived as a threat. The large *P. instabilis* nest described below and another moderately large one collected in 1973,

6 miles east of Riviera, Texas, were both extensively lacquered. This species is now rare or absent near Kingsville but was abundant in 1973, possibly an effect of the moist period initiated by hurricane Beulah, September 20, 1967.

Polistes is the most widespread of the 24 social wasp genera recognized by Snelling (1981), its range encompassing that of all others except some northern portions of the ranges of *Vespula* and *Vespa* (Akre and Davis 1978, Akre 1982). While often less abundant than Vespinae in the temperate and Polybiini in the tropics, the genus *Polistes* is dominant or the only genus present in many areas and widely pervasive in all, possibly in part attributable to the more generalized nature of the nest. *Polistes* nests are relatively small and inconspicuous and can develop in a great diversity of aboveground locations. Rau (1942) indicated the occupation of different territories by different species of *Polistes* in and around St. Louis, although in Kingsville the five species *Polistes apachus* Saussure, *P. bellicosus* Cresson, *P. carolina* L., *P. exclamans* and *P. major* Beauvois appear to intermingle almost indiscriminately. *Polistes carolina* is, however, often found in dim light, and *P. exlamans* is often found in the open.

Akre (1982) tabulated data concerning nest size for 15 species of *Polistes*, a total of 419 nests. The largest nest, that of *P. annularis* (L.), had 1,886 cells, and there were three other nests with over 1,000 cells, the average being 143 cells per nest. In the south Texas/northern Mexico area the three largest nests coming to my attention over an 18-year period were attributed to three species: *P. instabilis* collected in 1973 at Lake Corpus Christi near Mathis, Texas, 52 cm × 8.2 cm, 1,028 cells in 98 transverse rows with 900 pupal cells; a *P. annularis* from George West, Texas, 23 cm × 19 cm, 902 cells in 38 rows, with 633 pupal cells; and a *P. major* from Rancho del Cielo near Gomez Farias, Tamaulipas, Mexico, 23 cm × 19 cm, 867 cells in 37 rows, with 503 pupal cells.

Although total productivity of the *Polistes* nests cited above was not estimated, even with some of the pupal cells producing two or three wasps, it is not likely that more than 2,500 adults may be produced from the largest nests. West-Eberhard (1969) estimated 234 adults to have been produced in Michigan in one colony of *P. f. fuscatus* (F.) with 23 of 251 cells producing two adults and 46 producing none.

Polistes have been reported to reoccupy nests a second season (Starr 1976, J. E. Gillaspy, personal observation), which would offer an advantage toward early brood production in the spring. Although clean previous season nests are seen in North America, many nests are invaded by the pyralid moth *Chalcoela pegasalis* (Starr 1976). Starr (1976) hypothesized that wasps may adapt to these moths by dispersing and making new nests, often as single foundresses.

HOSTS AND FOOD RELATIONS

Environmental resources used by *Polistes*, other than prey, include nectar or other sources of carbohydrate, fiber, and water. The latter is used not only to meet immediate needs of the larvae and adults, but also for fiber moistening and evaporative cooling of the nest (Rau 1931).

Studies in North Carolina by Rabb and Lawson (1957), Lawson (1959), Rabb (1960), and Lawson et al. (1961) are the most thorough that have been made from the applied standpoint. Their studies were directed at the hornworm, *Manduca sexta* (L.), and the budworm, *Heliothis virescens* (F.). Based on over 2,000 prey pellets intercepted at the nest, Rabb (1960) found the pellets consisted of 16 families and 36 genera of Lepidoptera. In addition, Coleoptera comprised 19 pellets, Orthoptera 21 pellets, one pellet each for Diptera and Hemiptera, and 11 pellets of other Arthropoda. Larvae of all instars of Lepidoptera were taken, but early instars were consumed in the field and not evident in pellets returned to the nest. Intercepted prey from fields varied inversely both with distance and with caterpillar abundance. *Polistes exclamans* and *P. fuscatus fuscatus* were the principal wasp species involved although five other species were present. Gallego (1950) listed three Sphingidae, one Noctuidae, and one Pyralidae of tobacco as targets of *Polistes* augmentation in Colombia.

Published reports of lepidopterous larvae preyed on by *Polistes* involve those on many host plants other than tobacco, including cabbage, cassava, coffee, cotton, eggplant, garden vegetables, green gram, lawn and pasture grasses, sunflower, taro, and trees, including forest, fruit and shade. Rau (1929) noted predation of caterpillars on cabbage, *Brassica oleracea* L., by *P. metricus* Say, misidentified as *P. pallipes* (Krombein et al. 1979). Three studies by Morimoto (1960a, b, 1961) on cabbage caterpillars deal in turn with natural predation, wasps augmented by introduction into artificial nesting sites, and in screen wire field cages. Gallego (1950) cited A. Alfonso as reporting an instance of caterpillars on cabbage effectively controlled by *Polistes*, but only in sunlit portions of the field.

Cassava, *Manihot esculenta* Crantz, is the fourth most important source of food energy in the tropics (Cock 1982) and the cassava hornworm, *Erinnyis ello* (L.), is generally considered to be one of its most serious pests. *Polistes canadensis* (L.) and *P. erythrocephalis* Latreille appear to be the most effective predators of larvae (CIAT Annual Report 1974; Bellotti and Schoonhoven 1978a, b). There are other reports of *Polistes* predation of *E. ello* on cassava in Colombia by Gallego (1950) and Bellotti and Arias (1978), in Brazil by Farias et al. (1980) and in the Dominican Republic by Agudelo-Silva (1980). In Peru, Enriquez et al. (1976) regarded *P. peruviana* Beq. and *P. versicolor* Oliv. as important predators of a major coffee pest, the coffee leaf miner, *Leucoptera coffeella* Guer.-Men.

Ashmead (1894) reported *Polistes* predation of caterpillars on cotton, *Gossypium* sp., in Mississippi. Ballou (1909) found growers of cotton on the island of St. Vincent, Windward Islands, suffering relatively little from attacks of the leafworm, *Alabama argillacea* (Hubner) and he attributed it largely to the "Jack Spaniard," now regarded as being *Polistes dominicus* (Vallot) (Snelling 1983).

Experiments of Kirkton (1968, 1970) in Arkansas in an attempt to use *P. fuscatus, P. metricus* Say, *P. annularis, P. carolina* and *P. exclamans* against the bollworm, *Heliothis zea* (Boddie) on cotton, found the wasps were diverted to taking the fall armyworm *Spodoptera frugiperda* (Smith) from a dense population on adjacent soybeans. They appeared "fixed" on the armyworms, and would not take bollworms even when offered at the nest. Chinese investigators (Anonymous 1976) reported *Polistes* effective in controlling *Heliothis armigera* (Hubner) on cotton, and Gillaspy (1979b) found cabbage loopers, *Trichoplusia ni* (Hubner) on cotton to be the chief prey of *P. apachus* and *P. exclamans* introduced near a field in artificial shelters.

Other observations of *Polistes* as predators of pests have been reported for eggplant, *Solanum melongena* L., in Ghana (Frempong and Buahin 1977); garden vegetables (Peckham and Peckham 1905); grasses (Gillaspy 1979b); and green gram, *Vigna radiata* Wilczek in India (Rao 1980). Both *P. annularis* and *P. exclamans* have been reported as predators of *Chlosyne lacinia* Geyer, a nymphalid butterfly on sunflowers in the United States (Drummond et al. 1970, Rogers 1980), and other *Polistes* species attacking caterpillars of sunflowers in India (Garg and Sothi 1980); and the tobacco cutworm, *Spodoptera litura* F. in Japan infesting taro, *Colocasia esculenta* Schott (Nakasuji et al. 1976). Shiga (1979), in an eight-year study of population dynamics, provided life tables for *Malacosoma neustria testacea* in Japan on cherry, peach and pear trees with *Polistes* providing the chief mortality for fourth and fifth instar larvae for six years. Oliver (1964) found eggs and all larval instars of the fall webworm *Hyphantria cunea* (Drury) attacked by six species of *Polistes*, and Ito and Miyashita (1968), Gillaspy (1973), Morris (1972), and Schaeffer (1977) found *Polistes* an important predator of the webworm. Furuta (1968) placed eggs and larvae of *Dendrolimus spectabilis* Butler on pine trees and found predation by *P. japonicus* Saussure to occur at high prey density.

Taking of adult Lepidoptera has also been reported, a butterfly (Garcia, 1971) and a few adult moths by caged wasps (Gillaspy 1979b). Gillaspy (pers. obs.) has seen *P. bellicosus* capture an adult male *Mocis* sp. (Noctuidae) and prepare the thorax as a food pellet.

Records of Coleoptera prey include the Colorado potato beetle, *Lep-*

tinotarsa decemlineata (Say). Lawson et al. (1961) and Wheeler (1977) reported *P. fuscatus* taking larvae of the alfalfa weevil, *Hypera postica* (Gyl.), in New York. Gibo (1974, 1977) provided further evidence of the acceptability of Coleoptera as prey by using the mealworm, *Tenebrio molitor* L., to rear over 100 laboratory colonies of *Polistes* in box cages.

Hymenopterous prey has been reported by R. B. Friend, cited by Clausen (1940), who observed *P. f. fuscatus* cutting open the leaf mines of the birch leaf miner, *Fenusa pumila* Klug (Tenthredinidae) and feeding on the larvae. Iwata (1976) also cites use of Tenthredinidae as prey. Miscellaneous other records include Homoptera (aphids), and Orthoptera (nymphs of the Mantidae, *Tenodera aridifolia*), by Iwata (1976), and Thysanoptera by Dhaliwal (1975). A total of eight insect orders have been reported as prey, but only three are regarded of any significance.

In addition to insect prey, *Polistes* utilize carbohydrates mainly in the form of nectar, which may also contain amino acids (Baker and Baker 1973). Barrows (1979) noted use of *Kermes* scale honeydew, and along with Beckmann and Stucky (1981) found *Polistes* actively repelling other insects from the carbohydrate source the wasps were using as food. In the case of the latter authors, extrafloral nectaries were being used and protection by the wasps prevented damage to the plants by phytophagous insects. At Kingsville during periods of apparent nectar scarcity, *Polistes* have been noted in abundance in early spring around arbor vitae, *Thuja orientalis* L., infested by aphids, *Cinara tujafilina* (del Guer.), and in fall around live oak, *Quercus virginiana* Mill., infested with mealy oak gall, *Disholcapsis cinerosa* Bassett, a cynipid.

Rau (1928) discovered the storage of a honey-like liquid in *Polistes* cells throughout the nesting season but in greater abundance toward the end of the summer, where 75% of the cells contained some liquid. Honey was absent from cells containing mature larvae, but was found with eggs or newly hatched larvae.

Grinfel'd (1978) studied food storage in cells of *P. gallicus* L. in the Belgorod Province of the Soviet Union. This species, like those studied by Rau, often stored liquid with eggs or first instar larvae. Liquid droplets were placed on the side of the cell away from the nest center, with the egg placed on the other side. Food pellets brought to the nest were examined microscopically as were the crops of ten wasps. The crop was found to contain a thin "gruel" with scraps of integument, setae, crochets, fragments of spiracles, etc. Thus it was established that adults consume some of the food. Even wasps returning to the nest without a pellet had this gruel. This supplements the observations of Lawson (1959), which established that small larvae were eaten in the field and thus not detectable as prey pellets. To demonstrate that carbohydrates were fed to larvae, Grinfel'd (1978) furnished honey colored with neutral red dye to adult wasps and it was then found in the digestive tract of larvae.

Strassman (1975) found that *P. annularis* females store honey in their nests in the autumn and return on warm winter days to consume some of the honey and defend it from nonsisters. Honey deprivation decreased numbers surviving the winter, and females surviving without honey built smaller nests in the spring.

Polistes may play a role in pollination, and experiments have shown *P. exclamans* to be an effective pollinator of safflower, *Carthamus tinctorius* L. (Levin et al. 1967, Free 1970). Four tropical crops in Jamaica, cashew, mango, akee, and avocado, appear to require insect pollination, and were found to be visited by more *Polistes* than by honey bees from colonies maintained in the vicinity. The *Polistes* were, however, less efficient in transporting pollen (Free and Williams 1976). *Polistes crinitus* (Felton) were also numerous at flowers of coconut, *Cocos nucifera* L., in Jamaica, but were less effective in pollination than honey bees (Free et al. 1975). *Polistes* may result in reduced pollination because of their flower-guarding behavior (Free et al. 1975).

ENEMIES OF *POLISTES*

While *Polistes* suffer from various natural predators and parasites, their chief enemy may be man, who destroys colonies, or least knocks down nests wherever found. Moore et al. (1982) found that the majority of magazine articles about insects are unfavorable, even though entomologists estimate that less than one in a thousand can be regarded as harmful to man. Added to this generally negative perception of insects is the fear many people experience on seeing a hymenopterous insect such as *Polistes* that might sting. Few gardeners realize there could be an argument for coexisting with *Polistes* (Olkowski and Olkowski 1976). Gallego (1950) deplored the merciless destruction of *Polistes* colonies carried on daily in Colombia because it might be possible to transport them to fields suffering caterpillar damage. Quaintance and Brues (1905) likewise deplored wanton destruction of *Polistes* describing them as being in vast numbers in cotton fields, constantly in search of prey. In working with *Polistes* and honeybees over a period of years, it would appear to me (personal observation) that a majority of people could become tolerant and even in some cases protective on being educated to their benefits.

The remaining natural enemies of *Polistes* fall more or less into the categories of whole-nest predators, brood predators/parasitoids, and adult predators. Raccoons, opossums, and birds cause damage to both brood and adults (Gillaspy 1979b; Strassman 1981), while adults are apparently able to escape ants that are important in brood destruction in the tropics (Jeanne 1979a). The brood is also attacked by a number of parasitoids or predators. Little is known about the two Sarcophagidae that attack *Po-*

listes. *Sarcophaga polistensis* Hall has a yellowish adult coloration, similar to some species of *Polistes*. *Chalcoela* moths have a similar yellowish coloration, although there is no evidence that colonies of yellowish *Polistes* are more successfully attacked. Some of the nest destruction observed in nests attacked by parasitoids and predators may be done by the wasps in response to presence of the predators and parasitoids. Total nest destruction or removal may result in abandonment of the area in a manner resembling absconding (West Eberhard 1982) but vigorously developing colonies very often remain and renest on or near the old nest base.

Pyralid moths, *Chalcoela iphitalis* (Walker), indicated to be a junior synonym of *Chalcoela pegasalis* (Walker) by Richards (1978), are especially prevalent in wasp nests on buildings (Nelson 1968, Reed and Vinson 1979). Although Munroe (1972) provides a key to the species it is not clear to what extent he had variable populations available for analysis. Nelson (1968) and Rau (1941) accept the two-species concept. Ballou (1915, 1934) regarded *C. pegasalis* as important in eliminating *Polistes* populations on some islands. Rau (1941), Gillaspy (1971, 1973), and Strassman (1981) have discussed *Chalcoela* infestations, but there are questions concerning its biology. *Chalcoela* may well be predators rather than parasitoids (Gillaspy 1971). Strassman (1981) mentions only pupae and prepupae as being attacked, but larvae may also be consumed as evidenced by *Chalcoela* larvae and webbing in cells not previously capped. In other instances the wasp caps are present and modified or partially incorporated in construction of the moth cocoon. Early *Chalcoela* instars evidently feed on the meconium before attacking wasp brood, as occurs with the tineid moth attacking *Polistes canadensis canadensis* (L.) in Brazil (Jeanne 1979b). J. R. Lara (personal communication) noted that *Chalcoela* larvae emerging from *Polistes* nests without live brood doubled in size before dying at the end of the week. He also noted that moths emerge fully functional through the unsealed terminal slit of their cocoon and are thus able to evade wasp attack.

Elasmus polistis Burks, a chalcidoid, also attacks and consumes wasp brood as a gregarious ectoparasitoid. These small parasitoids produce a flat, black transverse partition or false bottom to cells of the nest by means of meconium and secretions prior to pupation. Reed and Vinson (1979) indicate *E. polistis* attacks the wasp brood in the prepupal and pupal stages. As in the case of *Chalcoela*, it is possible that larvae may also be attacked since partitions may be evident in uncapped open cells. Other enemies of wasp brood include Strepsiptera and also Ichneumonidae of the genus *Pachysomoides*, which are discussed by Nelson (1968). Predators attacking adult wasps include robber flies of the family Asilidae, spiders, and birds.

MANAGEMENT EXPERIMENTS

Biological control involving the manipulation of natural enemies (Rabb 1971) is older than "classical" biological control, in which exotic control agents are introduced (Caltagirone 1981). However, due to the development of pesticides and the success of some biological control introductions, the management of natural enemies has not received much attention. *Polistes* would appear more promising for manipulation than predatory ants, birds, small mammals, or toads that have been used or experimented with since 900 A.D. (Coppel and Mertens 1977). References to biological control derived from *Polistes* predation date to Ashmead (1894) and Peckham and Peckham (1905). Specific attempts have been made to exploit wasp populations for crop protection by the provision of nesting sites, supplemental feeding, and the reduction of parasites by culling of infested nests (Gillaspy 1979b).

Various structures of wood (Ballou 1910, 1915; Rabb and Lawson 1957, Kirkton 1970, Gillaspy 1971, Anonymous 1976), metal (Morimoto 1960b), plastic (Gillaspy 1979b), and other materials (Gallego 1950, Bellotti and Schoonhoven 1978a, b) have been offered as nesting sites. Ballou (1910) was first to report artificial measures to increase *Polistes* predation, that is, farmers on the island of St. Vincent building sheds on the edge of fields to provide nesting sites for wasps. Gallego (1950) reported the use of *Polistes* by the Colombian Tobacco Co. involving over 40 ha of tobacco where shelters were constructed from 1936 to 1944 for the wasps. Gallego (1950) reported that during a visit with an economic entomology class in 1937 he had difficulty in finding a single caterpillar.

About 40 years after Ballou's original report, augmentation with *Polistes* was undertaken by entomologists in the United States and there apparently existed a favorable attitude as expressed by Bishopp (1952), "They can be a nuisance about houses because they sting viciously when they are molested. The benefits derived from the predaceous habits of the *Polistes* outweigh their objectionable traits."

Augmentation experiments in North Carolina in 1955–1956 involved boxlike wooden shelters, open below but with a door for the wasps' entrapment. Placement was at a height of 1 m at the margins of fields. Some occupied shelters were relocated from as far as 200 km. In 1956 there were 468 such shelters in use. Tests involved two half-acre plots (1 protected, 1 control) separated by at least 275 m at each of five locations across the upper half of the state. A mean of 24 wasp colonies were active in the field margin shelters of each protected plot, but strength of the colonies in terms of adult female wasps as an index of predation intensity was not reported. Results were summarized in terms of integrated control: "Biological control with *Polistes* wasps is inexpensive and leaves no residue, but this method by itself will not give adequate control of hornworms under all conditions, nor will it prevent damage by budworms" (Lawson et al. 1961).

Rabb (1971) subsequently reported that changes in tobacco production practices had reduced hornworm abundance and that *Polistes* were diverted to more abundant alternate prey, of little economic importance. However, he cited *Polistes* as an instance of natural enemies tractable to manipulation and also as an example of habitat modification to enhance the beneficial effects of a native natural enemy. As problems in use of *Polistes* he mentioned failure to take boring and tunneling larvae, generalized predation resulting in taking of high-density prey rather than a particular prey species, declining activity in late summer, and foraging of some species in wooded areas rather than in croplands (Rabb 1971). Through research in equipment and methodology, some deficiencies may be overcome by better population control and deployment through research in equipment and methodology.

Kirkton (1968, 1970) tested 1,000 each of three types of small storeable and transportable nesting containers for *Polistes*. All were smaller — potentially less cumbersome than those of Lawson et al. (1961), although entrapment of wasps for movement was not provided for. Kirkton (1968, 1970) found 15 cm wooden boxes faced downward and at a height of 0.6 m from the ground were most acceptable to the wasps. At 1.2 m or when opening laterally in any compass direction wasp occupancy decreased, and those opening laterally were sometimes occupied by birds and mice. Wasps thrived, with 700–900 nests in one "wasp line" of 0.8 km. Kirkton (1968, 1970) appears to be the first to undertake an experimental approach to the problem of optimally meeting wasp nesting preferences and requirements.

Investigators in China (Anonymous 1976) reported *Polistes* to be effective in controlling *Heliothis armigera* on cotton. Wasps were moved at night to horizontal supports, 1.8 m from the ground beneath roofs. *Heliothis* control of 70–80% was reported five to seven days after introduction of wasps.

In South America, a program using natural *Trichogramma* egg parasitism plus *Polistes* predation of the cassava hornworm was reported (Bellotti and Arias 1978) to have been in operation at the Centro Internacional de Agricultural Tropical, Cali, Colombia, since 1973 on 50–60 ha of cassava with no worm outbreak during that period. These authors also reported introduction of *Polistes* on several farms in a cassava-growing region of Colombia, with biweekly evaluations of hornworm oviposition, egg parasitism, and larval and wasp populations. Control was most effective when a tentlike protective shelter was provided for the wasps in the center of cassava fields. However, wasps were attacked by parasites (CIAT 1975).

Gillaspy (1971, 1973, 1979b) conducted experiments using *P. apachus* and *P. exclamans* directed against bollworms and budworms on cotton. In these experiments 1279 wasp colonies were placed near fields of cotton in eight separate tests over a span of three seasons, 1971–73. Nests were at first mounted beneath rooflike shelters and then beneath flat table-like wooden shelters on stakes. Experiments have been carried out by Gillaspy (1982) to develop nesting receptacles and increase resident *Polistes* populations, rather than continuing to treat wasps as an environmental resource to be harvested from places of natural occurrence and transported to fields as suggested by Gallego (1950).

Half-gallon plastic containers were adopted in 1973 and are economical, readily available, relatively light, and readily modifiable. In some cases they have been resistant to weathering through two or more seasons when protected by structures or vegetation, but became very brittle on direct exposure to the sun for a year or two. Covers permit entrapment of wasps when desired.

1982–1983 MANAGEMENT STUDIES

Management expectations for *Polistes* center largely around their need for suitable nesting sites. The potential to manage *Polistes* for control purposes centers around the ability to provide suitable nesting containers. Such containers need to be simple and capable of stimulating spontaneous nest-founding. They also must facilitate the introduction of established nests for the initiation and maintenance of a managed wasp population and to control the species representation of wasps. Such containers should provide optimal conditions for colony development, offer protection from both environmental and biological factors that could harm the nests, and be easily moved with a minimum of problems to both the wasps and to the management personnel. Use of individual limited-access nest containers facilitates culling of parasitized nests and other caretaking operations.

Plastic one-half gallon ice cream containers were modified, inverted and attached to wooden stakes. A painted container lid was attached to the stake and served as a sun shade. To further reduce the heat when exposed to the sun, the containers were painted with two coats of fiber-aluminum paint, a corrugated cardboard disk was riveted with a single rivet to the bottom of the container (to serve as the ceiling where the nests attach), and three holes ¾ in in diameter were made about 1½ in from the cardboard in the sides of the container to provide ventilation. Tests have shown painted containers at least 5°F cooler inside than unpainted ones. The center of the container lid, which served as the floor of the nesting container, was replaced with ⅛-in hardware cloth to provide ventilation. A number of pins or paper clamps were inserted around the edge of the lid to discourage its removal by opossums, which previously proved destructive to nests.

The containers were attached to stakes about 1 m from the ground by a 2-in × ¼-in bolt and wing nut to permit easy removal or turning to permit inspection of nests. Plugs inserted into the holes could be used to trap wasps in the container for transport or inspection. Developed nests could be installed by inserting the pedicel into a small paper binder clip in the container ceiling. The clip, perforated by a drill hole, served in lieu of a washer for the rivet retaining the ceiling cardboard.

In a study using these containers in 1982 and 1983, 95 containers were placed on stakes about 2 m apart over a 1/5-ha tract of land surrounded by uncultivated land of weeds and brush. Of 85 containers placed in June 1983, nine were used by naturally occurring *Polistes*. Colonies were introduced into the remaining containers between late June and early August with the addition of 11 more wasp-containing containers in August. In addition to the 95 colonies, three natural colonies were present in the area.

Of the 95 colonies, 13 were destroyed by an opossum. Further, the carrying capacity of the area was exceeded rendering the establishment of transplanted colonies difficult. By the end of 1982 mature brood had been produced in 34 colonies (Table 13.1). Table 13.2 shows the population counts through 1983. Nests were left over winter and some hibernation in the containers was noted, although counts were not made. Nesting resumed in February and March 1983.

Various authors, Pardi (1942), Rau (1928, 1929, 1931, 1940), West-Eberhard (1969), Evans and West-Eberhard (1970), Gibo (1972) Gillaspy (1979b) Klahn (1979), and Pratte and Gervet (1980a, b) have noted a tendency for *Polistes* to return to previously occupied nest sites and to rebuild destroyed nests. Twenty-six containers were reoccupied the second year, plus one *P. bellicosus* nest and one nest of *P. major*, a species not observed in the area in 1982. The latter developed a nest with 12 cells, three of these capped, but disappeared on April 20. Three males emerged from the capped cells between May 6 and May 8, but attempts to hand rear (Michener and Michener 1951) the other nine apparently healthy larvae and eggs failed. The results indicated the containers are acceptable and could prove useful in *Polistes* management, if shaded to prevent rapid deterioration of the plastic.

PROBLEMS IN *Polistes* MANAGEMENT

Since *Polistes* do not appear to store protein food they have little buffer against varying prey densities. While supplemental feeding would seem to be important in a management program, the use of live insect prey

TABLE 13.1
Colonies that Produced Mature Brood Within the Experimental Area in 1982

	Exclamans		Apachus		Carolina	
	Nests	New cocoons	Nests	New cocoons	Nests	New cocoons
Introduced	12	286	12	184	1	32
Renesting	1	4	3	44		
Volunteer	3	38	1	15	1	3
	16	328	16	243	2	35

TABLE 13.2
Population Counts, 1982–1983 Experiment

	No. hives	Exclamans			Apachus			Carolina		
		n^a	a	c	n	a	c	n	a	c
Introduced	96	52	778	1003	24	160	452	1	4	19
After 1 week	96	39	310	489	20	111	232	1	3	13
October 10	96	24	246	96	17	191	42	2	12	3
February 25	62	8	16	0	6	8	0	1	1	0
March 11[b]	62	10	18	0	14	20	0	2	2	0

^{a}n = nests, a = adult wasps, c = cocoons.
[b]On March 11 there were also single females of p. bellicosus and p. major with small nests.

345

is probably not practical. The use of an artificial diet might be possible but much more information is needed as to how *Polistes* find their prey and the factors that stimulate them to feed.

Whether fiber material for nest construction is a limiting factor is unknown. Certainly more information on what constitutes good nest material would be useful. Also more information is needed in regard to nesting site choice and microhabitat attributes that stimulate nesting site choice and maximum colony growth. Edwards (1980) considered a number of factors important to nest location in the Vespinae and some of these are probably also important to *Polistes*. Basic research on nest founding in the laboratory such as that provided by Pratte and Gervet (1980a, b) may yield information that will prove useful in designing artificial nesting containers readily utilized by foundress wasps.

Understanding the conditions needed for overwintering sites and providing such conditions could be important in management. Much is also unknown concerning the distances over which *Polistes* forage or how far they may travel in seeking sites to initiate nests. Gillaspy (1979b and personal observation) released several hundred marked wasps on two occasions but was unable to locate any initiating nests in the area. Kasuya (1982) postulated *Polistes* as foraging in their immediate vicinity although capable of foraging 2 to 8 km. However, the distances over which they forage have not been established. Such information is essential to management. One of the more significant problems is the indication that *Polistes* tend to forage on the most abundant prey thus causing *Polistes* to be less effective in crop pest control when other nonpest species are more abundant or when the pest species population declines.

Another problem is that the *Polistes* tend to disperse when prey declines resulting in a decrease in the population of the area. In contrast, Gillaspy (1979b) noted nest initiation by subordinates leaving parental nests under favorable conditions of prey, etc.

Two additional problems are location of nest sites and nest destruction. The importance of nest sites is borne out by studies of Gillaspy (1979b) who found that plastic nest containers mounted beneath wooden shelters were not utilized by foundresses. The ideal height to place containers may be important not only with regard to the ideal height for nesting but to reduce vandalism. Some vandalism of nest containers has occurred during studies in Kingsville (Gillaspy 1979b). Although farmers and other personnel may be convinced not to harm the nests, other people encountering nests may not be as understanding. Gallego (1950) reported that fines and even the possible loss of a job were penalties inflicted on the Colombia Tobacco Company employees who damaged wasp nests. But damage is not just due to people, since at Kingsville opossums destroyed 13 of 46 colonies in 1982. Those nests destroyed were the largest and adults as well as brood were consumed. A coatimundi was also noted to consume adults (Gillaspy 1971) as do birds as well (Gillaspy 1971, Strassman 1981). Probably the biggest problem may be the spiders, ants, and parasites.

Three species of ants and a spider, *Phidippus audax* Hentz, were common in the 1982–1983 study (J. E. Gillaspy personal observation). The ants were *Solenopsis geminata* (F.), *Camponotus abdominalis transvectus* Wheeler and *Forelius foetidus* (Buckley). The spiders caused problems with nests in the containers while the ants interfered with attempts at supplemental feeding of the wasps. The parasitoids or nest predators may also become a problem particularly if nests are maintained in a particular area year after year. Another problem may be cannibalism. Kasuya (1982) found cannibalism to occur between conspecifics in Japan.

ACKNOWLEDGMENTS

Experimentation with improvement of nesting receptacles has received support from Texas A&I University Faculty Research Grants 2893 and 2867. Graduate students participating part of the time were Chris Pease and Larry Gamble.

REFERENCES

Agudelo-Silva, F. 1980. Parasitism of *Erinnyis ello* eggs by *Telenomus sphingis* in the Dominican Republic. *Environ. Entomol.* 9:233–235.

Akre, R. D., 1982. Social wasps. *In*: Social Insects, pp. 1–105 (Hermann, H. R., ed.), 4, xiii + 384 pp.

Akre, R. D. and H. G. Davis. 1978. Biology and pest status of venomous wasps. *Annu. Rev. Entomol.* 23:215–238.

Anonymous 1976. A preliminary study of the bionomics of hunting wasps and their utilization in cotton insect control. *Acta Entomol. Sinica* 19:303–308.

Ashmead, W. H. 1894. Notes on cotton insects found in Mississippi. *Insect Life* 7:240–274.

Baker, H. G. and I. Baker. 1973. Amino acids in nectar and their evolutionary significance. *Nature (London)* 241:543–545.

Ballou, H. A. 1909. Treatment of cotton insects in the West Indies in 1907. *W. Indian Bull.* 9:235–241, 10 figs.

Ballou, H. A. 1910. Introduction of the St. Vincent "Jack Spaniard" into Montserrat. *Agric. News* 9:378.

Ballou, H. A. 1915. West Indian wasps. *Agric. News* 14:298.

Ballou, H. A. 1934. Notes on some insect pests of the Lesser Antilles. *Trop. Agric. Trin.* 11:210–212.

Barrows, E. M. 1979. Polistes wasps show interference competition with other insects for *Kermes* scale insect secretions. *Proc. Entomol. Soc.* 81:570–575.

Beckmann, R. L., Jr. and J. M. Stucky. 1981. Extrafloral nectaries and plant guarding in *Ipomoea pandurata* (L.) G. L. W. Mey (Convolvulaceae). *Amer. J. Bot.* 68:72–79.

Bellotti, A. and B. Arias. 1978. Biology, ecology and biological control of the cassava hornworm (*Erinnyis ello*), pp. 227–232. In: Proc. Cassava Prot. Workshop (Brekelbaum, T., Bellotti, A., and Lozano, J. C., eds.). CIAT, Cali, Colombia, 7–12 November 1977.

Bellotti, A. and A. van Schoonhoven. 1978a. Cassava Pests and Their Control. Cassava Information Center, CIAT, Cali, Colombia, 71 pp.

Bellotti, A. and A. van Schoonhoven. 1978b. Mite and insect pests of cassava. *Annu. Rev. Entomol.* 23:39–67,

Bishopp, F. C. 1952. Insect friends of man, pp. 79–87. In: Insects. Yearbook of Agriculture 1952. U.S. Dept. Agri., Wash. D.C., 780 pp.

Caltagirone, L. E. 1981. Landmark examples in classical biological control. *Annu. Rev. Entomol.* 26:213–232.

CIAT. 1974. Annual Report, 1973. Centro Internacional de Agricultura Tropical, Cali, Colombia, 284 pp.

CIAT. 1975. Annual Report, 1974. Centro Internacional de Agricultura Tropical, Cali, Colombia, 260 pp.

Clausen, C. P. 1940. Entomophagous Insects. McGraw-Hill, New York, 688 pp.

Cock, J. H. 1982. Cassava: A basic energy source in the tropics. *Science* 218:755–762.

Coppell, H. C. and J. W. Mertens. 1977. Biological Insect Suppression. Springer Verlag, New York, 314 pp.

Dhaliwal, J. S., 1975. *Polistes hebraeus* preying upon Thysanoptera. *Curr. Sci.* 44: 368.

Drummond, B. A. III, G. L. Bush, and T. C. Emmel. 1970. The biology and laboratory culture of *Chlosyne lacinia* Geyer (Nymphalidae). *J. Lepidop. Soc.* 24:135–142.

Duncan, C. D., 1928. Plant hairs as building material for *Polistes* Pan-Pac. *Entomology* 5: 90.

Ebeling, W., 1978. Urban Entomology. University of California Press, Berkeley. 695 pp.

Edwards, R., 1980. Social Wasps. Rentokil, Ltd., East Grenstead, England. 398 pp.

Enriquez, E., S. Bejarano, and V. Vila. 1976. (1975). Observaciones sobre avispas predatoras de *Leucoptera coffeella* Guer.-Men. en el centro y sur del Peru. *Rev. Per. Entomol.* 18: 82–83.

Evans, H. E. and M. J. West-Eberhard. 1970. The Wasps. Univ. Michigan Press, Ann Arbor, 265 pp.

Faegri, K., and L. van der Pijl. 1979. The Principles of Pollination Ecology, 3rd Ed. Pergamon Press, New York.

Farias, A. R. N., F. N. Ezeta, and J. L. L. Dentas. 1980. Circular Tecnica, EMBRAPA, CNPMF, 5: 11 pp.

Fowler, H. G. 1983. Human effects on nest survivorship of urban synanthropic wasps. *Urban Ecol.* 7: 137–145.

Free, J. B. 1970. Insect Pollination of Crops. Academic Press, New York. 544 pp.

Free, J. B., Raw, A., and I. H. Williams. 1975. Pollination of coconut (*Cocos nucifera* L.) in Jamaica by honeybees and wasps. *Appl. Anim. Ethol.* 1: 213–223.

Free, J. B., and I. H. Williams. 1976. Insect pollination of *Anacardium occidentale* L., *Mangifera indica* L., *Blighia sapida* Koenig and *Persea americana* Mill. *Trop. Agric. Trin.* 53: 125–139.

Frempong, E. and G. K. A. Buahin. 1977. Studies on the insect pests of eggplant, *Solanum melongnea* L., in Ghana. *Bull. Inst. Fond. Afrique Noire* (Ser. A.), 39: 627–641.

Furuta, K. 1968. The relationship between population density and mortality in the range of latency of *Dendrolimus spectabilis* Butler. *Jap. J. Appl. Entomol. Zool.* 12: 129–136.

Fye, R. E. 1972. Manipulation of *Polistes exclamans arizonensis*. *Environ. Entomol.* 1: 55–57.

Gallego, F. L. 1950. Estudios entomologicos; el gusano de las hojas de la yuca. *Rev. Fac. Nac. Agron. (Medellin, Colombia)* 12: 84–110.

Garcia, C. 1971. *Polistes* spp. [Sic.] predatory on adult Lepidoptera. *Entomol. News.* 82, 274.

Garg, A. K. and G. R. Sothi. 1980. Some observations on wasps attacking sunflower crop. *Indian J. Entomol.* 42: 267–269.

Gibo, D. L. 1972. A introduced population of social wasps, *Polistes apachus*, that has persisted for 10 years. *Bull. South. Calif. Acad. Sci.* 71: 53.

Gibo, D. L. 1974. A laboratory study on the selective advantage of foundress association in *Polistes fuscatus*. *Can. Entomol.* 106: 101–106.

Gibo, D. L. 1977. A method for rearing various species of social wasps of the genus *Polistes* (Hymenoptera: Vespidae) under controlled conditions. *Can. Entomol.* 109: 1013–1015.

Gillaspy, J. E. 1971. Papernest wasps (*Polistes*): Observations and study methods. *Ann. Entomol. Soc. Amer.* 64: 1357–1361.

Gillaspy, J. E. 1973. Behavioral observations on paper-nest wasps. *Amer. Midl. Nat.* 90: 1–12.

Gillaspy, J. E. 1979a. Mass collection of *Polistes* wasp venom by electrical stimulation. *Southw. Entomol.* 4: 96–101.

Gillaspy, J. E. 1979b. Management of *Polistes* wasps for caterpillar predation. *Southw. Entomol.* 4: 334–352.

Gillaspy, J. E. 1982. The impact on man of *Polistes* wasps, with special reference to caterpillar suppression, pp. 129–133. *In:* The Biology of Social Insects (Breed, M. D., C. D. Michener, and H. E. Evans, eds.). Westview Press, Boulder, Colo., 419 pp.

Grinfel'd, E. K. 1978. The feeding of the social wasp, *Polistes gallicus*. *Entomol. Rev.* 56: 24–29.

Hermann, H. R. and M. S. Blum. 1981. Defensive mechanisms in the social Hymenoptera. In: Social Insects, pp. 77–197 (Hermann, H. R., ed.) Vol 2., Academic Press, New York, 491 pp.

Huffaker, C. B. 1971. Biological Control. Plenum Press, New York, 511 pp.

Hunt, J. H., I. Baker, and H. G. Baker. 1982. Similarity of amino acids in nectar and larval saliva. The nutritional basis for trophallaxis in social wasps. *Evolution* 36: 1318–1322.

Ito, Y. and K. Miyashita. 1968. Biology of *Hyphantria cunea* Drury in Japan. V.

Preliminary life tables and mortality data in urban areas. *Res. Popul. Ecol.* 10: 177–209.

Iwata, K. 1976. Evolution of Instinct. Comparative Ethology of Hymenoptera. Amerind, New Delhi (for Smithsonian Institution), 535 pp.

Jeanne, R. L. 1979a. A latitudinal gradient in rates of ant predation. *Ecology* 60: 1211–1224.

Jeanne, R. L. 1979b. Construction and utilization of multiple combs in *Polistes canadensis* in relation to the biology of a predaceous moth. *Behav. Ecol. Sociobiol.* 4: 293–310.

Kasuya, E. 1982. Central place water collection in a Japanese paper wasp, *Polistes chinensis antennalis*. *Anim. Behav.* 30: 1910–1914.

Kirkton, R. M. 1968. Building up desirable wasp populations. *Ark. Farm Res.* 17: 8.

Kirkton, R. M. 1970. Habitat management and its effect on populations of *Polistes* and *Iridomyrmex*. *Proc. Tall Timbers Conf.* 2: 243–246.

Klahn, J. E. 1979. Philopatric and nonphilopatric foundress associations in the social wasp *Polistes fuscatus*. *Behav. Ecol. Sociobiol.* 5: 417–424.

Knipling, E. F. 1977. The theoretical basis for augmentation of natural enemies. pp. 79–123 (Ridgway, R. L., and Vinson, S. B., eds.). In: Biological Control by Augmentation of Natural Enemies. Plenum Press, New York, 480 pp.

Krombein, K. V., P. D. Hurd, Jr., O. R. Smith, and P. D. Burks. 1979. Catalog of Hymenoptera in America North of Mexico, Vol. 2, Apocrita (Accleata). Smithsonian Inst. Press, Washington, D.C., pp. i-xvit, 1199-2209.

Lawson, F. R. 1959. The natural enemies of the hornworms on tobacco. *Ann. Entomol. Soc. Amer.* 52: 741–755.

Lawson, F. R., R. L. Rabb, F. E. Guthrie, and T. G. Bowery. 1961. Studies on an integrated control system for hornworms on tobacco. *J. Econ. Entomol.* 54: 93–97.

Levin, M. D., G. D. Butler, Jr., and D. D. Rubis. 1967. Pollination of safflower by insects other than honey bees. *J. Econ. Entomol.* 60: 1481–1482.

Michener, C. D. and M. H. Michener. 1951. American Social Insects. D. Van Nostrand Co., Inc., New York, 267 pp.

Moore, W. S., D. R. Bowers, and T. A. Granovsky. 1982. What are magazine articles telling us about insects? *Journalism Quart.* 59: 464–467.

Morimoto, R. 1960a, 1960b, 1961. *Polistes* wasps as natural enemies of agricultural and forest pests. I, II, III. (Studies on the social Hymenoptera of Japan X, XI, XII). *Sci. Bull. Fac. Agric.* (Kyushu Univ.) 18: 109–116; 117–132; 243–252.

Morris, R. F. 1972. Predation by wasps, birds and mammals on *Hyphantria cunea*. *Can. Entomol.* 104: 1581–1591.

Munroe, E. 1972–1973. Pyraloidea. In: The Moths of America North of Mexico, pp. 137–304 (Dominick, R. B., manag. ed.). E. W. Classey, London.

Nakasuji, F., H. Yamanaka, and K. Kiritani. 1976. Predation of larvae of the tobacco cutworm *Spodoptera litura* by Polistes wasps. *Kontyu (Tokyo)* 44: 205–213.

Naveh, Z. 1982. Landscape ecology as an emerging branch of human ecosystem science. *Adv. Ecological Res.* 12: 189–237.

Nelson, J. M. 1968. Parasites and symbionts of nests of *Polistes* wasps. *Ann. Entomol. Soc. Amer.* 61: 1528–1539.

Oliver, A. D. 1964. Studies on the biological control of the fall webworm, *Hyphantria cunea*, in Louisiana. *J. Econ. Entomol.* 57: 314–318.

Olkowski, H. and W. Olkowski. 1976. Entomophobia in the urban ecosystem, some observations and suggestions. *Bull. Entomol. Soc. Amer.* 22: 313–318.

Pardi, L. 1942. Richerche sui Polistini V. 12 polidinia iniziola in *Polistes gallicus* (L.). *Boll. Ist. Entomol. Univ. Bologna* 14: 1–106.

Peckham, G. W. and E. G. Peckham. 1905. Wasps, Social and Solitary. Houghton, Mifflin, Boston, 311 pp.

Pratte, M. and J. Gervet. 1980a. Relations entre conditions d'excercice de l'activite batisseuse et la duree de la phase de fondation chez la guepe *Polistes*. *Biol. Behav.* 5: 351–360.

Pratte, M. and J. Gervet. 1980b. Influence des stimulations sociales sur la persistance du comportement de fondation chez la guêpe poliste, *Polistes gallicus* L. *Insectes Soc.* 27: 108–126.

Quaintance, A. L. and Brues, C. T. 1905. The cotton bollworm. *Bull. Bur. Entomol.* 50: 155 pp.

Rabb, R. L. 1960. Biological studies of *Polistes* in North Carolina *Ann. Entomol. Soc. Amer.* 53: 111–121.

Rabb, R. L. 1971. Naturally-occurring biological control in the eastern United States, with particular reference to tobacco insects, In: Biological Control, pp. 294–311 (Huffaker, C. N., ed.). Plenum Press, New York, 511 pp.

Rabb, R. L. and F. R. Lawson. 1957. Some factors influencing the predation of *Polistes* wasps on the tobacco hornworm. *J. Econ. Entomol.* 50: 778–784.

Rao, P. S. 1980. A new record of predatory wasp, *Polistes hebreaus* Fabricius, associated with the pest complex of green gram, *Vigna radiata* Wilczek. *Indian J. Entomol.* 42: 278–279.

Rau, P. 1928–1942. See Krombein et al. 1979 and West-Eberhard 1969 for extensive lists of publications by Rau.

Reed, H. C. and S. B. Vinson, 1979. Observations of the life history and behavior of *Elasmus polistis* Burks. *J. Kans. Entomol. Soc.* 52: 247–257.

Richards, O. W. 1971. The biology of the social wasp. *Biol. Rev.* 46: 483–528.

Richards, O. W. 1978. The Social Wasps of the Americas. British Mus., London, 580 pp.

Rogers, C. E. 1980. Natural enemies of insect pests of sunflower: A world view. Texas Agric. Exp. Sta. MP-1457, 30 pp.

Schaffer, F. W. 1977. Attacking wasps, *Polistes* and *Therion*, penetrate silk nests of fall webworm. *Environ. Entomol.* 6: 591.

Shiga, M. 1979. Population dynamics of *Malacosoma neustria testacea*. *Bull. Fruit Tree Res. Sta.* (6): 59–168.

Snelling, R. R. 1981. Systematics of social Hymenoptera, pp. 369–453 In: Social Insects (Hermann, H. R., ed.), Vol. II, 491 pp.

Snelling, R. R. 1983. Taxonomic and nomenclatorial studies on American polistine wasps. *Pan-Pac. Entomol.* 59: 267–280.

Spradbery, J. P. 1973. Wasps. Univ. Washington Press, Seattle, 408 pp.

Starr, C. K. 1976. Nest reutilization by *Polistes metricus* and possible limitation of multiple foundress associations by parasitoids. *J. Kans. Entomol. Soc.* 49: 142–144.

Strassman, J. E. 1975. Honey catches help female paper wasps (*Polistes annularis*) survive Texas winters. *Science* 204: 207–209.

Strassman, J. E. 1981. Parasitoids, predators and group size in the paper wasp, *Polistes exclamans*. *Ecology* 62: 1225–1233.

Tweeten, L. 1983. The economics of small farms. *Science* 219: 1037–1041.

West-Eberhard, M. J. 1969. The social biology of polistine wasps. Misc. Pubs. Mus. Zool. Univ. Michigan No. 140: 1–101.

West-Eberhard, M. J. 1982. The nature and evolution of swarming in tropical social wasps. In: Social Insects in the Tropics (Jaisson, P., ed.), Vol. I, 280 pp.

Wheeler, A. G. 1977. Studies on the arthropod fauna of alfalfa. VII. Predaceous insects. *Can. Entomol.* 109: 423–427.

Chapter 14

BIOLOGY, ECONOMIC IMPORTANCE AND CONTROL OF YELLOW JACKETS

Roger D. Akre and John F. MacDonald

INTRODUCTION

Yellow jackets are small- to medium-size social wasps belonging to the genera *Dolichovespula* Rohwer and *Vespula* Thomson. Together with the closely related, and usually much larger, hornets (*Vespa* L.) and *Provespa* Ashmead, they comprise the subfamily Vespinae of the Vespidae (Akre and Davis 1978). This group of wasps probably originated in southeast Asia and then extensively invaded the temperate regions (Richards 1971). Yellow jackets are strictly temperate with the possible exceptions of *Vespula squamosa* (Drury) (Miller 1961), and *V. nursei* Archer found in the Philippines (Archer 1981a). There are about 18–19 species of *Dolichovespula* (Archer 1980a, 1981b; Eck 1980, 1981; Edwards 1980) and 21–24 species of *Vespula* (Archer 1981a, 1982, Edwards 1980, Yamane et al. 1980b). Most yellow jackets are predominantly yellow and black, but a few species, such as *Dolichovespula maculata* (L.), and *Vespula consobrina* (Saussure), are black and white. Yellow jackets are the dominant social wasps in the north temperate areas, and are responsible for most problems caused by wasps in these areas. This is particularly true for stinging incidents since colonies can be quite populous and stings are frequently multiple.

Several recent publications (Akre 1982, Akre et al. 1981, Edwards 1980) have extensively reviewed yellow jacket biology and control. Therefore this chapter deals only briefly with the history of control, and concentrates on more recent literature on biology, previously unpublished aspects of economic and medical damage caused by yellow jackets, and the importance of basic studies to possible management and control. Since scant information is available on vespine economic impact on a world basis (Akre and Davis 1978), the chapter is limited mostly to problems with yellow jackets in the United States, supplemented, whenever possible, with information on yellow jacket problems in other areas of the world.

NATURAL HISTORY

Taxonomy

Previously the Vespidae was restricted to the subfamilies Stenogastrinae, Polistinae, and Vespinae (Richards 1962). However, a newer classification scheme has reunited all diplopterous (wings folded longitudinally at rest) wasps into a single family, Vespidae, with six subfamilies (Carpenter 1982). Two subfamilies, the Vespinae (hornets and yellow jackets) and Polistinae (paper wasps, polybiines), include most social vespids. As previously stated, the Vespinae includes the true hornets (Vespa) with about 20 species, Provespa with three species, and the yellow jackets. Only the latter are of immediate concern to this chapter.

American researchers of yellow jackets have recognized only two genera, Dolichovespula and Vespula, based on morphology and biology. However, the species of Vespula are commonly further subdivided into two groups based, again, on morphology and biology (Table 14.1): the Vespula rufa (L.) species group (Vespula of European workers) and the Vespula vulgaris (L.) species group (Paravespula Blüthgen of European workers) (Bequaert 1931, Akre et al. 1981). There is now sufficient biological information (especially nest architecture, foraging behavior, and colony cycles) available to indicate that members of the V. vulgaris species group (Table 14.2) are closely related and distinct from other Vespula, and thus Paravespula should be accepted as a genus. Therefore, Paravespula is used for these species throughout the remainder of this chapter, while Vespula is retained for all other species of Vespula, including members of the V. rufa group (subgenus Vespula) and others (see Yamane et al. 1980b). Additionally, subgroupings of Paravespula have been proposed (Jacobson et al. 1978, Yamane et al. 1980b) for germanica-pensylvanica, and for flavopilosa-maculifrons-structor-vulgaris that indicate close morphological and biological relationships between these species (see also MacDonald and Matthews 1981). While these appear to be natural subgroups, conflicting evidence from immunological studies indicates they may not be as clear-cut as previously thought (Hoffman and MacDonald 1982).

Yellow jacket taxonomy is presently in a state of flux with new species being described and affinities within groups and subgroups being

TABLE 14.1

Distribution and Colony Characteristics of Yellow Jackets in North America (Modified from Akre et al. 1981)

Genus	Distribution	Colony parameters
Paravespula[a] (*V. vulgaris* species group)		Size: 500 to 5,000 workers
flavopilosa	Eastern	Nest: 3,500 to 15,000 cells
		Workers: predators plus scavengers
germanica	Palearctic, now eastern North America and Washington State	Decline: Late Sept.–Dec.
maculifrons	Eastern	
pensylvanica	Western	
vulgaris	Holarctic, transcontinental	
Vespula (*V. rufa* species group)		Size: 75 to 400 workers
		Nest: 500 to 2,500 cells
acadica	Transcontinental	Workers: strictly predators
atropilosa	Western	
consobrina	Transcontinental	Decline: late Aug.–early Sept.
vidua	Eastern	
intermedia	Transcontinental (Boreal)	
austriaca	Holarctic, transcontinental	
Uncertain (currently *Vespula*)		
squamosa[a]	Eastern, southeastern to Mexico	Size: 500 to 4,000 workers
		Nest: 2,500 to 10,000 cells
		Workers: predators plus scavengers
		Decline: Late Sept.–Nov.
sulphurea[b]	Western	Size: 100–1,100 workers
		Nest: 400–3,300 cells
		Workers: predators plus scavengers
		Decline: Sept.–Oct.

(*continued*)

355

TABLE 14.1 (continued)

Genus	Distribution	Colony parameters
Dolichovespula		Size: 100 to 700 workers
arctica	Transcontinental	Nest: 500 to 4,500 cells
arenaria[a]	Transcontinental	Workers: predators
maculata[a]	Transcontinental	Decline: July–Sept.
norvegicoides	Transcontinental	
norwegica (= albida)	Transcontinental (Boreal)	

[a]Species responsible for most stinging problems.
[b]Personal communication, R. E. Wagner, University of California, Riverside; see also Akre et al. 1981.

TABLE 14.2
Distribution of Species of *Paravespula* (= *V. vulgaris* Species Group) with References to Biology and Taxonomy of the Species (*koreensis* Radoszkowski and *orbata* Buysson are omitted from consideration as members of the subgenus *Rugovespula*, Archer 1982)

Species	Distribution	Reference
flaviceps (Smith)	Asia	Yamane et al. 1980b, Iwata 1976
flavopilosa (Jacobson)	Eastern U.S.	Jacobson et al., 1978, MacDonald and Matthews 1981
germanica (Fab.)	Eurasia, now introduced worldwide	Spradbery 1973b, Edwards, 1980, MacDonald et al. 1980a
maculifrons (Buysson)	Eastern U.S.	MacDonald and Matthews 1981
pensylvanica (Saussure)	Western U.S.	Akre et al. 1981
shidai (Ishikawa, Sk. Yamane, Wagner)	Asia	Yamane et al. 1980b
structor (Smith)	Asia	Yamane et al. 1980b
vulgaris (L.)	Holarctic, transcontinental	Edwards 1980, Akre et al. 1981

reexamined and redefined (Archer 1981a, 1982, Edwards 1980, Jacobson et al. 1978, Yamane et al. 1980b). However, most taxonomic problems, including differences of opinion on several species, occur with species remaining in the genus *Vespula*.

Geographical Distribution

Yellow jackets occur transcontinently in the temperate regions of Asia, Africa, America, and Europe. Additionally, three species have been accidentally introduced into areas where they did not occur naturally. Thus, *Paravespula germanica* (Fab.), a species native to Europe, northern Africa, and Asia, now occurs in numerous countries including New Zealand, Chile, Argentina, Australia (including Tasmania), South Africa, the United States, and Canada (summarized in MacDonald et al. 1980a, MacDonald and Akre 1984, Smithers and Holloway 1978, Willink 1980). Also, *P. vulgaris* (L.) was introduced into Australia (Spradbery 1973a) and Hawaii (Howarth 1975). *Paravespula pensylvanica* was introduced into the Hawaiian Islands, and by 1978–1980 was firmly established as a pest species (Nakahara and Lai 1981).

The distribution of yellow jackets in the Asian region had been little studied until Archer (1980a,b, 1982), Iwata (1976), and Yamane et al. (1980b) added distributional records and described several new species mostly from specimens collected for taxonomic purposes. Actual distributions of many of the Asiatic species are still very poorly known.

The distributions of yellow jackets in Europe are more completely known, and were summarized by Edwards (1980), Guiglia (1972), and Spradbery (1973b). Exact distributions for the British Isles have been worked out for the six species of yellow jackets and one hornet occurring there (Archer 1973).

North America has the greatest species diversity with five species of *Paravespula*, eight species of *Vespula*, and five species of *Dolichovespula* (Table 14.1). The 18 species include the introduced *P. germanica* and *V. squamosa* and *V. sulphurea* (Saussure). Although included as *Vespula*, the latter two species are of uncertain status and do not belong in the *V. rufa* species group (*Vespula*). *Dolichovespula albida* has been shown to be conspecific with *D. norwegica* (Fab.) (Eck 1981) and a new species of *Dolichovespula* will soon be described. Distributional maps for North American species are available in Akre et al. (1981) and Miller (1961).

Paravespula germanica, the German yellow jacket, was probably introduced into northeastern North America several times during the past century, but it only became permanently established in the 1960s. Since then it has rapidly spread westward, reaching Winnipeg, Manitoba and Minneapolis, Minnesota in 1979–1980 (MacDonald et al. 1980a). In October 1981 a colony of *P. germanica* was found in Nampa, Idaho and in September 1982 workers were collected in Puyallup, Washington (MacDonald and Akre 1984). Also, during the summer of 1984, workers were collected in Vancouver, British Columbia. Dispersal of this species will probably continue until it occurs in most of northern North America. However, to date, *P. germanica* has not spread into the southern United States.

The only other species increasing its distribution in the United States (Hawaii) is *P. pensylvanica*. First found on the island of Kauai in 1919, by 1978–1980 it had spread to all the major islands of the State, and was causing concern to the agriculture and tourism industries (Nakahara and Lai 1981).

Colony Cycle

The timing of the cycle and colony duration varies somewhat among the genera of yellow jackets. Most species of *Dolichovespula* have a relatively short colony cycle. In California, nest construction by queens of *D. arenaria* (Fab.) (aerial yellow jacket) begins as early as March with colonies dying by July (Duncan 1939) (Table 14.1). In more northern areas (Washington) queens become active in April and nests are initiated April–early June, with colonies entering decline in August. A few colonies of *D. arenaria* and *D. maculata* (bald-faced hornet) may endure until September –October. The species of *Vespula* also have a relatively short colony cycle. Nests are constructed April–early June and are usually in advanced stages of decline by late August–early September. *Vespula squamosa*, an atypical member of this genus that may ultimately be placed in a different genus or subgenus, has a longer colony duration, with colonies declining in late September–early November. A sister species, *V. sulphurea*, apparently has similar habits (Table 14.1). Similarly, all species of *Paravespula* have colonies that do not decline until late September– December. Colony duration is, of course, also related to colony size, and colonies of *Paravespula* and *V. squamosa* are much larger than those of *Dolichovespula* or other *Vespula* (Table 14.1). In warmer areas, such as California, Florida, and Hawaii, an occasional colony continues to expand and becomes perennial (Akre et al. 1981 (summarized), Akre and Reed 1981a, Ross and Matthews 1982).

Nesting Biology

Dolichovespula nests are usually aerial, on branches of trees and bushes, or frequently on the eaves, porches, or ceilings of building (Table 14.3). *Dolichovespula arenaria* nests are constructed a few centimeters above

TABLE 14.3
Nest Site and Architectural Characteristics of Nests of Yellow Jackets
(Modified from Akre et al. 1981)

Genus	Nest site	Nest architectural parameters
Paravespula	Subterranean, inside structures, rarely aerial	Large nest: 3,500–15,000 cells Several worker cell combs Reproductive cells may occur on worker cell comb Suspensoria cordlike Envelope fragile[a] Scalloped envelope paper
Vespula	Subterranean, inside structures, rarely aerial	Small nest: 500–2,500 cells One worker cell comb Reproductive cells never on worker comb Top suspensoria buttresslike Pliable envelope Laminar envelope paper
Dolichovespula	Aerial, sometimes subterranean	Small nest: 500–4,500 cells Several worker cell combs Reproductive cells frequent on worker comb[b] Suspensoria buttresslike Pliable envelope Laminar envelope paper
Uncertain V. squamosa	Subterranean, inside structures, rarely aerial	Large nest: 2,500–10,000 cells Several worker cell combs Reproductive cells may occur on worker cell comb Suspensoria cordlike Pliable envelope Laminar envelope paper

[a]Gray (P. germanica, P. pensylvanica) or tan (P. flavopilosa, P. maculifrons, P. vulgaris).
[b]D. maculata has one worker-cell comb; worker cells rarely, if ever, appear on the second and succeeding combs.

ground in grass clumps to the very tops of trees. They are also found on man-made structures, and sometimes inside wall voids. Nests have been reported as subterranean (Greene et al. 1976, Roush and Akre 1978), and are probably fairly common in mountainous areas of the West. Nests of D. norwegica (Fab.) and D. norvegicoides (Sladen) are usually aerial or built partly into the soil surface (Akre and Bleicher 1985). Most D. macu-

lata nests are constructed in exposed locations, usually a meter or more above the ground. Only one nest of *D. maculata* has been reported as subterranean, and another was inside a hollow tree (see Greene et al. 1976). The sole remaining species of *Dolichovespula*, *D. arctica* (Rohwer), is a social parasite that does not construct a nest, but usurps those of other yellow jacket species.

Dolichovespula use mostly sound, weathered wood for the combs and envelope, although a few nests may have strips of colored envelope from workers incorporating fibers from berries (e.g., barberry) or other colored sources. Envelope paper usually consists of laminar sheets, although the outer roof portion of some mature *D. maculata* nests may have a high degree of scalloping. The paper is stronger, more flexible, and more resistant to water damage than the brittle paper constructed by species of *Paravespula*. Queen nests also have a visible, shiny coating on the pedicel supporting the comb, which makes this support very flexible and tough (Greene et al. 1976, Jeanne 1977, Yamane et al. 1980a). Typical mature nests have two to six combs totaling 300–1,500 cells, although the largest *D. arenaria* nests may have up to 4,300 cells and vigorous *D. maculata* colonies have nests of approximately 3,500 cells. All combs are connected by stout buttress-like suspensoria. Peak worker populations range from 200 to 700 individuals (see Akre et al. 1981) (Table 14.3).

Nearly all *Paravespula* nests are subterranean or in a darkened enclosure (Table 14.3). Most species initiate their nests in rodent burrows or tunnels or in other preformed cavities in soil, logs, or trees. However, all species will also construct nests in wall voids and attics of houses (summarized in Akre et al. 1981). This is especially true of *P. germanica* in North America, which constructs nearly all nests in buildings (MacDonald et al. 1980a). Exposed, aerial, nests are rare for *Paravespula*. We have collected approximately 230 nests of *P. pensylvanica* and 55 of *P. vulgaris* in the Pacific Northwest, but only a single *P. vulgaris* nest was aerial. The nest was attached to the eaves of a house in Friday Harbor, Washington. The only other reported aerial nest of a *Paravespula* was a *P. maculifrons* (Buysson) nest constructed on an exposed building wall (Green et al. 1970).

Paravespula germanica and *P. pensylvanica* nests are grey as most fibers collected by the queen and by foraging workers are of sound, weathered wood. Occasionally the nest envelope is streaked with colors as workers will also collect fibers of painted wood, styrofoam, and even magazines and newspapers. *Paravespula flavopilosa* (Jacobson), *P. maculifrons*, and *P. vulgaris* use decayed wood as a fiber source and the envelope paper and combs are tan to reddish-brown in color. All *Paravespula* paper is brittle, but nests of the latter three species are extremely fragile and easily broken. Queen nests of all species have a flexible pedicel, but it is not visibly coated with a glossy material as in *Dolichovespula* (Akre et al. 1981). Typical mature nests have five to seven combs comprised of 3,500 to 12,000 cells, although some nests are considerably larger (MacDonald et al. 1974, 1980b, MacDonald and Matthews 1981, Akre et al. 1981). Unlike *Vespula* and *Dolichovespula* nests, all suspensoria connecting combs are cordlike (Table 14.3). Peak worker populations vary from 500 to 5,000.

Vespula nests are also primarily subterranean, initiated in rodent burrows, although a few species will also construct nests in wall voids or attics of buildings. However, nests in buildings are fairly rare. For example, 95 nests of *Vespula atropilosa* (Sladen) (prairie yellow jacket) were collected in the vicinity of Pullman, Washington from 1971–1977. Only three were in wall voids or buildings, while a single nest was aerial (Akre et al. 1981). Similarly, only five of 38 nests of *V. consobrina* (Saussure) (black jacket) collected in North America were built inside structures (summarized Akre et al. 1982). Although few nests of *V. vidua* (Saussure) have been collected, it also nests primarily in subterranean locations (MacDonald and Matthews 1976), as does *V. acadica* (Sladen) (forest yellow jacket) (Reed and Akre 1983a).

Vespula use mostly sound, weathered wood as a fiber source for nest construction, and nests are usually grey. *Vespula acadica* will also incorporate some decayed wood fibers into their envelope creating some red or tan stripes (Reed and Akre 1983a). In general, workers do not malaxate the fibers thoroughly, and this results in a loosely woven envelope with many air spaces between the fibers (Roush and Akre 1978, Akre et al. 1982, Reed and Akre 1983a). Therefore, while the envelope is very pliable, it is not very strong nor water repellent. Nests of mature colonies are usually quite small and have one to four combs with 300–2,700 cells (Akre et al. 1981, 1982; MacDonald and Matthews 1976). The suspensoria connecting the uppermost comb to the substrate are buttress-like, while suspensoria between combs are usually cordlike. A few nests may have a mixture of buttress-like and cordlike suspensoria between combs. Peak worker populations range from 53 to about 500 in exceptional colonies. Typical colonies have 100–150 workers. *Vespula consobrina* has somewhat smaller colonies than the other species.

Vespula squamosa is a facultative social parasite of *V. maculifrons*, and probably of *V. flavopilosa* (MacDonald and Matthews 1975, 1984) and *V. vidua* (Taylor 1939). About 80% of investigated *V. squamosa* colonies showed visible remnants of *V. maculifrons* nest construction indicating that the original nest was constructed by a *V. maculifrons* queen whose position was usurped. Gradually the host colony becomes solely populated by *V. squamosa* workers. *Vespula squamosa* nests are typically subterranean, but aerial nests have been reported (Tissot and Robinson

1954). In areas where V. maculifrons commonly nests in the walls of houses, V. squamosa nests are also found in these locations. Presumably, most of these nests were originally V. maculifrons.

Vespula squamosa use sound wood fibers to produce envelope paper and combs that are grey in color. Mature V. squamosa nests typically have 2,500–10,000 cells in three to seven combs, and a peak worker population of 500–4,000 (Table 14.3). All suspensoria are cordlike. Thus, this species has many nest and colony characteristics similar to those exhibited by Paravespula (MacDonald and Matthews 1984).

Colony and Population Dynamics

The most complete and accurate mathematical model of yellow jacket colony and population dynamics to date was constructed from information collected from colonies of Dolichovespula sylvestris (Scopoli) and P. vulgaris in the British Isles (Archer 1980b, 1981c). Dolichovespula, of course, have a short colony cycle, while Paravespula have a much longer cycle (Table 14.1). The model was used to compare the length of colony cycle, the size of the nest, the ratio of small (worker) to large (reproductive) cells, the number of new queens produced, and the number of workers needed to rear each queen. The latter is the ultimate measure of efficiency of the colony, and Dolichovespula were shown to produce more queens per worker than Paravespula, as is typical for small-colony species. The short cycle/long cycle strategies of these two groups of yellow jackets may have originally arisen as a means of separating the species temporally and reducing competition (Archer 1980b).

Models of this type may also someday be used to investigate subtle population cycles of abundance among the yellow jacket species, and may also be used to predict years of extreme abundance or "wasp" years that occur in irregular cycles in various geographical areas of the world (summarized Akre and Reed 1981b). This will require intensive investigation of the influence that parasites, pathogens, competition, and weather have on wasp colonies and populations. Thus far no definitive study has been made of the causes of this phenomenon, but several indirect lines of evidence indicate that weather may be one of the principal factors affecting populations (Akre and Reed 1981b, Madden 1981). This is particularly true in the Pacific Northwest in North America where outbreak populations occur only during those years with a warm, dry spring. On the other hand, in the Southeast, weather does not seem to have much influence on yearly abundance. Nest usurpation is a major mortality factor of early-season V. maculifrons colonies and competition may play a large role in species abundance in the Southeast during a particular year (MacDonald and Matthews 1981). However, there are no published reports showing that years of extreme abundance and relative scarcity occur anywhere else in North America to the extent that this occurs in the Pacific Northwest.

Mathematical models of colony growth may also be useful in yellow jacket control since a sample of a limited number of colonies can be used to predict the correct timing for control measures during a particular year. For example, probably the best time to control scavenging species by the use of toxic baits is when the colonies have entered the exponential growth phase, with the first 150–300 workers present, and surely before reproductive production starts. Analyses of a few colonies early in the season can be used to accurately predict when this period will occur for that year. The model could also be used to indicate if colonies undergoing a particular control treatment were arrested in development, and approximately how rapidly this occurred.

BEHAVIOR

General

Studies comparing the behavior of *Dolichovespula* (Greene et al. 1976), *Vespula* (Akre et al. 1976, 1981, 1982), and *Paravespula* (Akre et al. 1976) indicate a similar repertoire of behavior, with subtle, but significant, differences among the species. These studies have revealed that yellow jacket behavior is much more diverse than previously thought, and behavior has been used, in conjunction with traditional morphological studies, to propose a new phylogenetic sequence for the Vespinae. Yamane (1976) had previously proposed a sequence for hornets and yellow jackets based primarily on larval morphological characteristics. While there is general agreement that the ancestral yellow jacket evolved from true hornets (*Vespa*), the sequence from this point onward is unclear. Yamane suggested that *Paravespula* are closely related to hornets, at the bottom of the yellow jacket lineage, while *Dolichovespula* are the most recently evolved taxon. Greene (1979), however, proposed an alternative phylogeny, based mostly on behavior. His studies of *Dolichovespula arenaria* and *D. maculata* indicated that the behavior of *Dolichovespula* is more simple (primitive) than that observed in *Paravespula*. The foundresses of *Dolichovespula* maxalate prey throughout the season, new queens also exhibit this behavior, and they also trim cell caps after the occupants emerge, behavior reported only for species of *Vespula* (Akre et al. 1982). In addition, reproductive dominance in *Paravespula* seems to be completely pheromonal, and oophagy by the queen does not occur (Akre et al. 1976). Conversely, the queen of *D. maculata* competes with egg-laying workers for cells, and chases and mauls workers. These studies, and morphological similarities, led Greene (1979) to conclude that *Dolichovespula* are the ancestral yellow jacket group. The behavior of *Vespula* (Akre et al. 1982) indicates that they may be intermediate in the phylogenetic sequence, while the complex behavior of large-colony *Paravespula* indicates they should be at the apex as the most recently evolved genus (Akre 1982).

Foraging

All yellow jackets construct paper nests of masticated vegetable fibers, and all forage for this fiber, for carbohydrates and protein in the form of flesh to feed the brood, and for water used in construction, cooling, and drinking.

The foraging behavior of several European species of yellow jackets has been extensively investigated with regard to light, temperature, weather, and daily cycle (see Spradbery 1973b, Edwards 1980). Although workers scavenging for protein will respond to odors, foraging for arthropod prey is largely visual (Gaul 1952). Yellow jacket workers hunt independently and seem unable to communicate the location of a food source to other colony members (Kalmus 1954, Kemper 1961), although Maschwitz et al. (1974) reported that communication occurs in *Paravespula germanica* (Fab.) and *P. vulgaris*. Similar studies of *P. pensylvanica*, however, indicated no communication occurred among workers regarding food location (Akre 1982), and it has been suggested that the results indicating communication occurred are instead due to social facilitation (Parrish and Fowler 1983).

Foraging is highly dependent on light intensity, but yellow jackets will forage whenever light is above a minimum threshold. A definitive study of colonies of *P. vulgaris* in England showed workers started to forage about an hour before sunrise and continued to about an hour after dark (Potter 1964). In general, foraging peaked in the early morning, declined during the middle of the day, and reached a smaller peak in the late afternoon or evening. Foraging for prey was low in the morning and gradually increased throughout the day. Somewhat different results were observed for workers of *V. consobrina* that exhibited maximum foraging for prey from 8:00 to 12:00 AM and for fiber from 8:00 AM to 2:00 PM (Akre et al. 1982).

Just as workers hunt independently, they also forage for wood fibers independently and never share these loads with nest mates when they return to the nest. Rather, they attach them to the envelope and then only rarely will another worker assist in incorporating the fibers. Workers also use the silken cap remnants of reproductive cells to add to the envelope and cell walls, and to reinforce struts above the comb. Nearly all fibers used to construct or lengthen cells are obtained from the envelope. Fiber loads collected outside are rarely added directly to cells (Akre et al. 1982).

Most *Vespula* are predators foraging only for live arthropod prey. An exception is *V. squamosa*, with workers that scavenge for protein (Akre et al. 1981). *Paravespula*, of course, are scavengers on nearly any source of flesh, while *Dolichovespula* rarely scavenge (Greene et al. 1976). *Paravespula pensylvanica* (R. D. Akre, unpublished data) and *P. germanica* (Greene 1982) scavenge early in the season for both protein and vegetable tissues from fruits that are fed to the larvae, and this may be one reason colonies of these species are able to expand so rapidly. Workers of all species of yellow jackets are sometimes attracted to sweets such as fruit juices and other sweet beverages and may be somewhat pestiferous. This attraction to sweets is especially strong when the colony has entered decline.

Few studies have been made of foraging distances of yellow jackets. Arnold (1966), in England, determined that workers of *V. rufa* (L.) and *D. sylvestris* (Scopoli) foraged within 180–275 m of the nest. Similarly, a California study showed that *P. pensylvanica* and *P. vulgaris* workers foraged within 400 m of the nest and did not have any directional preferences (Rogers 1972a). An additional study of the foraging behavior of *P. pensylvanica* workers using small metal tags and a magnetic recovery system determined that 80% of all workers foraged within 340 m of the nest (Akre et al. 1975). These studies indicate that most foraging occurs quite close to the nest, even though workers are capable of foraging over longer distances.

An aspect of yellow jacket foraging behavior that has not been investigated is the possibility that some degree of vertical partitioning or stratification of the habitat occurs among the various species. For example, workers of *V. consobrina* (Akre et al. 1982) and *V. acadica* (Reed and Akre 1983a) hover while foraging, and this ability, rarely observed in other foraging yellow jackets, may aid them in searching dense vegetation for prey. Also, workers were usually observed searching low vegetation for prey and not 2–4 m up in trees as is commonly observed for *D. arenaria*. This suggests that some vertical partitioning of foraging areas may occur.

Mating

Mating behavior of yellow jackets has received little attention (see MacDonald et al. 1974). Few natural pairings have been reported, and mating descriptions are based primarily on caged individuals. A review of the mating behavior of Palearctic species was made by Spradbery (1973b) and Edwards (1980). A typical mating sequence in yellow jackets has been described for *P. pensylvanica* by MacDonald et al. (1974). Additional observations revealed that many matings of *V. consobrina* (Akre et al. 1982) and *V. acadica* (Reed and Akre 1983b) occurred inside the nest. This, coupled with the fact that few new queens are seen after leaving the parent nest (Spradbery 1973b), suggests that sibling matings may be the rule in yellow jacket colonies (Akre et al. 1982). Post (1980) observed five natural matings of *P. maculifrons* in the field so at least some do occur. However, since yellow jackets will mate more than once under correct conditions (MacDonald et al. 1974, Akre et al. 1982), it is possible these matings were of already-mated queens. The inseminated queens then seek a hibernation site, regardless of the time of year, and enter diapause.

Inter- and Intraspecific Competition

Since many yellow jacket queens are still actively searching for nest sites long after many queens have already established nests, inter- and intraspecies competition for these nests is probably frequent (Akre et al. 1977; MacDonald and Matthews 1975, 1981, 1984; Matthews and Matthews 1979). A foundress queen is highly intolerant of other queens and vigorously attacks invaders (Akre et al. 1976). Evidence that these conflicts occur include reports of dead queens found in the entrance tunnels of subterranean nests (Akre et al. 1976, MacDonald and Matthews 1975). The resident queen wins most confrontations (Matthews and Matthews 1979), and therefore most of these dead queens are probably the invaders. Direct evidence of competition is very rare except in cases like *V. squamosa*/*P. maculifrons* where different-colored building materials of the two species leaves architectural evidence of nest usurpation (MacDonald and Matthews 1975). However, some direct evidence for colony usurpation exists. Matthews and Matthews (1979), in a study using marked queens, determined that usurpation rates of embryo nests, mostly *P. maculifrons*, ranged from 23–37% annually over a 3-year period. What role this competition plays in abundance of the various species is totally unknown.

Yellow jackets are opportunistic foragers, and they will accept a wide range of protein and carbohydrate foods. Food has seldom been considered a limiting factor to their populations (Spadbery 1973b). However, with many species occurring in the same areas it seems logical to assume that there is some sort of resource partitioning among the various species. Inferences to partitioning have been made by several authors including Edwards (1980), who stated that *P. germanica* and *P. vulgaris* in England not only forage in slightly different habitats, but that the workers forage at different heights above the ground. Roush and Akre (1978) also noted that workers of *D. arenaria* and *D. maculata* tended to forage high in trees while workers of *P. pensylvanica* and *P. vulgaris* foraged near the ground. Keyel (1983) studied the foraging of *P. flavopilosa*, *P. germanica*, and *P. maculifrons* in New York State. He found these yellow jackets do indeed "partition" the resources, mostly through aggressive encounters at the food source. Workers of *P. germanica* are displaced from food by both *P. flavopilosa* and *P. maculifrons*, but they are still the dominant yellow jackets scavenging on human garbage. Since *P. germanica* workers are recruited to the food more rapidly by social facilitation, they literally "swamp" the opposition (Parrish and Fowler, 1983). In addition, workers of *P. flavopilosa* displaced workers of *P. maculifrons* from both protein and sugar foods. It was also concluded that *P. maculifrons* uses scattered food sources, while *P. flavopilosa* forages more for longer-lasting food sources, and therefore these resources were also partitioned. From this study, it can probably be safely concluded that similar partitioning of food also occurs among the other yellow jacket species in other areas of North America.

In addition to partitioning food resources, yellow jackets may also partition the habitat as to nest sites. It was previously mentioned that the gross distributions of many yellow jacket species overlap geographically. For example, the three species just mentioned, *P. flavopilosa*, *P. germanica*, and *P. maculifrons*, occur together in large sections of eastern North America. Within these areas, there seems to be a tendency to occupy different nesting sites. Therefore, while *P. germanica* nests nearly exclusively in buildings (MacDonald et al. 1980a), *P. flavopilosa* is a subterranean nester in open, disturbed habitats (MacDonald et al. 1980b), and *P. maculifrons* (MacDonald and Matthews 1981) will nest in all areas including disturbed forests and residential areas, but it also nests in undisturbed forests. The social parasite, *V. squamosa*, usurped fewer nests of *P. maculifrons* in undisturbed forest habitats, and no *V. squamosa* nests occurred in hardwood forests (MacDonald and Matthews 1984). The apparent tendency of species to nest in slightly different habitats may reduce competition, not only for nest sites, but for food in the areas immediately surrounding the colonies.

Similar nest site preferences occur among species in the West (Akre et al. 1982). In the areas of overlap, at least some competition for nests occurs, and *P. pensylvanica* queens usurp some *P. vulgaris* nests. Similar divisions of habitat apparently occur between *V. atropilosa*, which nests mostly in drier areas, and *V. acadica*, which is restricted nearly exclusively to dense forests (Akre et al. 1982). There are no studies of competition between these species.

Intracolonial Interactions and Communication

One of the principal activities observed inside yellow jacket nests is a nearly constant trophallaxis among colony members (Spradbery 1973b, Edwards 1980, Akre et al. 1982). This constant contact and exchange of food, in addition to its obvious function, probably plays a large role in maintaining colony odor.

Colony odor must be very important in large colonies for recognition of other individuals as belonging to the same colony. This is especially important for workers that return from foraging. Individuals from the colony are permitted to enter without challenge, while conspecifics from another colony or workers of other species are not permitted to enter and are often attacked. This may be one explanation for "mauling," a vigorous biting by one individual of another, as the recipient remains motionless. Mauling occurs in all studied yellow jacket species and may result when an individual "smells" slightly different (Akre 1982, Akre et al. 1982).

The paper carton nests of these wasps are excellent conductors of sound and there is a growing volume of evidence that sound is a major method of communication among colony members. Ishay and Brown (1975) analyzed sounds produced by the larvae of *P. germanica*, which scrape the sides of their cell walls when hungry, and concluded that this stimulates the adults to feed them. Similar work has been done on *D. sylvestris* and *P. vulgaris* larvae (Es'kov 1977). It is now known that the larvae of most yellow jacket species scrape the sides of their cells to produce an audible sound, and this seems to attract workers to attend them (Akre et al. 1976, 1982, see also Akre 1982). It has been postulated that the sound produced by this behavior functions to let larvae know food is available (Akre 1982).

Among the primitively eusocial wasps such as *Polistes*, stance or overt physical actions are used by the queen to maintain her position in the colony as the primary egg layer (West-Eberhard 1969). *Polistes* colonies usually consist of less than 100 individuals, and dominance hierarchies formed among the colony members by physical means, including fighting, are easily maintained. However, as colonies become more populous, it is probably increasingly difficult, or impossible, to maintain dominance by physical means. It is therefore likely in larger colonies that the dominant position of the queen is maintained chemically by means of a queen pheromone. One of the prime functions of this pheromone is to inhibit egg-laying by workers through supression of ovarian development. Indeed, a queen pheromone was isolated from the head of the Oriental hornet, *Vespa orientalis* F., that seems to serve this function (Ishay 1965, Ikan et al. 1969). Similar pheromones have been postulated for yellow jackets (Table 14.4). Also, in species with smaller colonies such

TABLE 14.4
Pheromones Suggested for Yellow Jackets (Modified from Akre 1982)

Type	Species	Sources	Reference
Alarm	P. germanica P. vulgaris	Venom	Maschwitz 1964, 1966, see also Edwards 1980
Nest marking (footprint)	P. vulgaris V. atropilosa	— 7th sternal gland (?)	Butler et al. 1969 See text
Pupal brooding	P. germanica P. vulgaris D. saxonica D. media	Pupae	Ishay 1972
Queen sex attractant	D. sylvestris V. atropilosa	— —	Sandeman 1938 MacDonald et al. 1974
	V. consobrina P. germanica	— —	Akre et al. 1982 Thomas 1960
Queen pheromone	P. vulgaris P. pensylvanica	— —	Potter 1964 Akre 1982, Akre and Reed 1983
	P. vulgaris V. atropilosa	— —	Landolt et al. 1977
Pacifying phero-mone	D. arctica	Dufour's gland (?)	Greene et al. 1978
Appeasement phero-mone	P. vulgaris	gaster	Franke et al. 1979
Wasp assembly	Vespula Paravespula Dolichovespula	Insemi-nated queens, also workers	Ishay and Perna 1979

as *Dolichovespula* and *Vespula*, queen control may be maintained by a combination of physical dominance and pheromones, while in large-colony species (*Paravespula*) it is probably solely pheromonal (Akre 1982). Additional pheromones have also been suggested. For example, Butler et al. (1969) reported a nest-marking (footprint) pheromone for *P. vulgaris* that was placed about the nest entrance. This pheromone also likely occurs in other yellow jacket species. Workers of *V. atropilosa* have been seen alighting up to 20 cm away from their nest entrance on vertical surfaces of buildings and following what appears to be a chemical trail to the entrance. Workers also drag the tip of their gaster along the surface. Covering or washing the trail thoroughly disrupts returning workers for a short time (unpublished observations). Investigations of *Vespa crabro* L., the European hornet, suggest that a possible source of a pheromone may be the seventh sternal gland (Wheeler et al. 1982). In addition to the pheromones listed in Table 14.4, a male wasp assembly, a building initiator, and a building depressor pheromone have been suggested (Ishay and Perna 1979). However, most reports of pheromones are tentative and are based primarily on behavioral evidence. None of these proposed pheromones has been isolated, chemically identified, or bioassayed.

Colony Defense

Most reports of vespine nest defense indicate that the inhabitants become alarmed by vibration of the nest or by the entry of intruders (Edwards 1980, Gaul 1953). Indeed, the slightest vibration of a yellow jacket nest results in an immediate outpouring of workers to investigate the cause of the disturbance. Then, unless the source of the disturbance is readily apparent, the aroused wasps fly in ever increasing circular paths in search of the intruder (Gaul 1953). Any movement is rapidly investigated, and the wasps concentrate their attack on fleeing nest invaders. In most instances these defensive flights are relatively short in both distance and duration. Although Gaul (1953) reported that defensive flights were never more than 7 m from the nest, this distance varies among species and individual colonies. Thirty meters has been reported as the pursuing distance for workers of *D. maculata*, the bald-faced "hornet" (Rau 1929), and workers of *V. acadica*, *P. pensylvanica*, and *P. vulgaris* will pursue vertebrate predators at least that far (unpublished observations). Apparently, duration of defensive flights increases with a rise in temperature and may last from 90 seconds to five minutes or longer on hot days (Gaul 1953).

Only wasps inside the nest at the time of the disturbance participate in defense, and returning foragers neither participate, nor communicate alarm to the nest inhabitants (Edwards 1980, Gaul 1953, Potter 1964, Spradbery 1973b). The number of wasps responding to a disturbance is roughly proportional to the extent of the disturbance (Gaul 1953), but each wasp apparently acts independently to locate the source of the disturbance (Gaul 1953, Spradbery 1973b). However, alarm pheromones have been reported for *P. germanica* and *P. vulgaris* (Maschwitz 1964, 1966), presumably associated with the sting (Edwards 1980). Conversely, experiments on North American yellow jackets have revealed no indications of an alarm pheromone (Akre 1982), but this area must be thoroughly investigated before any definitive conclusions can be drawn. If alarm pheromones exist among yellow jackets, they surely do not cause mass recruitment of workers. It seems more likely that sound (vibration), coupled with visual stimulation, is responsible for most defensive reactions.

Colony aggressiveness is related to colony size, with smaller colonies usually responding less aggressively (Spradbery 1973b). This is reflected somewhat by the defensive reactions of individuals on the nest at various stages of development. When a queen nest is jostled, the queen responds by immediately coming out on the nest envelope, and rapidly encircling it several times before reentering the nest (Yamane and Makino 1977). She rarely tries to sting a vertebrate predator, but instead flies away. Invertebrate predators or parasites, including other yellow jackets trying to usurp the nest, are vigorously attacked with both mandibles and sting. As the colony grows to the point where about ten workers are present, one or two serve as guards and investigate all returning workers and any nest disturbances (Akre et al. 1976). Eventually more workers serve as guards, until the nest entrance is completely surrounded, by as many as 20 individuals, in mature *P. pensylvanica* colonies. The number of guard workers is roughly proportional to colony size. For example, *Paravespula* colonies are more populous than *Vespula* colonies, and have many more guards. *Vespula consobrina* (Saussure), which has colonies consisting of less than 200 workers, averaged 2.5 guards in the entrance at any one time (Akre et al. 1982). The number of guards varied by time of day with the greatest number present from 8:00 to 10:00 PM. Large colonies of *P. vulgaris* have more guards, but timing is quite similar. The greatest number of guard workers are present in the morning before foraging begins, and in the late evening when foraging wanes (Potter 1964). Numerous guards are also present during inclement weather when few workers are able to forage.

The chemistry of yellow jacket venom indicates that it functions primarily as a defensive weapon against vertebrate predators (Schmidt and Blum 1979). The venom causes intense pain when injected by the sting apparatus, and this frequently causes the predator to cease the attack. In addition, workers of *D. arenaria* can spray venom which causes intense urtication to human eyes (Greene et al. 1976), and probably causes a similar reaction in most other vertebrates. This aspect of vespine defense has been little investigated, and it is possible other species also spray venom (see Maschwitz and Kloft 1971).

In spite of these defenses, yellow jacket colonies are attacked and eaten by bear (Bigelow 1922, Buckell and Spencer 1950, MacDonald and Matthews 1981), skunks (Preiss 1967), raccoons (MacDonald 1977), coyotes (Akre et al. 1981), badgers (Edwards 1980), and birds (Cobb 1979). Predation can be heavy, as 13 of 27 late-season *P. maculifrons* colonies were destroyed by raccoons near Athens, Georgia (MacDonald 1977) and

119 of 151 *P. maculifrons* colonies were eaten by skunks in Delaware over a two-year period (Preiss 1967). Published records of vertebrate predation are few, but predation in other sections of the country, such as in the Pacific Northwest, seems to be much lighter (Roush and Akre 1978).

The relatively low rate of parasitism of vespine colonies by invertebrates suggests that defense against these invaders is quite effective. Indeed, biological control agents for pestiferous yellow jackets are frequently proposed, but no effective agent exists, and none is forseen (MacDonald et al. 1976).

Although yellow jacket colonies must succumb to diseases, no studies have been made of possible pathogenic organisms or vespine defenses against them. Norepinephrine has been found in the nest envelope of *P. germanica* (Lecomte et al. 1976) and *P. vulgaris* (Bourdon et al. 1975). Perhaps this material serves as a protection against microbial invaders or fungi.

PHYSIOLOGY

Diapause

In midsummer or fall yellow jacket colonies produce reproductives. About a week after emerging from the pupal cocoon, the queens mate, and shortly thereafter seek sheltered locations in which to hibernate for the winter. During this time, the queens are in reproductive diapause. Their ovaries are totally undeveloped and nonfunctional. However, under certain conditions the new queens do not enter reproductive diapause after mating, but instead join the parent or another colony as functional queens. This sometimes results in perennial colonies with multiple queens. Queens become active again during the first warm days of spring and begin to forage for nectar from flowers and for arthropod prey. Reproductive diapause is soon broken, and the ovaries develop rapidly. Simultaneously, the queens begin to construct nests and to rear brood.

Perennial colonies occur only in the warmer temperate or subtropical areas. For example, perennial colonies of *P. germanica* have been reported from Algeria and Morocco (Vuillaume et al. 1969), Chile (Jeanne 1980), New Zealand (Thomas 1960, Spradbery 1973b), and Tasmania (Spradbery 1973a). Perennial colonies of *P. pensylvanica* and *P. vulgaris* occur in California (Duncan 1939, R. E. Wagner, personal communication), and huge perennial colonies of *V. squamosa* have been recorded from Florida (Akre et al. 1981, Ross and Matthews 1982, Tissot and Robinson 1954). In addition, perennial colonies of *P. pensylvanica* have been reported from Hawaii (Nakahara 1980, Nakahara and Lai 1981), and possibly from Vancouver, British Columbia (Spencer 1960). Also, a large, polygynous colony of *P. pensylvanica* was discovered near Prosser, Washington in 1979 (Akre and Reed 1981a). Some perennial colonies are exceptionally large, and one nest of *P. germanica* was estimated to weigh 454 kg (1,000 lb) (Spradbery 1973b).

There are no reports of perennial colonies of *Dolichovespula* nor *Vespula*. Perhaps this is because colonies are usually quite small, and species of *Vespula* (excluding *V. squamosa*, *V. sulphurea*) have nests with only one worker comb.

Paravespula germanica often has perennial colonies in New Zealand, with 1,000 or more functional queens. Spradbery (1973a) postulated that the causative factors for these perennial colonies is photoperiod coupled with a mild climate, and they are not due to a genetic, nondiapausing biotype. In normal colonies there is a diminished animosity toward foreign queens and workers as the foundress becomes older. In New Zealand, new queens that emerge when daylength is less than ten hours apparently do not enter ovarian diapause, but instead mate and rejoin the colony as functional queens. Since sufficient prey is available year around, the colony becomes perennial. Spradbery concluded, from comparisons of data from colonies in Europe and those in New Zealand, that new queens enter diapause only when subjected to daylengths of 10–14 hours. New queens of *D. arenaria* in California emerge and mate in late June–July (Duncan 1939) when photoperiod is longer than 14 hours, and they immediately enter reproductive diapause. Similarly, new queens of *Vespula atropilosa*, *V. consobrina*, and *V. acadica* in Washington State emerge August–early September. They also enter diapause after mating even though daylength exceeds 14 hours. This suggests that the photoperiod parameters set by Spradbery for queens entering diapause are probably too narrow or vary among species. The upper limit is probably 16 hours or more of light. Since colonies in the United States are typically initiated during April–June and reproductives are produced when daylength is decreasing, it is unlikely any perennial colonies result from new queens being exposed to daylengths of 16 hours or more. Perennial colonies more likely result from nondiapausing queens produced very late in the year (November–January) when photoperiod is less than 10 hours.

Nutrition

Yellow jackets feed on animal flesh, mostly in the form of arthropod prey, and on carbohydrates usually collected as nectar from flowers, honeydew from homopterans, and from plant saps (Spradbery 1973b). Hibernating queens, when they first emerge, visit flowers for nectar, and shortly thereafter start to hunt arthropod prey. Since adult yellow jackets have a very long, thin esophagus extending from just behind the head all the way through the thorax to the crop, it is impossible for them to ingest much solid food (see Edwards 1980). However, when the queens kill and malaxate prey, they imbibe most of the body fluids before discarding the dry remnant. Proteins are required for egg maturation, and prey is necessary to sustain egg production. Once a nest is initiated and larvae are present, the queen feeds them a diet of flesh, with some liquid protein, and, possibly, nectar.

When the first four to seven workers emerge in the nest, they soon take over all feeding and nest maintenance duties, and the queen no longer leaves the nest to forage. From this time onward, she is dependent on food brought into the nest by workers. However, the queen continues to share prey brought into the nest, and she also obtains nourishment from larval oral secretions (saliva). This gradually changes as the colony grows larger and more larvae are present, until the queen obtains nearly all her nourishment from larvae. Only queens of D. arenaria (and D. maculata, Greene, personal communication) continue to take prey from returning workers late into the colony cycle (Greene et al. 1976). Workers also solicit and receive secretions from larvae, and these contain many sugars and soluble proteins that should be highly nutritious (see Edwards 1980, Hunt 1982). All adult wasps have proteases (Spradbery 1973b, Grogan and Hunt 1977), and workers imbibe juices from the prey they capture before it is fed to the larvae.

When new queens and males emerge, late in the colony cycle, they are nearly exclusively dependent on larval secretions for nourishment, at least during the first few days (Akre et al. 1976, 1982). However, as more reproductives emerge, the demand on the larvae increases to a point where they are constantly solicited for secretions. The larvae respond less and less readily to these demands, and the reproductives then begin to vigorously solicit returning workers. Only new queens and males of D. arenaria (Greene et al. 1976), D. maculata (A. Greene, personal communication), and V. acadica (Reed and Akre 1983b) have been reported to take prey from workers. This is thoroughly malaxated before the completely dry remnant is discarded, or, in the case of D. arenaria, it may be fed to larvae. All other studied species rarely solicited prey, instead they were fed only liquids.

Workers are highly attracted to new queens and rarely refuse to feed them. However, males are fed readily only through the early stages of colony decline. Thereafter, they are mauled frequently, and ultimately vigorously attacked until they leave the nest.

While the usual prey items consist of various small arthropods, species of Paravespula also scavenge extensively. Nearly all types of flesh are collected, and P. germanica (Greene 1982) and P. pensylvanica will also scavenge for plant tissues, such as the pulp of apples and other fruits, to feed the larvae (R. D. Akre, unpublished observations). With a constant source of garbage available in most localities, this enables colonies of the synanthropic P. germanica to increase in size much more rapidly than colonies of yellow jackets that scavenge little or not at all (Greene 1982).

Sensory Physiology

The compound eyes of yellow jackets are comprised of a large number of ommatidia, suggesting that these wasps have a high degree of visual acuity. Whether this is true is largely unknown, but there is no question that vision is the primary sense used in foraging and in orienting to the nest (summarized, Edwards 1980, Spradbery 1973b). For example, when workers leave the nest for the first time, they circle around or fly back and forth in front of the nest entrance for a minute or more. They frequently reenter the nest, and repeat this procedure several times before flying any distance from the nest. During this orientation the positions of objects immediately around the nest entrance are learned. This type of learning also seems to extend to the learning of figures since workers of P. vulgaris can distinguish patterns placed beneath a feeding dish. However, they are apparently not as efficient as honeybees in this regard (Mazokhin-Porshnyakov and Kartsev 1979). However, their visual acuity is not as sharp in other situations, since workers of nearly all species will repeatedly attack nail heads on the sides of light-colored buildings. They are presumably mistaken for prey items. Nevertheless, vision plays a major role in prey location and capture. Foraging workers thoroughly investigate vegetation for prey, and nearly any anthropod within an acceptable size range is attacked. The slightest movement of the prey appears to trigger the attack, although odor may also play a role. Searching workers extend the antennae toward possible prey, and nearly touch or touch it.

Yellow jackets have at least some degree of color vision. Mazokhin-Porshnyakov (1960) showed that workers of P. vulgaris, D. sylvestris, and V. rufa can distinguish green, yellow, and orange, while Menzel (1971) determined that workers of P. germanica have three types of color receptors for ultraviolet, blue, and green. However, field tests of colored bait materials for yellow jacket control indicate that yellow jackets may be indifferent to colors, with the possible exception that lighter colors or shades are preferred (Edwards 1980). Of course, this may not be true for all species, as more workers of V. squamosa were attracted to traps painted two shades of yellow than were attracted to traps left white or painted other colors (Sharp and James 1979).

Yellow jackets are extremely perceptive to odors. The workers of Paravespula detect odors of flesh for several meters' distance, and if a slight breeze is blowing, they can be seen following the air currents to the source. Thus, they very rapidly find freshly killed animals or garbage to scavenge. They are also attracted to the odors of overripe fruit, and to specific chemical attractants such as 2,4-hexadienyl butyrate and heptyl butyrate (Davis et al. 1967, 1969). The latter material is especially attractive to workers and queens of P. pensylvanica and V. atropilosa in the western United States. Yellow jackets are also attracted to certain perfumes, hair sprays, and lotions, and it is recommended that the use of these materials be avoided by individuals hypersensitive to stings (Akre et al. 1981).

The principal sense receptors for odors are located on the antennae. Indeed, the antennae are literally covered with a variety of sense organs from special hairs to various types of peg and plate sensilla (Spradbery 1973b).

Insecticides

Few data are available on the susceptibility of yellow jackets to various insecticides, and then most mortality data are primarily based on whether treated colonies died or if a particular population was reduced in numbers to an acceptable level (Akre et al. 1981). In the only carefully controlled laboratory investigation of the toxicity of insecticides to yellow jackets, toxicants were applied to individual P. pensylvanica workers by means of a microapplicator (Johansen and Davis 1971). Nine materials were tested, and it was determined that methomyl was very highly toxic ($LD_{50} < 1\mu g/g$); malathion, diazionon, carbaryl, and methoxychlor were highly toxic ($< 10\mu g/g$); and chlordane, DDT, and toxaphene were moderately toxic ($< 100 \mu g/g$). Some of these materials are no longer available for use because of potential environmental hazards. Currently registered materials that can be used for colony control by application to the nest include chlorpyriphos (Caron 1974), propoxur, carbaryl, resmethrin (see Akre et al. 1981), malathion, diazinon, and pyrethrins. In addition, the treatment of garbage containers with an aqueous spray of DDVP has been shown to greatly reduce populations in the immediate area (Wagner 1961). Thus, a number of insecticides can be used for colony control when the location of the nest is known. Unfortunately, control of yellow jackets is frequently necessary when nest locations are unknown.

ECONOMIC AND SOCIAL IMPACT OF YELLOW JACKETS

The direct and indirect impact of yellow jackets on humans has been addressed by numerous authors (see Akre et al. 1981). The unifying theme in these reports is that although yellow jackets are serious nuisances and stinging threats and may cause economic losses to various industries, typical evaluation of impact that one associates with other economically important groups is extremely meager. Furthermore, these reports suggest that the yellow jacket problem is increasing in importance as a result of human encroachment into previously undisturbed areas and because of the emergence of several highly synathropic species. Our most significant pest groups are the *V. vulgaris* species group (*Paravespula*), and *D. maculata* and *D. arenaria*. Not only may these species be closely associated with human activities, but some attain high population densities in given areas during certain years.

We have attempted to quantify some aspects of yellow jacket impact, but few data exist and are difficult to obtain. We will focus on the impact of foragers, and colonies; discussing their nuisance value, the stinging threat, and physical damage to structures and products. First, though, we begin with a discussion of the benevolent attributes of yellow jackets.

Beneficial Aspects

Numerous authors (Akre et al. 1981, Duncan 1939, Edwards 1980, MacDonald et al. 1974, Spradbery 1973b) have suggested that yellow jackets are beneficial predators. Still, nearly all accounts are anecdotal, and rigorous evaluation of yellow jackets as natural biocontrol agents is yet to be conducted.

Workers of yellow jackets are general predators that remove enormous numbers of insect prey within their foraging range, the amount dependent on colony size and colony density within a given area. Large *Paravespula* colonies, with nests of over 15,000 cells, produce more than 25,000 workers during a season and in some very productive habitats colony density can be quite high. For example, MacDonald and Matthews (1981) reported 44 *Paravespula* colonies along a 300-m stretch of stream. Average nest size was about 8,000 cells, representing colonies producing 15,000 workers during a season. Based on these data, this 300-m stretch of stream was patrolled by nearly 700,000 workers. Assuming each worker captures 50 prey items per day for each day of its 20-day career, workers in this area remove nearly 7×10^8 prey items each season. These figures are gross estimations and do not indicate if the prey taken is of economic importance. Still, they provide some indication of the importance of yellow jackets as predators.

Medical Importance

The ability of yellow jacket workers to sting and the potential of the venom to induce sensitivity in some people, with possible severe allergic reaction on subsequent stinging episodes, constitutes the greatest impact of yellow jackets on humans (Akre et al. 1981). This section focuses on the accelerating new developments associated with this problem. Researchers and other interested individuals without a background in medicine and immunology will find the article by Lichtenstein et al. (1979) and the pamphlet on "Insect Allergy" by NIAID (1981) helpful.

Assessment

Based on old reports (Barnard 1973, Barr 1971, Parrish 1963), between 50 to 100 people in the United States die each year as a result of bee and wasp stings. While mortality seems extremely low (Rubenstein 1980), Parrish (1963) stated that many deaths purportedly due to heart attack, heat stroke and natural causes, especially those of people engaged in outdoor activities such as hiking, fishing, hunting, etc., could have very well been the result of hypersensitive reactions to Hymenoptera venom; most recent reports by various physicians reveal the same belief. Almost no additional estimates of deaths exist, but Ennik (1980) reported a death rate of 0.1 per 100,000 due to stings, based on death certificate data from California. Extrapolating from Ennik's estimate, over 200 people may die each year in the United States as a result of stings by wasps and bees. All reports of mortality (Ennik 1980, Barr 1971, Barnard 1973, Parrish 1963) indicate a median death age of over 40, with deaths among children extremely uncommon (Lichtenstein et al. 1979).

Morbidity associated with stings is much greater, with most estimates (Chaffee 1970, Settipane and Boyd 1970, Settipane et al. 1972, 1978) indicating that between 0.4% and 0.8% of our population has experienced a serious reaction following bee or wasp stings. Extrapolating from these figures, 1–2 million people are hypersensitive to Hymenoptera venom and many may be at risk of life-threatening reaction on subsequent stinging episodes. Lichtenstein et al. (1980) stated that morbidity probably is vastly greater than documented. They also commented on the drastic changes in life style that may result to individuals who have experienced a life-threatening reaction. Although data on mortality and morbidity are meager, we feel safe in stating that venom hypersensitivity constitutes a major public health problem in this country. Based on case histories, immunological surveys, and biochemical research, yellow jacket and honeybee venoms appear to be the two major causes of deaths and serious hypersensitive reactions (King et al. 1978, Lichtenstein et al. 1979). However, the imported fire ant is an increasing problem in the southern United States (see Chapter 8).

Venom

Honeybee venom has been thoroughly studied and its components characterized as to chemical structure and pharmacological activity (Habermann 1972, Schmidt 1982). Similar research has been conducted on the

Oriental hornet, *Vespa orientalis* (Edery et al. 1978), but detailed study of yellow jacket venoms has only recently begun (Geller et al. 1976, Hoffman et al. 1981, King et al. 1978).

Table 14.5 shows the major components of yellow jacket venoms, including their chemical classification and pharmacological activity. All

TABLE 14.5
Pharmacologically Active Components of Yellow Jacket (*Dolichovespula*, *Paravespula*, *Vespula*) Venoms (Compiled from Blum 1978, Geller et al. 1976, Habermann 1972, Klein et al. 1981, Schmidt 1982, Schmidt and Blum 1979, Schmidt et al. 1980, Tu 1977)

Type	Material	Effect
Enzymes (Macromolecules) (ca. 25%)	Phospholipase A Phospholipase B	Tissue hydrolysis by cell destruction, inhibit coagulation of blood, act strongly as antigens capable of inducing strong allergic response
	Hyaluronidase	Increased permeability of tissue, "spreading agent"
	Acid phosphatase	Important factor in allergic responses
	Histidine decarboxylase	Converts host tissue histidine to histamines (see Blum 1978)
Peptides (Moderate-sized Molecules) (ca. 75%)	Vespulakinin 1 and 2	Induces histamine release, intense pain
	Neurotoxin (see Schmidt and Blum 1979)	Paralysis in invertebrates, unknown for vertebrates
	Mastoparan	Degranulates mast cells, destroys blood cells
Biogenic amines (Micromolecules) (small fraction)	Histamine	Pain, itch, vasodilation
	5-hydroxytryptamine (Serotonin)	Pain, vasodilation
	Dopamine	Vasoconstriction
	Epinephrine	Vasoconstriction
	Norepinephrine	Vasoconstriction
Free amino acids (Micromolecules)	Many, see Klein et al. 1981	Unknown, may be excretory products

materials found in the investigated species are listed, and any single species may not have one or more of these materials in its venom. Earlier reports (Feingold 1973, Habermann 1972) indicated that yellow jacket venom was a complex mixture of proteins and biogenic amines that may cause contraction of smooth muscle, stimulate glands of external secretion, increase vascular permeability, cause hypotension, destroy normal tissue barriers, and induce immediate and intense pain. However, their most significant role is as allergens, some of which may induce hypersensitivity in certain people. King et al. (1978) reported that yellow jacket venom consisted almost entirely of proteins (about 75% peptides and 25% enzymes) with a small proportion (by weight) of biogenic amines. They found three major enzymatic proteins in yellow jacket venoms, a hyaluronidase, a phospholipase, and an undetermined protein they named antigen 5 (Ag5); the latter appears to be the major allergen. However, more recent evidence strongly suggests Ag5 is a fragment of phospholipase (Alagon et al. 1982). Significantly, there is no allergenic or antigenic cross-reactivity between these yellow jacket proteins and similar proteins found in honeybee venom; however, cross-reactivity between venoms and individual allergins existed among the tested yellow jacket species (King et al. 1978). As Lichtenstein et al. (1979) point out, these interactions have not been fully determined and are of the greatest clinical importance.

King et al. (1978) focused on proteins while peptides and amines were discussed by Edery et al. (1978) and Schmidt (1982). The most noteworthy are the kinin or kinin-like peptides which, thus far, have only been found in the venom of vespid wasps. These compounds, known as vespakinins and vespulakinins, are pharmacologically active and more than likely allergenic, but their role as allergins has not been evaluated.

Reactions to Venoms

Injected venom produces immediate, intense pain and local inflammation, but subsequent symptomatology is quite variable and dependent on an individual's immunological status. Nearly all fatalities resulting from stinging are the result of a hypersensitive reaction to the venom, but very few people suffer such a reaction. Although the allergic reaction is responsible for most deaths, relatively uncommon complications such as neurological and vascular pathology are known (Light et al. 1977); however, deaths due to the toxic effects of massive amounts of venom injected in multiple sting episodes are extremely rare.

Table 14.6 summarizes the various categories of reactions to yellow jacket venoms. Local reactions typically are self-limiting and may be ameliorated by application of a cold compress. Such local reactions are dangerous only if occurring in the neck region. These reactions are more

TABLE 14.6
Reactions Associated with Yellow Jacket Stings (Modified from Frazier 1976, Lichtenstein et al. 1979, Light et al. 1977, NIAID pamphlet 1981)

Type of reaction	Cause	Onset	Symptoms
Normal (local)	Venom itself	Immediate	Pain, redness, swelling, itching at site
Toxic	Venom itself (multiple stings)	Immediate	Same as above, plus muscle cramps, headache, fever, drowsiness
Allergic	IgE-mediated (one or more stings)	Immediate to delay of hours	*Local:* swelling and itching with hives at site and involving extensive additional swelling *Systemic:* complications away from sting site Slight: cutaneous involvement with general itching and possible outbreak of hives, feeling of malaise and anxiety Moderate: respiratory distress, abdominal pain, dizziness, general swelling Serious: labored breathing and swallowing, affected speech, confusion, feeling of impending disaster Anaphylactic (shock): hypotension, cyanosis, vascular collapse, unconsciousness
Delayed	?	Hours to several days after stings	Serum sickness-like symptoms (painful joints, fever, hives, swollen lymph glands) Neurological Nephrotoxic Encephalopathic

381

manifest in some individuals, and may persist for a few days and involve substantial swelling around the sting site; however, unless symptoms appear elsewhere, these reactions are nonallergic. Allergic or systemic reactions vary in intensity and are mediated by IgE (Lichtenstein et al. 1979). Such a reaction may only involve the skin (cutaneous systemic) and is characterized by itching away from the sting area itself and possibly the outbreak of hives. More serious systemic reactions involve certain internal systems including smooth muscle, lungs, and heart, and the person may experience feelings of impending disaster. Most deaths result from anaphylactic shock and symptoms include hypotension, respiratory failure, and loss of intravascular volume. Such reactions are considered explosive and death typically results within minutes to an hour (Barnard 1973).

Emergency Treatment

There exists general agreement that the initial step in the emergency treatment of anaphylaxis is the injection of epinephrine. Antihistamines and corticosteroids are ineffective in combating this condition (Lichtenstein et al. 1979). Therefore, immediate injection of epinephrine following a stinging episode and prior to onset of any symptom is the first step, barring a history of cardiac condition. Emergency treatment should also include application of a cold compress to retard absorption of the venom. Epinephrine will not reverse all anaphylactic reactions, and more rigorous therapy may be required.

Individuals with known hypersensitivity should wear a Medic Alert identification bracelet that states the allergic condition so that emergency therapy can be administered. This is particularly important if the victim should faint or become unconscious following stinging. Medic Alert tags can be ordered from the Medic Alert Foundation, Box 1009, Turlock, California 95380.

Prophylaxis

Nearly all physicians advise individuals with known hypersensitivity to wasp, bee, or ant venom to carry a sting emergency kit that contains, among other ameliorative materials, a syringe of epinephrine. The kit should be checked periodically to ensure it is in working order.

Hypersensitive people should incorporate a few life-style changes to preclude possibilities of receiving stings. For example, such individuals should avoid outdoor activities during the yellow jacket season. They are further advised to avoid use of sweet smelling lotions and the wearing of brightly colored clothing. The use of shoes, long pants, and long-sleeved shirts are also recommended.

Venom Immunotherapy

Until the past few years, desensitization of people with allergic reactions to hymenopteran stings was based on whole-body extracts as the therapeutic agent. Diagnosis of these people was also based on whole-body extracts. Although some allergists questioned this approach, it was not discredited until Hunt et al. (1978) conducted a controlled experiment. This study revealed the ineffectiveness of whole-body extract desensitization and demonstrated the efficacy of pure venom immunotherapy. Pure venom products are now commercially available, although they are expensive.

Pure venom immunotherapy involves the serial injection of pure venom to induce the production of venom-specific blocking antibody. The injection regimen typically employed, known as the modified "rush" regimen (Golden et al. 1980), involves a relatively rapid buildup of dosages over a short period of time compared to other allergy desensitization programs (Lichtenstein et al. 1979).

Unfortunately, diagnosis of insect sting hypersensitivity is presently imperfect, and since anaphylaxis is mediated not only by immunological mechanisms but also by physiologic, neurologic, and hormonal mechanisms, it is unlikely that any simple immunological test will ever provide a definitive answer as to who is at risk (Lichtenstein et al. 1979, 1980). Although three different immunological tests are available, including skin test, in vitro histamine release test, and radioallergosorbant test (RAST), the former is believed to be the best diagnostic tool, although equivocal tests are not uncommon.

Economic Impact

The economic impact directly associated with the stings of yellow jackets is substantial, and we have attempted to gather data as to the actual costs per person and then extrapolate this cost to a national level. Our intention is to provide an indication only, and while our projections are only estimates, they certainly suggest a major cost associated with sting hypersensitivity. Most insurance policies now cover the cost associated with the pure venom immunotherapy program.

One of the costs associated with desensitization is for pure venom. For example, honeybee venom costs $0.07 (U.S. dollars are used in this chapter) per mg, Polistes venom $40 per mg, and mixed yellow jacket venom $35 per mg (Sigma Chemical, St. Louis). However, only very small amounts of venom are used in a treatment program, so most costs must be associated with medical personnel and clinical procedures.

Using a modified "rush" regimen, which employs 1.345 mg of venom to attain maintenance level, yellow jacket immunotherapy costs over

$1,000 the first year, and nearly $300–$500 a year thereafter to maintain protection. If half of the 1–2 million people in this country with known hypersensitivity are engaged in immunotherapy, then the national costs the first year would be approximately $75 million and maintenance costs would approximate $25 million per year. Perhaps the best "buy" is the sting emergency kit which is available for about $12.

Residential and Recreational Areas

Although the greatest threat to individuals associated with yellow jackets is that of being stung, overall the greatest economic and social impact is probably due to nuisance value alone. Most people are either terrified of or at least not willing to tolerate yellow jacket foragers nearby or at one's food, and/or the existence of a colony in one's yard or home.

Quantifying these factors and establishing nuisance thresholds for yellow jackets may not be possible. However, we and other authors (Edwards 1980, Fluno 1961, Spradbery 1973b) suggest that yellow jackets adversely affect millions of people every summer and fall. Ascerno (1981) analyzed the 20 most frequent inquiries pertaining to insects by homeowners received by the Insect Information Service, University of Minnesota. The report revealed that from 1976 to 1979, inquiries about wasps ranked first three years and second the other year, averaging over 1,000 per year and accounting for about 10% of all inquiries. Of course, "wasps" include a number of different groups, but yellow jackets likely were responsible for most requests. Similarly, extension offices in New York State receive more calls about stinging insects than about any other pest (Heath, 1982). Akre and Reed (1981b) reported receiving about 20 calls per day in late summer and fall in 1979, an outbreak year for yellow jackets in the Pacific Northwest; the same year they received over 7,000 requests for an information bulletin on yellow jackets. The Department of Entomology, Purdue University, received about 10 to 15 calls per week during late summer and fall every year since 1977 from homeowners with colonies inside their houses, park personnel, and representatives of food-processing industries. An information bulletin (MacDonald 1980) was requested so often within a month of its initial release that a charge had to be implemented to cover mailing and new reproduction. Finally, the comprehensive bulletin by Akre et al. (1981) was sent to over 30,000 individuals and institutions the first year.

It is difficult to evaluate the impact of yellow jacket foragers in and around human dwellings, and at outdoor gatherings. We know that for many people outdoor activities, particularly those associated with eating, become either unpleasant or completely impossible in late summer and fall. Numerous reports are available from the Pacific Northwest and midwest of recreational areas such as parks, resorts, campgrounds, and amusement parks experiencing substantially reduced visitation and even closures during outbreak years (Akre et al. 1981, Akre and Reed 1981b). For example, the latter developed an index of forager activity, and concluded that a forager density of three to seven per square meter constituted intolerable levels. As alluded to earlier, similar outbreaks are less obvious in the eastern United States (MacDonald et al. 1976). However, the upper midwest appeared to experience a greater than "normal" yellow jacket year in 1982. The "outbreak" involved both *P. germanica* and *P. maculifrons*. One of us (JFM), working with park personnel in Lafayette, Indiana, witnessed the impact of intolerable numbers of *P. germanica* foragers on park visitors in late August to late September 1982. Park utilization remained high, but concession stands closed, people changed their eating habits, some refused to remain or return, and on peak weekends between 10 and 15 stings were reported.

While foraging yellow jackets cause nuisance, anxiety, and some stinging episodes, presence of a yellow jacket colony in one's yard or house evokes immediate fear in some and intolerance in nearly everyone. Establishment and range extension by the German yellow jacket (Mac-Donald et al. 1980a) has led to an additional problem in urban and residential areas. A subterranean nester in other countries, *P. germanica* in this country is primarily a structural nester (Fig. 14.1), but subterranean nests have recently become common in the Lafayette area (MacDonald and Akre 1984) (Table 14.7). Structural colonies not only constitute a stinging threat, but two additional problems may surface later in the seasonal cycle (MacDonald et al. 1980a): (1) nests filling their structural void may be expanded into the interior of the building through the efforts of workers chewing through walls; and (2) newly emerging queens and males may emerge into the building interior. Both situations may result in a home filled with hundreds of adult yellow jackets and consequent difficult control operation. The former may also result in substantial damage to interior walls and expensive repairs (Fig. 14.1).

Consideration of the number of yellow jacket colonies per year in a given area not only provides some measure of yellow jacket abundance but also some indication of economic impact associated with control of such colonies. Interestingly, using similar types of data and employing the same assumptions, we calculated nearly identical economic impact related to yellow jacket colony destruction in the three regions for which we have the most information and experience.

Table 14.7 presents data on numbers of yellow jacket colonies in the Lafayette, Indiana area (approx. pop. 75,000) from 1977 to 1982. In addition, a survey of two (out of ten) pest control companies in the city revealed that 25 to 30 structural yellow jacket colonies were destroyed

FIGURE 14.1

Paravespula germanica nest in poultry shed (a) with envelope partially removed, and (b) all but uppermost combs removed exposing damage to the wall.

TABLE 14.7

Nest Sites of *Vespula* Species in the Lafayette, Indiana (Pop. 75,000) Area, 1977 to 1982[a]

Species	1977 G	S[b]	1978 G	S	1979 G	S	1980 G	S	1981 G	S	1982 G	S	Total G	S
P. germanica	0	5	0	12	0	13	2	12	1	19	10	19	13	70
P. maculifrons	41	7[c]	7	2	8	2	2	2	5	2	15	1	78	16
P. flavopilosa	0	0	0	0	0	0	0	0	0	1	2	0	2	1
V. squamosa	2	2	1	0	0	2	0	0	1	0	0	0	4	4

[a]Requests for yellow jacket colonies were made in 1977 via newspaper ads. No such requests were made the following years; colonies represent unsolicited call-ins and some casual search by one of us (JFM).

[b]G = underground colonies; S = structural colonies.

[c]3 of these colonies were in unprotected sites (MacDonald and Matthews 1981).

by each company every year since 1978. The average charge ranged from $50 to $75 per operation. These two companies were moderate in size (six to eight employees) and probably represent the average. Assuming each of the ten companies destroys 25 colonies per year, about 250 to 300 colonies are treated each year in the Lafayette area. In addition, each year about 20 colonies are destroyed by researchers. Employing these figures we conclude that elimination of yellow jacket colonies in the Lafayette area costs over $16,000 per year. This works out to be $0.22 per person per year. Similarly, in Pullman, Washington, from 1971 to 1982, 577 colonies of yellow jackets were removed, an average of 73 colonies per year, 1976–1979 (Table 14.8). An additional 30 colonies per year were probably destroyed by pest control operators, at a cost of about $1800 yearly. Estimated total cost of destroying all colonies during this period by professionals would have been $35,000–$40,000 or about $6,000 per year, a large expense for a population center of only 25,000–30,000 people.

Comparative data are not available for many urban areas, but we mention several to support our contention that the economic impact is substantial. For example, MacDonald and Matthews (1981) reported 208 *Paravespula* and *Vespula* colonies in Athens, Georgia (pop. 45,000) from 1974 to 1975. Using only those colonies studied and assuming a charge of $60 per control operation, colony destruction would have been $6,240 per year, or over $0.14 per person. Colorado Springs, Colorado experienced an outbreak year in 1982 for *D. maculata* according

TABLE 14.8

Colonies of Yellow Jackets Studied in the Pullman, Washington (Pop. 30,000) Area, 1971–1982 (No records of *Dolichovespula* colonies were kept prior to 1976)

Species	Year												Total
	'71	'72	'73	'74	'75	'76	'77	'78	'79	'80[a]	'81	'82	
P. pensylvanica	4	9	47	41	48	18	51	15	23	2	6	7	271
V. atropilosa[b]	6	8	27	19	28	5	4	5	16	6	10	3	137
D. arenaria	—	—	—	—	—	27	14	39	37	5	7	3	132
D. maculata	—	—	—	—	—	5	5	24	3	0	0	0	37
													577

[a]Mt. St. Helen's erupted, and volcanic ash blanketed Pullman about 1 cm deep. Few yellow jackets survived.

[b]Although *V. atropilosa* is usually a nonpest species, these colonies were in lawns and similar areas and posed a stinging hazard to homeowners.

to one pest control operator who reported destroying over 100 colonies at an average charge of $75.

Economic impact in large urban areas is proportionately greater. For example, working in Seattle, Washington, one pest control operator reported destruction of 153 colonies in 1980, and 28 colonies through June of 1981. In nearby Tacoma, Washington, a pest control firm destroyed 76 yellow jacket colonies in 1977. In Portland, Oregon, another pest control operator destroyed 139 yellow jacket colonies in 1976; 52 inside structures and 87 in yards and on houses. Also, T. Nixon (1982, personal communication) reported the destruction of 100 colonies of *P. maculifrons*, 81 *P. germanica*, and 11 *V. squamosa* in the Washington, D.C., area during 1976–1982. Again, these companies revealed an average charge of $60 per operation. Assuming such large metropolitan areas support at least 20 exterminator companies each, and each destroys 100 colonies per year at $60 per treatment, then yellow jacket colony control costs about $120,000 per year or approximately $0.24 per person for an urban area of 600,000 people.

Using estimations of $0.20 per person and considering that at least half of our population lives in areas subject to yellow jacket infestation, colony control results in an expenditure of over $22 million.

Economic impact as a result of adverse effects of foraging yellow jackets (Figs. 14.2, 14.3) is unavailable, and probably is not measureable. Akre et al. (1981) discussed this topic and reported such affects as closure

FIGURE 14.2
Workers of *Paravespula pensylvanica* on ham in Pullman, WA.

FIGURE 14.3
Paravespula maculifrons workers on decaying melon. The 0.5-acre field in West Lafayette, Indiana was covered with abandoned melons and hordes of yellow jackets.

of campgrounds and parks in the Pacific Northwest during outbreak years in 1973 and 1979, and other more or less anecdotal reports. However, they also reported more objective indicators of economic impact such as an approximate loss of $5,000 per resort per year in parts of California (Poinar and Ennik 1972), and devotion of 6% of the annual budget of a California Mosquito Abatement District near San Francisco to yellow jacket control. Another indicator of potential economic impact of yellow jacket foragers occurred in a resort development northeast of Atlanta, Georgia (MacDonald and Matthews 1981). Intolerable numbers of workers were present around food preparation facilities, and walkways constructed from untreated timbers served as a source of wood fibers to pulp foragers. Passing along the walkways was frightening to prospective property buyers. While no data were obtainable, the resort sales manager believed that substantial losses in sales occurred as a direct result of the yellow jacket outbreak. Similar economic effects are probably being experienced wherever yellow jacket outbreaks occur.

Agriculture

This topic was addressed by Akre et al. (1981), and relatively little new information on economic losses has become available. What new data do exist are incorporated into a review of the problem. We conclude this section with a detailed accounting of the establishment and spread of the western yellow jacket, P. pensylvanica, in Hawaii, and report heretofore unavailable dollar losses directly attributable to the yellow jacket problem.

Most problems associated with yellow jackets in agriculture have occurred in Christmas tree plantations in the Pacific Northwest, in fruit orchards and vineyards, in association with lumbering, and in depredation of honeybee hives. The most serious problem is that of stinging when colonies are inadvertently disturbed. Other animals also receive stings as revealed in a report of 15 horses in El Cajon Valley, California being stung, and then developing abcesses that attracted foraging workers that cut flesh from the wound area. However, just the presence of foragers in a work area may result in work stoppage and economic loss. In addition, some fruits, including certain varieties of grapes, may be damaged due to activities of foragers, resulting in direct loss due to damage and subsequent problems of rot.

Christmas tree plantings appear heavily infested by yellow jackets. Workers attracted to honeydew on foliage (resulting from aphid infestations) often interact with pruners. Also, some species of Paravespula and Vespula nest in the sandy soil of plantings, and Dolichovespula build nests in the trees. Fanning (1979) reported that accidents attributed to yellow jackets account for about 16% of all accidents incurred in an industry cited as the most accident-prone in agriculture. Many such accidents arise from knife wounds inflicted by pruners attempting to ward off yellow jackets.

The original report of yellow jacket impact on fruit orchards in Oregon (Davis et al. 1973) revealed the necessity for control of yellow jacket workers precluding harvesting of the crop. Peaches appear most affected, but other orchard crops such as pears and apples may be affected also (Akre et al. 1981). Several reports of yellow jacket damage to grapes have been received from Oregon, Washington, Pennsylvania, and New York. According to Willard B. Robinson (personal communication) of Cornell University, yellow jackets penetrate the skin of some early maturing varieties, and account for up to 80% loss in yield. Doughty et al. (1981) indicated that yellow jackets are not direct pests of blueberries, but during outbreak years in the Pacific Northwest density of nests in highbush plots may preclude picking.

In a similar manner, yellow jackets affect the lumber industry. Several species of *Paravespula* and *Vespula* nest in forested areas and lumberjacks (also fire fighters) may be at a relatively high risk of stepping into a colony. Indeed, nearly one third of the *P. maculifrons* colonies reported by MacDonald and Matthews (1981) in forested areas were inadvertently discovered by foresters contacting them. Putnam (1977) reported that over a five-year period stings accounted for more than 5% of all Forest Service medical treatment and lost time. More recent information covering the period 1974–1981 revealed that 2,123 of a total of 2,268 lost-time accidents in the USDA due to insect bites or stings occurred in the Forest Service (L. Pestaner, USDA, Office of Finance and Management). Sixty-three percent of these accidents occurred in California, Idaho, Oregon, and Washington. Unfortunately, accidents due to stings rather than bites cannot be separated. However, since most of these lost-time accidents occurred in states with histories of yellow jacket outbreak years and large forest fires, we assume that many of these accidents were due to stings by yellow jackets.

Beekeepers throughout the world have reported losses due to depredation of hives by social wasps (Edwards 1980), and the German yellow jacket appears to be the most troublesome (Walton and Reid 1976). Due to the recent spread of *P. germanica* in North America, MacDonald et al. (1980a) suggested beekeepers should expect to experience problems in areas in which this species becomes established. According to J. L. Madden (Univ. Tasmania, personal communication), *P. germanica* only affects weak hives of honeybees in Tasmania. Still, DeJong (1979) feels yellow jackets pose a real threat to apiculture. Unfortunately, supporting data are lacking.

The establishment and spread of *P. pensylvanica* in Hawaii is revealing because it illustrates the exploitation of a new area by a yellow jacket species, similar to that of *P. germanica* in New Zealand (Thomas 1960). Also presented, for the first time, are actual dollar losses directly attributable to a yellow jacket problem. Our account is based on a report by Nakahara and Lai (1981), as well as personal communication with Mr. George Mills, Vector Control Supervisor, Hawaii, and Dr. V. Chang, Hawaiian Sugar Planter's Association.

Paravespula pensylvanica was first discovered on Kauai in 1919 and Oahu in 1936, but it remained obscure until 1978 when, according to Nakahara and Lai (1981), a more adaptable strain (probably introduced in the mid-1970s) spread throughout the islands. Due to increasing reports of stinging episodes and adverse effects on tourism and farming, three governmental agencies combined efforts to assess and attempt to remedy the problem. Recently, the Hawaii Sugar Planters Association joined the effort.

The problem with *P. pensylvanica* first manifested itself in 1978, and in 1980, 177 nests were destroyed on the island of Hawaii where the problem became the most serious. Between January and August 1981, 424 additional nests were destroyed. Although the number of nests is significant, the enormous size of *P. pensylvanica* colonies in Hawaii is remarkable. For example, 30% of 27 nests examined in July 1981 averaged 51 in wide by 49 in high with 47 combs (see Fig. 14.4); in comparison large nests from Washington State average 14 inches wide with six to nine combs. Nearly all colonies (98.1%) were subterranean, and most (54%) were found in agricultural situations. However, 82 colonies have been discovered in sugarcane fields, and because of the relatively long harvesting cycle (1½ to 4 years) more, and larger, colonies may be encountered in the future. Other agricultural areas supporting colonies include macadamia plantings, as well as coffee, avocado, guava, banana, and experimental farms. In addition to agricultural situations, *P. pensylvanica* colonies have been reported in residential areas (134 nests), parks and other recreational areas (80 nests) and along roadsides (33 nests).

Most of the economic losses resulting from this yellow jacket problem have been related to colony destruction. Cost to tourism, agricultural production, and food processing is unevaluated. The cost resulting from colony destruction is substantial and includes primarily salaries for employees assigned to the task ($53,261.20). Other costs included vehicle operation ($2,982.97), chemical insecticides ($8,798.25), and protective gear for employees ($825). Thus, direct costs for the 1½ year operation were nearly $66,000.

Akre et al. (1981) alluded to the impact of foraging yellow jackets on industries processing certain forms of sweets and other foods for human consumption. The problem is caused by foragers seeking sources of carbo-

FIGURE 14.4
Nest of perennial colony of *Paravespula pensylvanica*: Island of Hawaii, Laupahoe Hilo Forest Reserve; on dead or dying *Ohia lehua*, 8 ft from ground, nest height 8 ft.
(*Source*: C. Wakida, Division of Forestry and Wildlife, Honolulu, Hawaii)

hydrates and manifests itself relatively late in the seasonal cycle of yellow jackets, a period that coincides with the processing of fresh fruits such as pears, apples, strawberries, etc. A Del Monte™ pear processing plant near Vancouver, Washington experienced enough of a yellow jacket problem to have employed a number of techniques for abatement. Also, Smucker's™ jam-making operation in Ohio in 1980 experienced a work stoppage due to intolerable numbers of *P. germanica* in strawberry flats. Similarly, we have received pleas for assistance from confectionary, syrup, and ice cream companies in the upper Midwest. To remedy the situation, sanitation is essential to remove attractive foodstuffs, and changes in plant operations may be necessary—for example, switching some operations to after dark. However, these suggestions are disappointing to management and may be infeasible.

We are forced to conclude that yellow jacket problems will increase. We are particularly concerned by the propensity for some species of *Paravespula* to expand their range, and explode in certain situations. Examples include the spread and establishment of *P. germanica* throughout a large portion of temperature regions of both the northern and southern hemispheres, and the success of *P. pensylvanica* in Hawaii. On a less dramatic scale is the relatively rapid spread of *P. germanica* across the northern tier of eastern United States, and its recent detection in the Seattle, Washington area (MacDonald and Akre 1984). These special problems are in addition to basic ones associated with the emerging synanthropy of the *Paravespula* species in particular, but also of *D. arenaria* and *D. maculata*. These species have adopted human dwellings and their environs for nesting and food discards provide a rich food supply for the *Paravespula* species. In addition, our continued encroachment into previously natural areas opens new areas for nesting by yellow jackets and increases our exposure to them.

ABATEMENT

Yellow jacket workers are beneficial predators, and generally not a stinging threat unless their colony is disturbed. Control is warranted, however, when colonies are situated inside structures or in outdoor sites subject to human disturbance, and when numbers of foraging workers become intolerable in recreational and industrial areas.

Yellow jacket control has been reviewed several times (Akre et al. 1981, Akre and Davis 1978, Edwards 1980, MacDonald et al. 1976, 1980a, Spradbery 1973b). Spraying of a general area of infestation is completely ineffective. Worker yellow jackets are not easily killed by such application and those that may be killed are quickly replaced. The best control is to destroy yellow jacket colonies, and two approaches are employed: destruction of individual colonies by direct application of an insecticide and deployment of poisoned meat baits that are picked up by scavenger species of *Paravespula*, carried back to the nest, and distributed among colony members and brood.

Destruction of Individual Colonies

This method is used to quickly solve an existing problem, and remains the only abatement technique for many species of yellow jacket, especially those in the genera *Dolichovespula* and *Vespula*. With adequate caution, the application of an approved insecticide directly to a yellow jacket colony may be performed by homeowners, park personnel, foresters, etc., but generally is best left to a professional. Control is performed most safely after dark (see below) and while the applicator is protected by a bee suit, zip-on bee veil, and leather gloves (Wagner and Reierson 1975). Prior to attempting control, each colony should be examined for possible multiple entrances (underground colonies) or areas of incomplete envelope (aerial colonies), sites from which workers could escape. Colonies located inside structures should be carefully investigated to determine the safest and most effective approach (see below). Application of insecticides can be conducted with the aid of a red light; all other lights should be off, although a single light nearby will attract any escaping workers, which can then be killed.

(1) Subterranean colonies are the safest and easiest to destroy. An approved insecticide is applied into the entrance tunnel and the hole plugged with an insecticide-treated cotton wad. The latter step will control the few returning workers that were not inside the nest the night of application.

(2) Aerial colonies are more difficult and dangerous to control. Adequate protective gear is imperative since workers are easily aroused when control is attempted. The most effective insecticide formulations contain a quick "knockdown" pesticide such as synergized pyrethrins in conjunction with a residual insecticide such as propoxur. The spray is directed into the nest entrance and a six- to ten-second burst quickly kills the nest inhabitants. The nest can then safely be removed, and even if it is not, the residual material will kill workers emerging from capped cells.

(3) Structural colonies may be difficult to control, and require special application equipment and possibly a combination of treatments. The difficulty results from variable and unpredictable location of colonies within a structure. Figure 14.5 illustrates some of the possible nest locations inside a typical home. Also, some colonies may be situated 2–3 m from the entrance into the structure. Inappropriate control applications not only fail, but may drive the yellow jackets into the interior of the building. Typically, control of structural colonies is best left to professionals; readily available insecticide bombs and spray cans will not work in controlling these colonies. Under no circumstances should the entrance hole into the structure be plugged without first applying an approved insecticide.

Most structural colonies are destroyed by employing a pressurized aerosol generator equipped with a thin plastic wand and charged with a fast-acting insecticide with some residual action; resmethrin typically is used. The wand is inserted into the entrance hole and insecticide discharged for 30 to 45 seconds. Following this, the hole is plugged with steel wool dusted with 5% carbaryl. This plug precludes escape from the interior and kills returning workers attempting to chew through the plug from the outside. Some colonies may require a follow-up application to kill workers emerging from capped cells; resmethrin typically does not affect individuals in capped cells.

Colonies situated deeply within a structure may require an additional step, that is, the forcible blowing of 5% carbaryl dust into the entrance

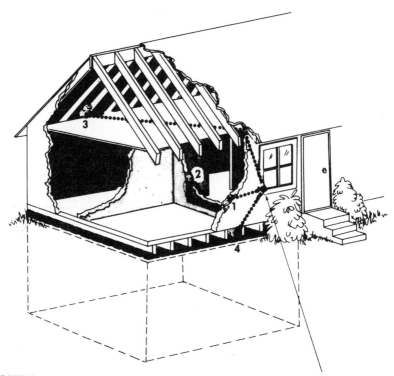

FIGURE 14.5

Schematic representation of possible structural nest sites of yellow jackets. Assuming a single hole at the window frame, four nest locations are depicted: (1) in an outside wall void; (2) in an interior wall void; (3) in the attic; (4) between the floor and the drop-ceiling of the basement. Only one situation is likely in a given year.

(*Source*: MacDonald 1980)

hole just before applying the resmethrin. The dust penetrates deeply into the void and even around corners; agitated workers crawl through the dust, repeatedly entering and leaving the nest proper, and thus tracking the insecticide into the nest interior where it contacts brood and the queen.

An inexpensive and simple technique employing a small piece of insecticide strip (DDVP) is useful for killing structural colonies situated in inopportune sites. For example, German yellow jacket colonies commonly are situated in wall voids of brick buildings often at second story or higher levels. Control requires the use of ladder trucks that cost over $300 if used at night. To save money, an exterminator, during the day, gently places a piece of insecticide strip into the entrance hole, taking care to allow passage of workers in and out. Colony demise typically occurs in three to four hours due to workers tracking in residues plus gradual penetration of insecticide vapor into the nest. Similarly, a small piece of insecticide strip tacked to the end of a 20- to 30-foot pole has facilitated the safe destruction of *Paravespula* colonies situated at dangerously high levels in homes. The pole is placed so that the insecticide strip rests against the entrance.

Not uncommonly, German yellow jacket nests are completely exposed in dark basements or suspended from ceilings or floor joists. Careful application of aerosol resmethrin (at night) to the outside of the envelope induces workers to enter the nest proper. After this, the aerosol generator wand may be gently inserted through the nest envelope for final application of about 20 seconds. Conducted in this manner, application does not elicit worker defense but some adult males and queens may escape. If this occurs, then immediate application of Wasp Freeze® (or other quick-knockdown materials) is necessary for their control.

Destruction of individual colonies may solve a local problem, but also may have little effect on the number of foraging workers in an area. In fact, even if all known colonies in an area could be located and destroyed, this may not effect abatement since foragers exhibit a long flight range (Akre et al. 1976, Rogers 1972a). Clearly, yellow jacket abatement requires a different approach, namely area-wide abatement.

Area-Wide Abatement

Area-wide abatement techniques have exploited the foraging behavior of scavenger yellow jacket species by incorporating an insecticide into some form of meat bait that foragers carry back to their colony, or by trapping workers attracted to synthetic lures. These techniques are environmentally sound, but have thus far enjoyed limited success, and then only against *P. pensylvanica*, and to a lesser extent *P. vulgaris* (MacDonald et al. 1976).

The most successful approach is the toxic bait abatement program. The approach is attractive because it is based on sound control principles: it is highly specific for species of *Paravespula*; it involves very limited risk of environmental contamination since the target organism itself actually delivers the insecticide; and it is modifiable to fit local situations and to be compatible with novel insecticides. Limited success rests with our inability to develop baits that are attractive to the various species or because some species may be responsive in some regions but not in others. This approach has not worked against eastern species of *Paravespula*.

Several explanations can be advanced to explain the failures: (1) differences in foraging behavior among species render some baits unattractive; (2) reduced attractiveness of a bait following addition of the insecticide; (3) short "field life" of the bait due to dehydration and spoilage; (4) lack of bait competitiveness with discarded foods in infested areas; and (5) reduced bait attractiveness resulting from commercial processing.

Differential attractiveness of various meat baits may never be surmounted and new baits may have to be developed for the different regions and yellow jacket species. The problem in the eastern United States may have been overcome when Ross et al. (1984) discovered that a horse-meat-based carnivore food, Nebraska Brands Feline Food (NBFF) was about seven times more attractive to P. germanica, P. maculifrons, and P. flavopilosa than the current standard, Puss n' Boots® fish-flavored cat food.

Initial success of the poison meat programs in California was based on the use of nonrepellent, slow-acting insecticides such as mirex and chlordane; other available insecticides rendered meat baits unattractive (Wagner and Reierson 1969). In anticipation of the loss of mirex and chlordane, Ennik (1973) ran tests and discovered that encapsulated diazinon (Knox Out®) and Rabon® did not reduce bait attractancy. Baits prepared with these toxicants abated P. pensylvanica and P. vulgaris in central California coastal areas. However, according to D. Caron (Univ. Delaware, personal communication), in the eastern United States, addition of insecticides to baits exhibiting some attractancy rendered them unattractive.

A limitation of meat baits is that they may lose their attractancy after placement in the field as a result of dehydration and associated crusting. This problem is pronounced when visitation rates are low and seems accentuated in the eastern United States where Paravespula species are less attracted to meat baits.

An effective program of sanitation has long been known to prevent and even reduce numbers of foragers in a park area, and also facilitates success of poison bait abatement programs (Wagner 1961). The presence of attractive food stuffs often results in entrainment of foragers to rewarding sites, leading to greater concentrations of workers as the season progresses. This may preclude success of poison bait abatement as baits may have difficulty competing with an unavoidable build-up of foodstuffs in refuse cans in a treatment area.

Finally, commercial preparation of a poison meat bait may result in its complete loss of attractancy. Rogers (1972b) reported that failure of Yellowjacket Stopper® to control P. pensylvanica stemmed from some aspect of its preparation since the original recipe, developed by Wagner and Reierson (1969), was effective. Indeed, any number of minor alterations to a meat bait may affect the success of an abatement program. For example, G. M. Reid and J. F. MacDonald (unpublished) discovered that while ground NBFF was as attractive as whole zoo meat few workers actually attempted to gather the bait, and almost none flew away with portions. Thus, high attractancy and bait acceptancy may not necessarily be correlated. Development of highly attractive and acceptable meat baits is the single most important, immediate research need. Furthermore, use of more novel approaches, incorporating insect growth regulators and possibly pheromones or fluorlipids (see Prestwich et al. 1981), will not be possible unless we develop a vehicle with which to dispense them by means of foraging workers of species of *Paravespula*.

Synthetic Chemical Attractants

The attraction of *Paravespula* workers to various synthetic organic compounds (heptyl butyrate, octyl butyrate, 2,4-hexadienyl butyrate and others) was first reported by Davis et al. (1967, 1969, 1973) and McGovern et al. (1970). Placed in a suitable trap, these compounds resulted in the capture of large numbers of *P. pensylvanica* workers, but no beneficial insects such as honeybees or bumblebees. This specificity for target species coupled with safe and convenient use in the field led to widespread belief that these lures could be used in depletion trapping (see MacDonald et al. 1976). However, although some local successes were reported (Davis et al. 1973), the original promise of synthetic lures has not been realized; indeed, no advances have occurred since the original discoveries of the early 1970s.

The greatest limitation is that of the over 300 compounds tested, the few with the greatest attractancy have worked only in the western United States, and almost exclusively against the western yellow jacket, *P. pensylvanica*. These lures attract some spring queens and workers throughout the season, but are not attractive to fall queens or males. The best synthetic lures have demonstrated extremely weak attractancy to all eastern species of *Paravespula* including the German yellow jacket (Grothaus et al. 1973, Howell et al. 1974, Reierson and Wagner 1978), and virtually no attractancy for *Dolichovespula* (same authors; Fluno 1973, MacDonald et al. 1974, Roush and Akre 1978). Even in the western United States where these lures work best, Roush and Akre (1978) discovered how weakly attracted *P. vulgaris* was to heptyl butyrate, reporting that while this species represented about 50% of the total yellow jacket catch in Malaise traps it constituted only 0.5% of the catch in heptyl butyrate traps.

Several studies have revealed that species of *Vespula* are more strongly attracted than *Paravespula* (MacDonald et al. 1973, 1974, Roush and

Akre 1978, Reierson and Wagner 1978). The latter authors do not believe this to be true, but scrutiny of their data supports the contention that *Vespula* species are more attracted than sympatric and synchronic species of *Paravespula*. Further confirmation of our contention stems from casual trapping with heptyl butyrate over four seasons in Lafayette, Indiana (J. F. MacDonald unpublished) during which time almost no *Paravespula* were collected (despite their overwhelming abundance), but three *V. vidua* queens were. The significance becomes apparent since in over seven years of collecting and examining student insect collections no other specimens of *V. vidua* have been encountered.

Two points of importance are associated with the strong attractancy of synthetic lures for members of *Vespula*: (1) misapplication of depletion trapping in areas supporting *Vespula*, most especially in the spring, could result in removing large numbers of these nontarget species, as pointed out by MacDonald et al. (1973); and (2) synthetic lures constitute a sensitive method of detecting existence of species of *Vespula* in areas in which they might not otherwise be detected, particularly rarer species (MacDonald et al. 1974; Roush and Akre 1978; Reierson and Wagner 1978). Indeed, all these authors employed synthetic lures to sample for *Vespula* and *Paravespula* in previously unstudied areas, and MacDonald et al. (1974) as well as Roush and Akre (1978) used attractant traps to monitor *Vespula* and *Paravespula* population dynamics. Of course, this latter use is applicable only in the western United States.

To date, depletion trapping has only effectively abated *P. pensylvanica* workers in some areas of the Pacific Northwest, and then only under certain conditions as in roadside rests and around small orchards of about 30 acres or less (Davis et al. 1973). Similar use in larger orchards, general recreational areas, or food-processing operations has failed to abate worker populations. For example, Reierson and Wagner (1975) reported the capture of over 500,000 *P. pensylvanica* workers from a site near Santa Monica, California, but, based on concurrent bait removal by workers in treated and control areas, attractant trapping had no detectable influence on the number of foragers.

Another use for synthetic lures was reported by Wagner and Reierson (1969) and Ennik (1973) and involved the addition to or "splashing" around, of heptyl butyrate and heptyl crotonate next to meat baits, which enhanced bait removal by *P. pensylvánica* workers. Perhaps additional workers attracted to the lures results in still other workers visiting the immediate area as a result of social facilitation. Such a phenomenon at meat baits has been shown by Ross (1983) and at carbohydrate baits by Parrish and Fowler (1983). These lures do not appear to be cues associated with the gathering of meat baits, since nonscavenging species of *Vespula* are more strongly attracted than are the scavenging *Paravespula* species. MacDonald (1977) suggested that the attractiveness of synthetic compounds like heptyl butyrate may somehow be related to cues associated with their subterranean nest site.

In conclusion, use of synthetic lures for depletion trapping is limited to the western United States and is completely ineffective in the East. Their use should be limited to those situations described by Davis et al. (1973).

Biological Control

Vertebrate predators of *Paravespula* and *Vespula* colonies include a variety of previously mentioned species. However, while these predators may destroy numerous colonies in a given area, predation typically occurs too late in the seasonal cycle to influence worker production. Similarly, most invertebrates are found with mature or declining colonies (Akre et al. 1981, Spradbery 1973b). Relatively few species actually invade active yellow jacket colonies (Akre et al. 1981).

Perhaps the best controlling agents of some yellow jacket species are social parasitic species. For example, the southern yellow jacket, V. *squamosa*, usurps nearly 40% of the young *P. maculifrons* colonies (MacDonald and Matthews 1975; 1981), and thus effects a good deal of control. But, of course, no one would advocate use of V. *squamosa* as a biological control agent.

Recommendations

The concept of integrated pest management can be applied to yellow jacket abatement. The various applicable measures include cultural, chemical, and physical control, and some common sense.

The proper management of garbage and other yellow jacket food sources is the single most important step in preventing a yellow jacket problem and facilitating abatement of an existing problem. Personnel in sites prone to yellow jacket problems should develop and conduct sanitation programs all summer and fall. The program must include both efficient, daily policing of discarded foods and deployment of well-maintained refuse cans fitted with can liners and tight-fitting lids.

The use of appropriate refuse cans helps in a number of ways. First, garbage does not collect and so workers do not become entrained to the site. Liners facilitate pick-up for employees and reduce the risk of stings; also their use precludes residues of food remaining in cans which may serve to attract workers. Tight-fitting lids prevent worker entry, and may also serve for attachment of an insecticide strip containing DDVP (Wagner 1961).

Success of a poison meat abatement program not only depends on use of an attractive bait, but also on elimination of competing food resources. Presently, the most effective abatement programs involve use of cat food mixed with Knox Out® or Rabon® against *P. pensylvanica* and *P. vulgaris* in the West. We suggest that the use of NBFF mixed with Knox Out® (preferably the washed formulation) may effect abatement of *P. maculifrons* and *P. germanica* in the eastern United States. Bait enhancement with the addition of synthetic attractant has been reported in the West (Wagner and Reierson 1969, Ennik 1973), but will probably be of no use in the East (see MacDonald et al. 1976, Ross 1983).

Deployment of a poison bait program should be initiated as soon as the problem manifests itself, but should not necessarily be implemented yearly since the yellow jacket problem may differ from year to year (Akre and Reed 1981b, Archer 1980b). The number of poison bait stations required in a given area is not yet established, but in general one per hectare is needed. However, greater densities may be necessary under severe outbreaks. Similarly, positioning of bait stations has not been established, but perimeter saturation coupled with placement of some stations within a treatment area should be effective. Charging of bait stations will depend on visitation and removal rate, but one should expect the best results with daily replenishment.

Destruction of individual colonies as located will not only eliminate an immediate risk, but, if enough are found and destroyed, number of foragers may be reduced.

A number of ameliorative and circumventive measures are advised. For example, people with known hypersensitivity to yellow jacket venom should avoid outdoor activities, eating in particular, during outbreak periods. Attractive perfumes, hair sprays, and lotions should be avoided, and certain brightly colored clothing not be worn. Homeowners should carefully search their premises for presence of yellow jacket colonies prior to lawn mowing, plant trimming, etc., and painting. Game and fish preparation may be accomplished inside screen enclosures in parks and campgrounds, but construction of such shelters is expensive. One of the oldest and most effective control measures for these areas involves suspending a raw fish above a bucket filled with water to which a wetting agent has been added. Yellow jacket workers cut pieces of flesh (from slits made in the flesh), and, if too heavy, they drop as they attempt to fly and thus fall into the water. Jackets impregnated with chemicals that are repellent to flies are completely ineffective.

Traps baited with synthetic lures such as heptyl butyrate may capture west coast specis of *Paravespula*, but unless used in a manner similar to that reported by Davis et al. (1973), depletion trapping is not an effective abatement technique for pest species, and may actually attract greater numbers of nonpestiferous species (MacDonald et al. 1973, Roush and Akre 1978). These lures do not attract eastern species of *Paravespula* nor any *Dolichovespula* species.

ACKNOWLEDGMENTS

We are deeply indebted to the many individuals who furnished information for this chapter. Indeed, the task would have been impossible without their unstinting efforts to render us assistance.

We thank Byron Reid and Donald Ross, Department of Entomology, Purdue University for sharing unpublished research results. Dr. R. L. Wagner, University of California, Riverside; Jay Nixon, American Disinfectant Company, Washington, D. C.; Dr. J. L. Madden, University of Tasmania; and Dr. Willard B. Robinson, Department of Food Science and Technology, Cornell University, are also thanked for the use of unpublished information collected during their studies of yellow jackets.

Sincere appreciation is extended to Jim Hanstra, Reliable Exterminators, West Lafayette, Indiana; Dr. Terry Whitworth, Whitworth Pest Control, Tacoma, Washington; Robert Stidham, Stidham Honey Farm, Auburn, Washington; William Ruhl, Jr., Portland, Oregon; and Jay Nixon, for providing detailed information on colonies destroyed by their companies. Mr. Satoru Togashi, Del Monte Corporation, Toppenish, Washington, furnished us with information on yellow jacket problems at some of the fruit packing plants owned by his company.

Dr. Donald Clayton, M.D., Arnett Clinic, Lafayette, Indiana is thanked for sharing his knowledge about venom hypersensitivity and immunotherapy. Similarly, Dr. William C. Mannschreck, M.D., Lewiston, Idaho furnished much needed information on costs of immunotherapy and on the number of patients undergoing treatment under his care.

Special thanks are due to Mr. George Mills, Vector Control Supervisor, Hawaii Department of Health, and to Dr. Vincent Chang, Hawaiian Sugar Planters' Association, for details and costs of the population explosion of the western yellow jacket in Hawaii. Charles K. Wakida, Division of Forestry and Wildlife, Hawaii, generously let us use his photograph of a huge perennial nest of *P. pensylvanica*.

L. C. Pestaner, program analyst, USDA, Office of Finance and Management, spent many hours collating lost-time accidents due to stings and bites. Our sincere appreciation is extended to her.

We thank A. Greene, J. Nixon, H. C. Reed, and R. Zack for their comments regarding manuscript improvement.

REFERENCES

Akre, R. D. 1982. Social Wasps. Vol. 4, Chapt. 1, pp. 1–105. In H. R. Hermann (Ed.), Social Insects. Academic Press, New York. 385 p.

Akre, R. D. and D. P. Bleicher. 1985. Nests of *Dolichovespula norwegica* and *D. norvegicoides* in North America (Hymenoptera: Vespidae). *Entomol. News* 96: 29–35.

Akre, R. D. and H. G. Davis. 1978. Biology and pest-status of venomous wasps. *Annu. Rev. Entomol.* 23: 19–42.

Akre, R. D., W. B. Garnett, J. F. MacDonald, A. Greene, and P. Landolt. 1976. Behavior and colony development of *Vespula pensylvanica* and *V. atropilosa* (Hymenoptera: Vespidae). *J. Kans. Entomol. Soc.* 49: 63–84.

Akre, R. D., A. Greene, J. F. MacDonald, P. J. Landolt, and H. G. Davis. 1981. The yellowjackets of America north of Mexico. USDA Agric. Handbook 552. 102 p.

Akre, R. D., W. B. Hill, J. F. MacDonald, and W. B. Garnett. 1975. Foraging distance of *Vespula pensylvanica* workers (Hymenoptera: Vespidae). *J. Kans. Entomol. Soc.* 48: 12–16.

Akre, R. D. and H. C. Reed. 1981a. A polygynous colony of *Vespula pensylvanica* (Saussure) (Hymenoptera: Vespidae). *Entomol. News* 92: 17–31.

Akre, R. D. and H. C. Reed. 1981b. Population cycles of yellowjackets (Hymenoptera: Vespinae) in the Pacific Northwest. *Environ. Entomol.* 10: 267–274.

Akre, R. D. and H. C. Reed. 1983. Evidence for a queen pheromone in *Vespula*. *Can. Ent.* 115: 371–377.

Akre, R. D., H. C. Reed, and P. J. Landolt. 1982. Nesting biology and behavior of the blackjacket, *Vespula consobrina* (Hymenoptera: Vespidae). *J. Kans. Entomol. Soc.* 55: 373–405.

Akre, R. D., C. F. Roush, and P. J. Landolt. 1977. A *Vespula pensylvanica*/*Vespula vulgaris* nest (Hymenoptera: Vespidae). *Environ. Entomol.* 6: 525–526.

Alagon, A. C., J. Kuan, and T. P. King. 1982. Venom allergens of *Vespula germanica*, *V. maculifrons* and *V. vulgaris*. *J. Allergy Clin. Immunol.* 69: 114.

Archer, M. E. 1973. The social and solitary wasp and solitary bee distribution maps scheme. *Entomol. Gaz.* 24: 361.

Archer, M. E. 1980a. A new species of *Dolichovespula* and subspecies of *D. pacifica* (Hymenoptera: Vespidae) from China. *Entomon.* 5: 341–344.

Archer, M. E. 1980b. Population dynamics. Chapt. 8, pp. 172–207. In R. Edwards, Social Wasps: Their biology and control. Rentokil, Sussex, England, 398 p.

Archer, M. E. 1981a. The Euro-Asian species of the *Vespula rufa* group (Hymenoptera, Vespidae) with descriptions of two new species and one new subspecies. *Kontyû* 49: 54–64.

Archer, M. E. 1981b. Taxonomy of the *sylvestris* group (Hymenoptera: Vespidae, *Dolichovespula*) with the introduction of a new name and notes on distribution. *Entomol. Scand.* 12: 187–193.

Archer, M. E. 1981c. A simulation model for the colonial development of *Paravespula vulgaris* (Linnaeus) and *Dolichovespula sylvestris* (Scopoli) (Hymenoptera: Vespidae). *Melanderia* 36: 1–59.

Archer, M. E. 1982. A revision of the subgenus *Rugovespula* nov. of the genus *Vespula* (Hymenoptera, Vespidae). *Kontyû* 50: 261–269.

Arnold, T. S. 1966. Biology of social wasps: Comparative ecology of the British species of social wasps belonging to the family Vespidae. M. Sc. Thesis, Univ. London. 119 p.

Ascerno, M. E. 1981. Diagnostic clinics: More than a public service. *Bull. Entomol. Soc. Amer.* 27: 97–101.

Barnard, J. H. 1973. Studies of 400 Hymenoptera sting deaths in the United States. *J. Allergy Clin. Immunol.* 52: 259–264.

Barr, S. E. 1971. Allergy to Hymenoptera stings – Review of the world literature: 1953–70. *Ann. Allergy* 29: 49–66.

Bequaert, J. 1931. A tentative synopsis of the hornets and yellowjackets of America. *Entomol. Amer.* 12: 71–138.

Bigelow, N. K. 1922. Insect food of the black bear (*Ursus americanus*). *Can. Entomol.* 54: 49–50.

Blum, M. S. 1978. Biochemical defenses of insects, pp. 465–513. *In* M. Rockstein (Ed.), Biochemistry of Insects. Academic Press, New York. 649p.

Bourdon, V., J. Lecomte, M. Leclercq, and J. Leclercq. 1975. Présence de Noradrenaline conjuguée dans l'enveloppe du nid de *Vespula vulgaris* Linné. *Bull. Soc. Roy. Sci. Liège.* 44: 474–476.

Buckell, E. R. and G. T. Spencer. 1950. The social wasps (Vespidae) of British Columbia. *Proc. Entomol. Soc. Brit. Col.* 46: 33–40.

Butler, C. G., D. J. C. Fletcher, and D. Walter. 1969. Nest-entrance marking with pheromones by the honeybee *Apis mellifera* L. and by a wasp *Vespula vulgaris* L. *Anim. Behav.* 17: 142–147.

Caron, D. M. 1974. Evaluation of chloropyrifos for hornet and wasp control. *Down to Earth* 30: 10–12.

Carpenter, J. M. 1982. The phylogenetic relationships and natural classification of the Vespoidea (Hymenoptera). *Syst. Entomol.* 7: 11–38.

Chaffee, F. H. 1970. The prevalence of bee sting allergy in an allergic population. *Acta Allergol.* 75: 292–293.

Cobb, F. K. 1979. Honey buzzard at wasps' nest. *Brit. Birds* 72: 59–64.

Davis, H. G., G. W. Eddy, T. P. McGovern, and M. Beroza. 1967. 2-4 Hexadienyl butyrate and related compounds highly attractive to yellowjackets (*Vespula* spp.). *J. Med. Entomol.* 4: 275–280.

Davis, H. G., G. W. Eddy, T. P. McGovern, and M. Beroza. 1969. Heptyl butyrate, a new synthetic attractant for yellowjackets. *J. Econ. Entomol.* 62: 1245.

Davis, H. G., R. W. Zwick, W. M. Rogoff, T. P. McGovern, and M. Beroza. 1973. Perimeter traps baited with synthetic lures for suppression of yellowjackets in fruit orchards. *Envir. Entomol.* 2: 569–571.

DeJong, D. 1979. Social wasps, enemies of honey bees. *Amer. Bee J.* 119: 505–507, 529.

Doughty, C. C., E. B. Adams, and L. W. Martin. 1981. Highbush blueberry production in Washington and Oregon. Coop. Ext., Washington State Univ. Pacific Northwest 215. 25 p.

Duncan, C. P. 1939. A contribution to the biology of North American vespine wasps. *Stanford Univ. Publ. Biol. Sci.* 8: 1–271.

Eck, R. 1980. *Dolichovespula loekenae* n. sp., eine neue soziale Faltenwespe aus Skandinavien. *Reichenbachia* 18: 213–217.

Eck, R. 1981. Zur Verbreitung und Variabilität von *Dolichovespula norwegica* (Hymenoptera, Vespidae). *Entomol. Abh.* 44: 133–152.

Edery, H., J. Ishay, S. Gitter, and H. Joshua. 1978. Venoms of Vespidae. *In* Rettini, S. (Ed.), Arthropod Venoms. Vol 48, New Series, Handbook of Experimental Pharmacology. Springer-Verlag: New York. 977 p.

Edwards, R. 1980. Social Wasps: Their biology and control. Rentokil, Sussex, England, 398 p.

Ennik, F. 1973. Abatement of yellowjackets using encapsulated formulations of diazinon and Rabon. *J. Econ. Entomol.* 66: 1097–1098.

Ennik, F. 1980. Deaths from bites and stings of venomous animals. *West. J. Med.* 133: 463–468.

Es'kov, E. K. 1977. The structure of acoustic signals in the larvae of the wasp *Dolichovespula sylvestris* and the hornet *Vespa crabro*. *Zh. Evol. Biokhim. Fiziol.* 13: 371–375. [English translation.]

Fanning, P. K. 1979. Christmas tree farm accidents and what to do about them. Coop. Ext. Coll. Agric. Wash. State Univ. EM 4512. 9 p.

Feingold, B. F. 1973. Introduction to Clinical Allergy. C. C. Thomas, Springfield, Ill. 380 p.

Fluno, J. A. 1961. Wasps as enemies of man. *Bull. Entomol. Soc. Amer.* 7: 117–119.

Fluno, J. A. 1973. Chemical proof of validity of the taxonomic separation of *Vespula* subgenus *Vespula* from *Vespula* subgenus *Dolichovespula* (Hymenoptera: Vespidae). *Proc. Entomol. Soc. Wash.* 75: 80–83.

Franke, W., F. Hindert, and W. Reith. 1979. Alkyl-1, 6-dioxaspiro (4.5)-decanes- a new class of pheromones. *Naturwissenschalten* 66: 618–619.

Gaul, A. T. 1952. Addition to Vespine biology X. Foraging and chemotaxis. *Bull. Brooklyn Entomol. Soc.* 47: 138–140.

Gaul, A. T. 1953. Addition to Vespine biology XI. Defense flight. *Bull. Brooklyn Entomol. Soc.* 48: 35–37.

Geller, R. G., H. Yoshida, M. A. Beaven, Z. Horakova, F. L. Atkins, H. Yamabe, and J. J. Pisano. 1976. Pharmacologically active substances in venoms of the bald-faced hornet, *Vespula (Dolichovespula) maculata* and the yellowjacket, *Vespula (Vespula) maculifrons*. *Toxicon* 14: 27–33.

Golden, D. B. K., M. D. Valentine, A. K. Sobotka, and L. M. Lichtenstein. 1980. Regimens of Hymenoptera venom immunotherapy. *Ann. Intern. Med.* 92: 620–624.

Green, S. G., R. A. Heckman, A. W. Benton, and B. F. Coon. 1970. An unusual nest location for *Vespula maculifrons* (Hymenoptera: Vespidae). *Ann. Entomol. Soc. Amer.* 63: 1197–1198.

Greene, A. 1979. Behavioral characters as indicators of yellowjacket phylogeny (Hymenoptera: Vespidae). *Ann. Entomol. Soc. Amer.* 72: 614–619.

Greene, A. 1982. Comparative early growth and foraging of two naturally established vespine wasp colonies, pp. 85–89. *In* Breed, M. D., C. D. Michener, and H. E. Evans (Eds.), The Biology of Social Insects. Westview, Boulder, Colorado. 419 p.

Greene, A., R. D. Akre, and P. J. Landolt. 1976. The aerial yellowjacket, *Dolichoves-*

pula arenaria (Fab.): Nesting biology, reproductive production, and behavior (Hymenoptera: Vespidae). *Melanderia* 26: 1–34.

Greene, A. and D. M. Caron. 1980. Entomological etymology: The common names of social wasps. *Bull. Entomol. Soc. Amer.* 26: 126–130.

Grogen, D. E. and J. H. Hunt. 1977. Digestive proteases of two species of wasps of the genus *Vespula*. *Insect Biochem.* 7: 191–196.

Grothaus, R. H., H. G. Davis, W. M. Rogoff, J. A. Fluno, and J. M. Hirst. 1973. Baits and attractants for east coast yellowjackets, *Vespula* spp. *Environ. Entomol.* 2: 717–718.

Guiglia, D. 1972. Les guepes sociales (Hymenoptera: Vespidae) d'Europe occidentale et septentrionale. Faune de l'Europe et du bassin mediterranéen No. 6. Masson et Cie, Paris. 181 p.

Habermann, E. 1972. Bee and wasp venoms. *Science* 177: 314–322.

Heath, J. 1982. Stinging insects. *Pest Control* 50: 22, 26.

Hoffman, D. R. and C. A. MacDonald. 1982. Allergens in Hymenoptera venom. VIII. Immunologic comparison of venoms from six species of *Vespula* (yellowjackets). *Ann. Allergy* 48: 78–81.

Hoffman, D. R., J. S. Miller and J. L. Sutton. 1981. Hymenoptera venom allergy: a geographical study. *Ann. Allergy* 45: 276–279.

Howarth, F. G. 1975. Reports of *Vespula vulgaris* on Maui, Hawaii. *Proc. Hawaiian Entomol. Soc.* 22: 11–12.

Howell, J. O., T. P. McGovern and M. Beroza. 1974. Attractiveness of synthetic compounds to some eastern *Vespula* species. *J. Econ. Entomol.* 67: 629–630.

Hunt, J. H. 1982. Trophallaxis and the evolution of eusocial Hymenoptera, pp. 201–205. *In* Breed, M. D., C. D. Michener, and H. E. Evans (Eds.), The Biology of Social Insects. Westview, Boulder, Colorado. 419 p.

Hunt, K. J., M. D. Valentine, A. K. Sobotka, A. W. Benton, F. J. Amodio, and L. M. Lichtenstein. 1978. A controlled trial of immunotherapy in insect hypersensitivity. *N. Eng. J. Med.* 299: 157–161.

Ikan, R., R. Gottlieb, E. D. Bergmann, and J. Ishay. 1969. The pheromone of the queen of the Oriental hornet, *Vespa orientalis*. *J. Insect Physiol.* 15: 1709–1712.

Ishay, J. 1965. Observations and experiments on colonies of the Oriental wasp. XX Internat. Beekeeping Jubilee Congress. Bucharest, Romania. pp. 140–145.

Ishay, J. 1972. Thermoregulatory pheromones in wasps. *Experientia* 28: 1185–1187.

Ishay, J. and M. B. Brown. 1975. Patterns in the sounds produced by *Paravespula germanica* wasps. *J. Acoust. Soc. Amer.* 57: 1521–1525.

Ishay, J. S. and B. Perna. 1979. Building pheromones of *Vespa orientalis* and *Polistes foederatus*. *J. Chem. Ecol.* 5: 259–272.

Iwata, K. 1976. Evolution of Instinct: Comparative ethology of Hymenoptera. Amerind: New Delhi. 535 p.

Jacobson, R. S., R. W. Matthews, and J. F. MacDonald. 1978. A systematic study of the *Vespula vulgaris* group with a description of a new yellowjacket species in eastern North America (Hymenoptera: Vespidae). *Ann. Entomol. Soc. Amer.* 71: 299–312.

Jeanne, R. L. 1977. A specialization in nest petiole construction by queens of *Vespula* spp. (Hymenoptera: Vespidae). *J. N. Y. Entomol. Soc.* 85: 127–129.

Jeanne, R. L. 1980. Evolution of social behavior in the Vespidae. *Annu. Rev. Entomol.* 25: 371–396.

Johansen, C. A. and H. G. Davis. 1972. Toxicity of nine insecticides to the western yellowjacket. *J. Econ. Entomol.* 65: 40–42.

Kalmus, H. 1954. Finding and exploitation of dishes of syrup by bees and wasps. *Brit. J. Anim. Behav.* 2: 136–139.

Kemper, H. 1961. Nestunterschiede bei den sozialen Faltenwespen Deutschlands. *Z. Angew. Zool.* 48: 31–85.

Keyel, R. E. 1983. Some aspects of niche relationships among yellowjackets (Hymenoptera: Vespidae) of the northeastern United States. Ph.D. Dissertation, Cornell Univ., Ithaca, NY. 161 p.

King, T. P., A. K. Sobotka, A. Alagon, L. Kochoumian, and L. M. Lichtenstein. 1978. Protein allergens of white-faced hornet, yellow hornet, and yellowjacket venoms. *Biochemistry* 17: 5165–5174.

Lecomte, J., V. Bourdon, J. Damas, M. Leclercq, and J. Leclercq. 1976. Presence de noradrenaline conjuguée dans les parios du nid de *Vespula germanica* Linne. *C. R. Séance. Soc. Biol.* 170: 212–215.

Lecomte, J., and M. Leclerq. 1973. Sur la mort provoquée per les piqûes d'Hyméenopteès Aculéates. *Bull. Acad. Méd. Belg.* 128: 615–693.

Lichtenstein, L. M., A. K. Sobotka, D. B. K. Golden, and M. D. Valentine. 1980. Once stung, twice shy. When should insect sting allergy be treated. *J. Amer. Med. Assoc.* 244: 1683–1684.

Lichtenstein, L. M., M. D. Valentine, and A. K. Sobotka. 1979. Insect allergy: The state of the art. *J. Allergy Clin. Immunol.* 64: 5–12.

Light, W. C., R. E. Reisman, M. Shimizu, and C. E. Arhesman. 1977. Unusual reactions following insect stings. Clinical features and immunologic analysis. *J. Allergy Clin. Immunol.* 59: 391–397.

MacDonald, J. 1977. Comparative and adaptive aspects of vespine nest construction. Proc. 8th Int. Congr. IUSSI, Wageningen, The Netherlands. p. 169–172.

MacDonald, J. F. 1980. Biology, recognition, medical importance and control of Indiana social wasps. Purdue Univ. Coop. Ext. Bull. E-91. Lafayette, Ind. 24 p.

MacDonald, J. F. and R. D. Akre. 1984. Range extension and emergence of subterranean nesting by the German yellowjacket, *Vespula germanica*, in North America (Hymenoptera: Vespidae). *Entomol. News* 95: 5–8.

MacDonald, J. F., R. D. Akre, and W. B. Hill. 1973. Attraction of yellowjackets (*Vespula* spp.) to heptyl butyrate in Washington State (Hymenoptera: Vespidae). *Environ. Entomol.* 2: 375–379.

MacDonald, J. F., R. D. Akre, and W. B. Hill. 1974. Comparative biology and behavior of *Vespula atropilosa* and *V. pensylvanica* (Hymenoptera: Vespidae). *Melanderia* 18: 1–66.

MacDonald, J. F., R. D. Akre, and R. Keyel. 1980a. The German yellowjacket (*Vespula germanica*) problem in the United States. *Bull. Entomol. Soc. Amer.* 26: 436–442.

MacDonald, J. F., R. D. Akre, and R. W. Matthews. 1976. Evaluation of yellowjacket abatement in the United States. *Bull. Entomol. Soc. Amer.* 22: 397–401.

MacDonald, J. F. and R. W. Matthews. 1975. *Vespula squamosa*: A yellow jacket wasp evolving toward parasitism. *Science* 190: 1003–1004.

MacDonald, J. F. and R. W. Matthews. 1976. Nest structure and colony composition of *Vespula vidua* and *V. consobrina* (Hymenoptera: Vespidae). *Ann. Entomol. Soc. Amer.* 69: 471–475.

MacDonald, J. F. and R. W. Matthews. 1981. Nesting biology of the eastern yellowjacket, *Vespula maculifrons* (Hymenoptera: Vespidae). *J. Kans. Entomol. Soc.* 54: 433–457.

MacDonald, J. F. and R. W. Matthews. 1984. Nesting biology of the southern yellowjacket, *Vespula squamosa* (Hymenoptera: Vespidae): Social parasitism and independent founding. *J. Kans. Entomol. Soc.* 57: 134–151.

MacDonald, J. F., R. W. Matthews, and R. S. Jacobson. 1980b. Nesting biology of the yellowjacket, *Vespula flavopilosa* (Hymenoptera: Vespidae). *J. Kans. Entomol. Soc.* 53: 448–458.

Madden, J. L. 1981. Factors influencing the abundance of the European wasp (*P. germanica* [F.]). *J. Aust. Entomol. Soc.* 20: 59–66.

Maschwitz, U. W. 1964. Alarm substances and alarm behavior in social Hymenoptera. *Nature* (London) 204: 324–327.

Maschwitz, U. W. 1966. Alarm substances and alarm behavior in social insects. *Vit. Horm.* 24: 267–290.

Maschwitz, U., W. Beier, I. Dietrich, and W. Keidel. 1974. Futterverstandigung bei Wespen der Gattung *Paravespula*. *Naturwisseuschaften* 61: 506.

Maschwitz, U. W. J. and W. Kloft. 1971. Morphology and function of the venom apparatus of insects—Bees, wasps, ants, and caterpillars. Chapter 44, pp. 1–60. In Bucherl, W. and E. Buckley (Eds.), Venomous Animals and Their Venoms. Vol. III. Venomous Invertebrates. Academic Press, New York. 537 p.

Matthews, R. W. and J. R. Matthews. 1979. War of the yellowjacket queens. *Nat. Hist.* 88: 56–65.

Mazokhin-Porshnyakov, G. A. 1960. Evidence of the existence of color vision in wasps (Vespidae). *Zool. Zh.* 39: 553–557. [In Russian, English summary.]

Mazokhin-Porshynyakov, G. A. and V. M. Kartsev. 1979. A study of the sequence of flight by insects to several equal food subjects. (Strategy of their visual search). *Zool. Zh.* 58: 1281–1289. [In Russian, English summary].

McGovern, T. P., H. G. Davis, M. Beroza, J. C. Ingangi, and G. W. Eddy. 1970. Esters highly attractive to *Vespula* spp. *J. Econ. Entomol.* 63: 1534–1536.

Menzel, R. 1971. Über den Farbensinn von *Paravespula germanica* F. (Hymenoptera): ERG und selektive adaptation. *Z. Vergl. Physiol.* 75: 86–104.

Miller, C. D. F. 1961. Taxonomy and distribution of Nearctic *Vespula*. *Can. Entomol.* (suppl. 22) 52 p.

N.I.A.I.D. 1981. Insect Allergy. National Institute of Health. Building 31-7A32, Bethesda, Maryland 20014.

Nakahara, L. M. 1980. First record of aerial *V. pensyl.* nest in Hawaii, prob. perennial. Cooperative Plant Pest Report USDA APHIS 5: 270.

Nakahara, L. M. and P. Lai. 1981. Hawaii Pest Report. Hawaii Dept. Agric., Plant Pest Control Branch 3: 1–4.

Nixon, J. 1982. Yellowjackets in houses—Research and control. *Pest Control* (August): 24–25.

Parrish, H. M. 1963. Analysis of 460 fatalities from venomous animals in the United States. *Amer. J. Med. Sci.* 245: 129–141.

Parrish, M. D. and H. G. Fowler. 1983. Contrasting foraging related behaviors in two sympatric wasps (Vespula maculifrons and V. germanica). Ecol. Entomol. 8: 185–190.

Poinar, G. O., Jr. and F. Ennik. 1972. The use of Neoplectana carpocapsae (Steinernematidae: Rhabditoidea) against adult yellowjackets (Vespula spp., Vespidae: Hymenoptera). J. Invert. Pathol. 19: 331–334.

Post, D. C. 1980. Observations on male behavior of the eastern yellowjacket, Vespula maculifrons (Hymenoptera: Vespidae). Entomol. News 91: 113–116.

Potter, N. B. 1964. A study of the biology of the common wasp, Vespula vulgaris L., with special reference to the foraging behavior. Ph.D. Dissertation, Univ. Bristol, England. 75 p.

Preiss, F. J. 1967. Nest site selection, microenvironment and predation of yellowjacket wasps, Vespula maculifrons (Buysson) (Hymenoptera: Vespidae) in a deciduous Delaware woodlot. M.A. Thesis, Univ. Delaware. 81 p.

Prestwich, G. D., M. E. Melcer, and K. A. Plavcan. 1981. Flurolipids as targeted termiticides and biochemical probes. J. Agric. Food Chem. 29: 1023–1027.

Putnam, S. E. Jr. 1977. Controlling stinging and biting insects at campsites. Project Record ED & T 2689. Control of stinging insects in Forest Service Camps. USDA Forest Service Equip. Dev. Center, Missoula, MT. 22 p.

Rau, P. 1929. The nesting habits of the bald-faced hornet, Vespula maculata. Ann. Entomol. Soc. Amer. 22: 659–675.

Reed, H. C. and R. D. Akre. 1983a. Nesting biology of a forest yellowjacket Vespula acadica (Sladen) (Hymenoptera: Vespidae). Ann. Entomol. Soc. Amer. 7C: 582–590.

Reed, H. C. and R. D. Akre. 1983b. Comparative colony behavior of the forest yellowjacket, Vespula acadica (Sladen) (Hymenoptera: Vespidae). J. Kans. Entomol. Soc. 56: 581–606.

Reierson, D. A. and R. E. Wagner. 1975. Trapping yellowjackets with a new standard plastic wet trap. J. Econ. Entomol. 68: 395–398.

Reierson, D. A. and R. E. Wagner. 1978. Trapping to determine the sympatry and seasonal abundance of various yellowjackets. Environ. Entomol. 7: 418–422.

Richards, O. W. 1962. A revisional study of the masarid wasps (Hymenoptera, Vespoidea). British Museum, London. 294 p.

Richards, O. W. 1971. The biology of the social wasps (Hymenoptera, Vespidae). Biol. Rev. 46: 483–528.

Rogers, C. J. 1972a. Flight and foraging patterns of ground-nesting yellowjackets affecting toxic baiting control programs. Proc. Calif. Mosq. Cont. Assoc. 40: 130–132.

Rogers, C. J. 1972b. Field testing of "Yellowjacket Stoppers" and population depletion trapping for the control of ground-nesting yellowjackets. Proc. Calif. Mosq. Cont. Assoc. 40: 132–134.

Ross, D. 1983. The role of volatiles and social facilitation in bait discovery by yellowjackets. MS Thesis, Purdue University. 96 p.

Ross, D. R., R. H. Shukle, and J. F. MacDonald. 1984. Meat extracts attractive to scavenger Vespula in eastern North America (Hymenoptera: Vespidae). J. Econ. Entomol. 77: 637–642.

Ross, K. G. 1982. Gastral vibrations in laboratory Vespula vulgaris and V. maculi-

frons colonies (Hymenoptera: Vespidae). *Fla. Entomol.* 65: 187–188.

Ross, K. G. and R. W. Matthews. 1982. Two polygynous overwintered *Vespula squamosa* colonies from the southeastern U.S. (Hymenoptera: Vespidae). *Fla. Entomol.* 65: 176–184.

Roush, C. F. and R. D. Akre. 1978. Nesting biologies and seasonal occurrence of yellowjackets in northeastern Oregon forests (Hymenoptera: Vespidae). *Melanderia* 30: 57–94.

Rubenstein, H. S. 1980. Allergists who alarm the public. *J. Amer. Med. Assoc.* 243: 793–794.

Schmidt, J. O. 1982. Biochemistry of Insect Venoms. *Annu. Rev. Ent.* 28: 1–61.

Schmidt, J. O. and M. S. Blum. 1979. Toxicity of *Dolichovespula maculata* venom. *Toxicon* 17: 645–648.

Schmidt, J. O., M. S. Blum, and W. L. Overal. 1980. Comparative lethality of venoms from stinging Hymenoptera. *Toxicon* 18: 469–474.

Settipane, G. A. and G. K. Boyd. 1970. Prevalence of bee sting allergy in 4992 boy scouts. *Acta Allergol.* 25: 286–291.

Settipane, G. A., D. E. Klein, and G. E. Boyd. 1978. Relationship of atopy and anaphylactic sensitization: A bee sting allergy model. *Clin. Allergy* 8: 259–265.

Settipane, G. A., C. J. Newstead, and G. K. Boyd. 1972. Frequency of Hymenoptera allergy in an atopic and normal population. *J. Allergy Clin. Immunol.* 50: 146–150.

Sharp, J. L. and J. James. 1979. Color preference of *Vespula squamosa. Environ. Entomol.* 8: 708–710.

Smithers, C. N. and G. A. Holloway. 1978. Establishment of *Vespula germanica* (Fabricius) (Hymenoptera: Vespidae) in New South Wales. *Aust. Entomol. Mag.* 5: 55–60.

Spencer, G. J. 1960. On the nests and populations of some vespid wasps. *Proc. Entomol. Soc. Brit. Col.* 57: 13–15.

Spradbery, J. P. 1973a. The European social wasp, *Paravespula germanica* (F.) (Hymenoptera: Vespidae) in Tasmania, Australia. VII Internat. Congr. IUSSI. pp. 375–380.

Spradbery, J. P. 1973b. Wasps: An account of the biology and natural history of solitary and social wasps. Univ. Wash. Press, Seattle. 408 p.

Taylor, L. H. 1939. Observations on social parasitism in the genus *Vespula* Thomson. *Ann. Entomol. Soc. Amer.* 32: 304–315.

Thomas, C. R. 1960. The European wasp (*Vespula germanica* Fab.) in New Zealand. *NZ Dept. Sci. Ind. Res. Inf. (Ser. 27)*: 1–74.

Tissot, A. N. and F. A. Robinson. 1954. Some unusual insect nests. *Fla. Entomol.* 37: 73–92.

Tu, A. 1977. Venoms: Chemistry and Molecular Biology. John Wiley and Sons, New York. 506 p.

Vuillaume, M., J. Schwander, and C. Roland. 1969. Note préliminaire sur l'existence des colonies pérenées et polygynes de *Paravespula germanica. C. R. Acad. Sci. (Ser. D)* 269: 2371–2372.

Wagner, R. E. 1961. Control of the yellowjacket *Vespula pensylvanica* in public parks. *J. Econ. Entomol.* 54: 628–630.

Wagner, R. E. and D. A. Reierson. 1969. Yellowjacket control by baiting. 1. Influence of toxicants and attractants on bait acceptance. *J. Econ. Entomol.* 62: 1192–1197.

Wagner, R. E. and D. A. Reierson. 1971. Yellowjacket control with a specific mirex-protein bait. *Calif. Agric.* 25: 8–10.

Wagner, R. E. and D. A. Reierson. 1975. Clothing for protection against venomous Hymenoptera. *J. Econ. Entomol.* 68: 126–128.

Walton, G. M. and G. M. Reid. 1976. The 1975 New Zealand European wasp survey. *N. Z. Beekeeper* 38: 26–30.

West-Eberhard, M. J. 1969. The social biology of polistine wasps. Misc. Publ. Mus. Zool. Univ. Michigan No. 140. 101 p.

Wheeler, J. W., F. O. Ayorinde, A. Greene, and R. M. Duffield. 1982. Citronellyl citronellate and citronellyl geranate in the European hornet, *Vespa crabro* (Hymenoptera: Vespidae). *Tetrahedron Lett.* 23: 2071–2072.

Willink, A. 1980. Sobre la presencia de *Vespula germanica* (Fabricius) en la Argentina (Hymenoptera: Vespidae). *Neotropica* 26: 205–206.

Wilson, E. O. 1971. The Insect Societies. Belknap Press of Harvard Univ. Press, Cambridge, Mass. 548 p.

Yamane, Sk. 1976. Morphological and taxonomic studies on vespine larvae, with reference to the phylogeny of the subfamily Vespinae (Hymenoptera: Vespidae). *Insecta Matsumurana N.S.* 8: 1–45.

Yamane, Sk., and S. Makino. 1981. Embryo nest architecture in three *Vespula* species (Hymenoptera, Vespidae). *Kontû* (Tokyo) 49: 491–497.

Yamane, Sk., S. Makino, and M. Toda. 1980a. Nests of *Dolichovespula albida* from the Arctic Canada (Hymenoptera: Vespidae). *Low Temp. Sci.* (Ser. B) 38: 61–68.

Yamane, Sk., R. E. Wagner, and S. Yamane. 1980b. A tentative revision of the subgenus *Paravespula* of eastern Asia (Hymenoptera: Vespidae). *Insecta Matsumurana N.S.* 19: 1–46.

INDEX

acidopore, 276
Acromyrmex, 165
Acromyrmex landolti, 169
African driver ant, 303
aggressive behavior, in ants, 318
agricultural crops: immunity to leaf-cutting ants, 168; termite control, 87
aldrin, 87
alkyl ammonium compounds (AAC), 117
allergy, to fire ant venom, 230
altitude, and termite distribution, 3
Amblyomma americanum, see lone star tick
Amdro, 247
amidinohydrazone, 117, 247
Amitermes neogermanus, 112
Amitermitinae, 106
Anacanthotermes, 82; damage to building, 83
Anacanthotermes mounds, as nitrate fertilizer, 148
ant communities, interactions, 317
ant-Homoptera relationship, 315
ant manipulation, 317; alteration of ant segregation, 320; compatibility with farming, 329; and crop interplanting, 322; ecological constraints, 325–327; experiments, 318; fungus growing, 295; habitat alteration, 320; influence on pests, 322; introduction from other regions, 318; introduction with-in the region, 319; pesticides role in, 328; selective chemical treatment, 319; selective mechanical removal, 320; side effects, 328; sustainability, 327

ant mosaic, 317
ant mounds, 294
ants, 193; aggressive behavior, 318; arthropods as food, 300; beneficial economic role, 290; beneficial effects, 314–331; development as pest control agent, 315; as human food, 306; interaction with man, 290; in medicine, 308; in nutrient cycling, 295; in pest control of coconuts, 304; as pollinator, 293, 297; as predators, 300; as predators of fruit fly, 304; as predators of leaf-cutting ants, 304; selection as pest control agents, 314; social insects, 290; and soil, 293; as sugar collectors, 298; in suturing wounds, 308; and termites, 151
ants, predatory: in agroecosystems, 303; characteristics, 315; colony size, 315; in date pest control, 301; food requirement, 315; in forest ecosystems, 305; habitat range, 315; in pest control, 301; queens, 315; relationship with Homoptera, 315; size, 315; trophic biology, 300
Apicotermes, nest building, 31, 146
Apicotermitinae, autothysis, 12
apple, damage by termites, 79
army ants, 300; in pest control, 302
arsenic trioxide, 94
Atta capiguara, 169
Australian termites: control measures, 113; damage to plastic, 113; damage to buried cables, 113; drywood, 109; economic importance, 103–129; eucalypt forests, 111; forest tree, 111; grass-eating, 112; pasture damage by, 112; pest

Australian termites (*continued*)
control cost, 120; termite families,
104
autothysis, 12

bacteria, cellulolytic, 38
bees, as pollinator, 298
bendiocarb, 265
benzalkonium chloride, 117
benzene hexachloride, 87
Bifiditermes beesoni, 79
blowflies, 303
boric acid, 117
buildings: damage by subterranean
termites, 83; on formerly forested
areas, 133; poor construction and
termite infestation, 133;
postconstruction termite control,
92; preconstruction termite control, 91, 134; termite damage, 82;
termite management, 91

cables, underground, termite control
cost, 124
calcium arsenate, 176
Camponotus pennsylvanicus, see
carpenter ants
carbaryl, 88, 376
carbon disulfide, 177
carpenter ants, 272–289; alarm
pheromone, 276; colony life cycle,
272; control costs, 282; development time, 274; digestive glands,
276; division of labor in, 275;
economic aspects, 281; forest loss,
283; fungi as food, 276; individual
specialization, 275; lack of
metapleural glands, 279; larvae,
277; larval food, 275; males, 272;
mating, 273; metapleural glands,
279; nesting behavior, 278; nest
location, 279; neurosecretory products, 274; nutrition, 275; orientation, 277; as pest, 280; pesticides
for, 284; polyethism, 275; polymorphism, 275; postpharyngeal glands,

276; as predator of insects, 276;
predators, 278; prevention, 284;
protein demand of colony, 276;
proventriculus of, 277; queen, 274;
satellite nests, 274; structural
damage to buildings, 283; trail
pheromone, 277; trees infestation,
279; as urban pests, 282; venom
apparatus, 276; vibrational communication, 279; worker's food,
275
cashew, damage by termites, 79
cell biology, 146
cellular division of labor, 145
cellulases, 40
cellulose acetate butyrate, 113
cellulose digestion, in termites, 39
chemical communication, 209; in termites, 25
chemical defense in termites, 15
chemical sex attractants, 17
chili crop, termite damage, 75
chlordecone, *see* kepone
chlorpyrifos, 135
chlorosulfonated polyethylene, 96
chrysanthemums, termites damage,
75
citrus damage: leaf-cutting ants, 169;
termites, 79
coal strip mine soil, 148
cocoa, damage by leaf-cutting ants,
169
coconut palm: damage by termites,
76; termite control, 89
coffee, damage by termites, 76
colony foundation, of termites, 17
color vision, of yellow jackets, 375
Coptotermes: attacking *Pinus radiata*,
111; damage to timber, 106; food
selection, 33
Coptotermes acinaciformis, damage
to timber, 109
Coptotermes curvignathus, 79
Coptotermes formosanus, damage to
rice, 73
Coptotermes heimi, 83

cotton: damage by termites, 75; termite control, 88
crops, termite control cost, 123
Cryptotermes acinaciformis, control of, 118
Cryptotermes bengalensis, 84
Cryptotermes brevis, 109

demethoxylation, 39
Diatraea saccharalis, 239
diazionon, 284, 376
p-dichlorobenzene, 176
dimorphism, in termite soldiers, 16
Dorylinae, 302
dragonflies, 150
Drepanotermes rubriceps, 149
drione, 138, 284
Dufour's gland, 260
dung beetles, 148
Dursban, 116, 284

ecosystem, termites beneficial effects, 147
Ekphysotermes, 104
endosulfan, 135
ethospecies, 146
eucalyptus: damage by leaf-cutting ants, 171; termite attack, 80

face fly, 243
faranal, 261
fenitrothion, 265
Ficam, 284
fire ant colony: development, 202; founding, 199; number, 200; site selection, 200; starvation, 206
fire ant damage: to citrus, 237; to corn, 238; to eggplants, 238; to okra plants, 238; to soybeans, 237
fire ants: attack on cattle, 239; behavior, 204; beneficial aspects in cotton fields, 242; beneficial effects, 239; biological control, 248; biomass, 197; brood pheromone, 209; chemical communication, 209; chemical toxicants for, 248; competition with other ants, 214; control cost, 228; control with boric acid, 247; control with mirex bait, 246; damage (*see* fire ant damage); digestion, 207; distribution, 195; ecology, 212; economic importance of, 227–256; effect on agriculture, 231; egg production, 202; feeding, 204; feeding glands, 207; female reproductive system, 201; flight muscle, 199; food distribution, 205; food exchange, 205; food habits, 236; food particle size, 206; foraging territories, 207; fumigant treatment of mounds, 246; glycogen changes in, 200; and ground nesting birds, 243; human response to venom, 229; introduction in United States, 195; invasion of homes, 244; invasion of telephone pedestals, 244; larval development, 203; larval excretory system, 208; life span, 203; male reproductive system, 201; mating behavior, 198; mating flights, 198; mound development, 203; mound temperature, 204; nector collection, 205; nutritional needs, 205; and parasites, 214, 249; physiology, 204; plant oils, fondness of, 204; predation on lizard eggs, 243; as predator of lone star tick, 243; as predator of tobacco budworm, 242; and predators, 213; predators of queen, 249; proteinaceous allergens in venom, 230; and public health, 229; pupal stage, 203; queen, 199; queen pheromone, 210; range of infestation, 196; reproduction, 197, 201; reproduction inhibition, 248; spreading of, 197; sterile males, 201; stinging statistics, 230; sugar feeding, 204; taxonomy, 195; temperature and distribution, 213; trail pheromone, 209; venom, 212; as weed species,

fire ants (continued)
228; and wildlife, 243; workers, 202
forest trees: damage by leaf-cutting ants, 170; damage by termites, 80; termite control cost, 123
forestry crops, termite control, 90
Formica obscuripes, 296
Formica schauffusi, 299
fruit trees, termite control cost, 123
fungi, woodrotting, 33
fungus comb, 19

β-glucosidase, 40
Glypotermes, 78
Glyptotermes dilatatus, damage to tea, 76
Gnathamitermes, seasonal variation in food selection, 33
Gnathamitermes tubiformans, 140
grapevines, damage by termites, 79
groundnut: damage by termites, 73; seed treatment for termite control, 88
guava, damage by termites, 79

Haematobia irritans, see hornfly
Heliothis virescens, see tobacco budworm
hemicellulose, 39
heptachlor, 135
Hesperotermes, 104
Heterotermes, attack on Pinus radiata, 111
Heterotermes aureus, nest size, 28
Heterotermes gertrudae, 83
Heterotermes indicola, 83
2,4-hexadienyl butyrate, 376
honeypot ants, 308
hornfly, 243
hornworm, 336
horticultural crops, termite control, 90
houseflies, 303
humus, 31

imported fire ants, see fire ants

Incisitermes, 104
insects, nutritional value, 306
invertase, 208
Iridomyrmex humilis, 196
isofenphos, 135
isoptera, 2

Jack-fruit, damage by termites, 79
jute, damage by termites, 75
juvenile hormone, 23

Kalotermes, number of ovarioles, 11
Kalotermes flavicollis, development, 21
Kalotermitidae, 105
kepone, 264

laminated wood, termite control, 96
Lasius alienus, 294
lead arsenate, 180
leaf-cutting ant damage: to agricultural crops, 167; to citrus, 169; to forest trees, 170; to horticultural crops, 168; to pastures, 169
leaf-cutting ants: biological control, 175; chemical attractants as baits, 182; control techniques, 172; chemicals for destroying, 177; destroying nests, 176, 177; economic importance, 165; gasoline treatment of nests, 178; kerosene treatment of nests, 178; killing foraging workers, 176; nest dusting with toxicants, 178; nest fumigation, 178; organization of control, 184; physical defenses of plants, 174; poison baits, destroying with, 179; preventing access to crops, 174; resistant crops, growing of, 172; toxic baits, 179, 181; toxic smoke treatment of nests, 179; types, 167
leather, damage by termites, 87
legumes: damage by termites, 74; insecticidal termite control, 87
lignin, 38; degrading enzymes, 40

lime, immunity to leaf-cutting ants, 168
lindane, 88, 135
linolenic acid, 204
lipase, 208
lone star tick, 243

Macrotermes: juvenile hormone, 23; royal cell, 28
Macrotermitnae: colony survival, 19; food consumption, 36; food storage, 35; fungi feeding, 32; sex and caste development, 24; soldier caste development, 22; soldiers, 22; symbiosis with fungi, 38
maize, damage by termites, 73
malathion, 284, 376
mandibular gland, in termites, 6
mango, damage by termites, 79
Mastotermes, sternal glands, 10
Mastotermes darwiniensis: control, 118; damage to fruit trees, 112; damage to timber, 106; development, 21
Mastotermitidae, 104; wings, 10
mathematical model, of yellow jacket population dynamics, 362
medical cost, for fire ants stinging, 231
medicinal plants, termite control, 88
mercuric chloride, 176
Metarhizium anisopliae, 118
methomyl, 376
methoprene, 265
methoxychlor, 376
methyl bromide, as anti-termite chemical, 116
Microcerotermes: cellulose synthesis, 40; damage to rice, 73
microclimate regulation, in termite nests, 29
Microtermes, population biology, 49
Microtermes obesi, damage to wheat, 71
millet, insecticidal termite control, 87
mirex, 237

Monomorium pharaonis, see Pharaoh's ant
moths, 340
Mulberry, damage by termites, 79
Musca autummalis, see face fly
Myrmicinae, 257

Nasutitermes: colony growth, 20; distribution, 47; food selection, 34
Nasutitermes exitiosus: attack on eucalypts, 111; food selection, 33
Nasutitermitinae, 106
Neotermes, colony growth, 19
Neotermes assmuthi, 79
Neotermes greeni, 79
Neotermes tectonae, damage to teak trees, 82
nests, termites: building and climate, 31; composition, 27; dimensions, 28; microclimate, 29; moisture, 31; soil selection for, 27; system, 26
noncellulose material, damage by termites, 85

oak, damage by termites, 82
ocelli, 9
Odontotermes obesus, 71, 79
oil loving fire ant, 204
oil palm, damage by termites, 74
Oriental Region: tea damaging termites, 76; termite distribution, 71; termite species, 69
oviposition, 19
oxymonad, cellulolytic, 37

Panicum maximum, 170
Paraneotermes simplicornis, 140
Paravespula, see yellow jackets
Parrhinotermes, 105
Pasteurella pestis, 258
pastures: damage by termites, 82; termite control, 91; termite control cost, 123
peanut, see groundnut
pearl millet, damage by termites, 73
pentachlorophenol, 135
permethrin, 135–136

Pharaoh's ant, 257–271; age polyethism of worker, 260; baits against, 265; biology, 259; boric acid baits, 266; colonies, 260; colony foundation, 261; developmental period, 260; food source, 261; food storage, 261; hospital infestation of, 259; insecticidal control, 264; liver baiting, 266; males, 260; mating, 260; migration, 261; nests, 259; oviposition rate, 262; poison gland, 262; queen, 260; sex pheromone, 260; spraying with chlordane, 264; temperature effect on colony, 262; trail pheromone, 261; in transmission of pathogenic bacteria, 258; workers, 260; world distribution, 257
Pharaoh's ant control, 263; with bendiocarb, 265; with fenitrothion, 265; with methoprene, 265; with sodium fluoride, 264
Pheidole megacephala, 139
pheromones: and caste regulation in termites, 23; in fire ants, 209
phorid fly, 278
physogastry, 11
pinenes, 33
Pinus radiata, 111
pipelines, termite control cost, 124
2,6-piperidines, 229
planthopper, 205
plants: ant-detering structures, 298; chemical defense system, 167
plastic resistance against termites, 119
Pogonomyrmex occidentalis, 301
Polistes wasps, *see* wasps
pollination biology, 297
polyethism, 6; in carpenter ants, 275
polyethylene tubes, damage by termites, 85
polymorphism: in carpenter ants, 275; in termites, 6
polyvinyl chloride, 96
plywood, termite control, 96
Ponerinae, in pest control, 302

poplar, attack by termites, 81
population ecology: sampling methods, 42; of termites, 42
Porotermes adamsoni: damage to timber, 109; species susceptibility, 111
Porotermitinae, 105
potassium cyanide, as bait for leaf-cutting ants, 180
potatoes, damage by termites, 75
Procryptotermes, 104
protozoan-termite relationship, 38
Psammotermes, 79
pyrethrum, 264
pyrrolidines, 262

rain infiltration, 149
red imported fire ant, see *Solenopsis invicta*
red tree ant, 291
red wood ants, 305
reforestation, 149
Rhinotermitidae, 105; tibial-tarsal exocrine gland, 10
rice: damage by termites, 73; termite control, 87
rubber: damage by termites, 78; termite control, 89
runoff, 149

Sesamum indicum, 179
sex attractants, 17
silica aerogel, 138
social organization, in termites, 25
sodium fluoride, 264
soil: and ants, 293; movement, 294; porosity, 148; sandy-loam, 75; treatment termiticides, 135
soil feeding termites: digestive process, 40; symbiosis with microorganisms, 38
soil fertility, 296; and termites, 147
soil genesis, 296
Solenopsis geminata, 194
Solenopsis invicta, 193
Solenopsis xyloni, 194
Sphaerotermes sphaerothorax, 38

Spirochaetes, 37
stable fly, see *Stomoxys calcitrans*
sternal glands, in termites, 10
Stolotermitinae, 105
Stomoxys calcitrans, 243
subterranean termites: control cost, 135; distribution in United States, 131; preventive measures, 133
sugarcane: damage by termites, 74; insecticidal termite control, 88
sulfuryl fluoride, 117
supraorganismic concept, 145

tea: insecticidal termite control, 89; termite damage, 76
teak trees, termite damage, 82
telegraphic cables, damage by termites, 85
Termitariophiles, 155
termite baiting system, 117
termite biomass, 37
termite colony: as ecological model, 145; growth, 18, 19; integration, 25; odor, 26; size, 20
termite colony foundation, 17; by budding, 18; by burrowing, 18; by social fragmentation, 18
termite damage, 71; to coconut palm, 76; to coffee, 76; to cotton, 75; to dams, 85; to fruit trees, 78–80; to groundnut, 73; to human dwellings, 82; to jute, 75; to legumes, 74; to maize, 73; to mango plants, 79; noncellulose material, 85; to oak, 82; to oil palm, 74; to pearl millet, 73; to plantation crops, 76–78; to rubber, 78; to stored wood, 84; to sugar-cane, 74; to tea crop, 76; to vegetables, 75; to water reservoirs, 85; to wheat crops, 71
termite detector, 94
termite development, 20; as incomplete metamorphosis, 21
termite mounds: and birds nests, 156; and honey bee hives, 156; and human constructions, 155; importance of, 155; mineral concentration, 150; as model ecosystem, 145; as shelter for other organisms, 155; use by snakes, 156
termite nest structure, and behavior theory modeling, 145
termite-protozoa relationship, 38
termite soldier caste, 15–16; compound eyes, 16; defense mechanism, 15; dimorphism, 16; mandibulate type, 15; masutoid type, 15
termite workers: alimentary system, 12; antennal segments, 12; buccal cavity, 12; defense mechanism, 12; form and function, 11; mandibles, 12
termites: abdomen, 10; antennal segments, 9; and ants, 151; attack by termites, 75; attack on chrysanthemum, 75; attacking eucalyptus, 80; bacterial symbiosis, 38; beneficial activities, 71; beneficial aspects, 144–164; biological control, 118; and carbohydrate decomposition, 147; carbon to nitrogen ratio, 40; caste and colony nutrition, 25; caste determination, 22; caste differentiation, 20; caste elimination, 25; caste regulation, 23; cellulases, 40; cellulose digestion, 38; chemical communication, 25; chemical composition of nests, 28; classification, 2; as classroom teaching tools, 146; climate and distribution, 47; competitors, 48; compound eyes, 9; control cost, 119; control in pastures, 119; crops attack, 71; daily foraging periods, 35; developmental pathways, 20; distribution affecting factors, 45; distribution and other animals, 47; division of labor, 6; as drug removal agents, 148; economic importance, 69–102; egg-laying capacity, 11; endocrine system, 16; environmental impact on population, 50; exocrine glands, 10; flight,

termites (continued)

17; fontanelle, 6; food consumption, 36; food digestion, 37; as food for birds, 152; as food for predators, 48, 150; food galleries, 34; food location, 34; food resources, 31; food selection, 32; food selection and stage of decomposition, 34; food size, 34; food storage, 35; food utilization, 37; foraging, 34; as forestry pests, 80; fossil species, 3; and fungi dissemination, 147; and fungus comb, 19; grassfeeding, 36, 112; hemicellulose digestion, 39; human consumption of, 154; injurious to pasture, 82; internal foraging, 35; lignin digestion, 39; as lizard food, 152; as mammals' food, 153; mandibles, 6; molar plates, 6; morphology, 4; nervous system, 16; nest architecture, 29; nests (see nests, termites); niche exploitation, 44; nitrogen content, 40; nitrogen fixation, 41; nutrient recycling, 150; panoistic ovarioles, 11; pheromones, 26; physogastry in, 11; population ecology, 42; and potatoe damage, 75; predators, 47; primary reproductives, 6; as primates diet, 153; and reforestation, 149; reproductive castes, 6; reproductive morphology, 6; reproductive system, 11; resistance to timber, 33; role in geochemical prospecting, 150; role in rain infiltration, 149; seasonal foraging, 35; secondary reproductives, 6, 21; sensory system, 16; sex attractant, 17; social organization, 25; soil, beneficial effect on, 147; soil feeding, 32; soil type and distribution, 45; soldier caste development, 22; soldiers to workers ratio, 22; spatial organization, 26; specialized feeding, 32; sperm viability, 11; swarming and weather, 17;

symbiosis with microorganisms, 37; thorax, 9; timber damage cost, 119; in tropical regions, 144; underground damage by, 85; use in cell biology, 146; vegetation, beneficial effects on, 149; vegetation and distribution, 45; veins, 10; wings, 10; wood feeding, 32; workers development, 22; world distribution, 2

termites, dry-wood, 84; of Australia, 109; control, 95, 138; fumigation with methyl bromide, 139; prevention, 138; remedial control, 138; Wood Treatment TC control, 139

termites in Australia, 103–129; control, 113

termites in United States, see United States, termites in

Termitidae, 105

Termitomyces, 19; cellulose degradation, 40

Termopsidae, 105

thallium sulfate, 264

timber, protection against termites, 117; termite damage, 106

tobacco budworm, 242

toxic baits, for fire ants control, 245

trichomonads, 37

trichonympha, 147

Trinervitermes geminatus, 36

trophallaxis, 42

tropical rain forests, termite distribution, 3

trail pheromone, in five ants, 209

United States, housing units, 132

United States, termites in, 130–143; agricultural crops damage, 139; damage cost, 134; damage to buried cables, 140; grapevine damage, 140; economic impact, 134; ornamental plants damage, 139; pesticide usage data, 135; range grass damage, 140; sugarcane damage, 139; tree damage, 139; wooden products, damage to, 132

uricolysis, 41

vegetables: damage by termites, 75; termite control, 88
venom, of fire ants, 212
venom, yellow jackets: active component, 379; as allergen, 380; chemistry, 371; immunotherapy, 383; reactions to, 380
venom immunotherapy, 383
Vespula, see yellow jackets
Vespula squamosa, 353
vibration communication, among carpenter ants, 279

Wasmannia auropunctata, 315
wasps: anatomy, 334; biological control, 341; biology, 332; birds as predator, 339; carbohydrate utilization, 338; enemies of, 339; food relations, 336; food storage, 338; forage distance, 346; honey storage, 338; hosts relations, 336; human enemy, 339; management, 341; management problems, 344; management studies, 343; nest material, 346; nesting containers, 343; nests, 333; nest site location, 346; predation of caterpillars, 337; as predator of pests, 337; productivity of nests, 335; role in pollination, 339; supplemental feeding, 344. See also yellow jackets
wheat: seed treatment, 87; soil treatment, 87; termites attack on, 71; termite control, 87
wind erosion, 149
wood, pressure treated, 138
wood-feeding termites, food selection, 33
wood preservatives, 117
wood treatment, 113
wood Treatment TC, 139

yellow jacket colony, 362; aggressiveness, 371; cycle, 358; defense, 370; odor, 368

yellow jackets, 353–412; area-wide control, 397; attack by bears, 371; behavior, 363; beneficial aspects, 377; biological control, 401; chemical attractants, 376; color vision, 375; communication, 368; competition for nest site, 367; compound eyes, 375; control, 394; damage to Christmas tree plantation, 390; defensive flights, 370; diapause, 372; economic impact of stings, 383; emergency treatment after stinging, 382; foraging, 364; foraging distance, 635; foraging on human food, 392; geographical distribution, 357; impact on agriculture, 390; impact on fruit orchards, 391; impact on lumbar industry, 391; individual colonies destruction, 394; insecticides for, 376; inter-species competition, 366; intracolonial interactions, 368; intra-species competition, 366; light and foraging, 364; mathematical model for population dynamics, 362; mating behavior, 365; medical importance, 377; mortality statistics of stinging, 378; nest color, 360; nest construction, 361; nest location, 361; nesting biology, 358; nutrition, 373; odor perception, 375; partitioning food resources, 367; pheromones, 369; physical dominance, 368; physiology, 372; population dynamics, 362; in recreational areas, 384; in residential areas, 384; sensory physiology, 375; social impact, 376; synthetic chemical attractants, 399; taxonomy, 354; toxic bait abatement program, 397; venom (see venom, yellow jackets); workers, 374

zinc chloride, 117
zootermopsis: endocrine caste control, 24; flight, 17

ABOUT THE EDITOR

S. Bradleigh Vinson is Professor of Entomology at Texas A&M University, College Station, Texas. Until 1969 he was associate professor of Entomology at Mississippi State University.

Dr. Vinson has published widely with over two hundred research articles and dozens of reviews. He is internationally recognized for his work in the physiology, behavior and ecology of parasite-host relationships and social insects (specifically ants). In 1977 he coedited a book "Biological Control by the Augmentation of Natural Enemies" with Dr. R. L. Ridgway.

Dr. Vinson holds a B. S. from the Ohio State University, Columbus, Ohio and an M. S. and Ph.D. from Mississippi State University, State College, Mississippi.